Microbial Based Land Restoration Handbook

Microorganisms are a good indicator of soil health. Plant growth-promoting microorganisms protect plants from the stresses of water, salt, metal, biotic, and so on, and are well known for strategically modulating the plant mechanisms to defend and mitigate environmental stresses. Taking a multidisciplinary approach, this volume explores the role of plant microorganisms in ecological and agricultural revitalization beyond normal agriculture practices and offers practical and applied solutions for the restoration of degraded lands to fulfill human needs with food, fodder, fuel, and fiber. It also provides a single comprehensive platform for soil scientists, agriculture specialists, ecologists, and those in related disciplines.

Features

- Presents cutting-edge microbial biotechnology as a tool for restoring degraded lands
- Explores the aspects of sustainable development of degraded lands using microorganism-inspired land remediation
- Highlights sustainable food production intensification in nutrient-poor lands through innovative use of microbial inoculants
- Explains the remediation of polluted land for regaining biodiversity and achieving United Nations Sustainable Development Goals
- Includes many real-life applications from South Asia offering solutions to today's agricultural problems

This book will be of interest to professionals, researchers, and students in environmental, soil, and agricultural sciences as well as stakeholders, policy makers, and practitioners with an interest in this field.

Microbial Based Land Restoration Handbook

VOLUME 2

Soil and Plant Health Development

Edited by
Vimal Chandra Pandey and Umesh Pankaj

CRC Press
Taylor & Francis Group
Boca Raton London New York

CRC Press is an imprint of the
Taylor & Francis Group, an **informa** business

First edition published 2023
by CRC Press
6000 Broken Sound Parkway NW, Suite 300, Boca Raton, FL 33487-2742

and by CRC Press
4 Park Square, Milton Park, Abingdon, Oxon, OX14 4RN

CRC Press is an imprint of Taylor & Francis Group, LLC

ISBN: 9780367702243 (hbk)
ISBN: 9780367705855 (pbk)
ISBN: 9781003147077 (ebk)

DOI: 10.1201/9781003147077

Typeset in Times
by KnowledgeWorks Global Ltd.

Dedication

This book is dedicated to our beloved mothers.

—Vimal Chandra Pandey and Umesh Pankaj

Contents

Foreword

The growing human population in the world is expected to create an alarming situation in the future regarding food security. Due to disproportionate application of chemical fertilizers and pesticides, agriculture is facing severe provocation like deterioration of land, reduced productivity and increased susceptibility to several biotic and abiotic stresses. Salinity and drought stress are major abiotic stresses that limit sustainable agriculture production. Nature has bestowed upon us the gift of soil microflora that has the potential to cope with these stresses by improving the nutrient uptake, antioxidant system and phytohormone regulation in plants. These microorganisms and their biosynthetic activities under different environmental conditions make them suitable candidates for application in agriculture and other allied fields. Since soil biodiversity is an essential driver of ecosystem functions, these microorganisms tend to improve the soil-plant health, thus playing a pivotal role in agriculture. Beneficial microorganisms and arbuscular mycorrhizal fungi help in ecological restoration and sustainable agriculture through nitrogen fixation, siderophore production, mineral solubilization, bioremediation of metal(loid)s and salts, and so on.

The importance of beneficial microorganisms in soil-plant health development under degraded lands has created a great interest in sustainable agriculture. The present book, *Microbial Based Land Restoration Handbook: Soil and Plant Health Development*, describes such promising and potential tools to enhance plant performance with microbial intervention. This book covers how plant growth is influenced under various soil abiotic stresses by the application of beneficial soil microorganisms and their responses in plants through physiological, biochemical, and molecular approaches to mitigate the stresses.

I congratulate the editors Dr. Vimal Chandra Pandey and Dr. Umesh Pankaj for bringing out this valuable book published by a renowned publisher CRC Press/ Taylor & Francis. This book consists of 13 chapters covering various aspects of soil-plant health development. I am sure that this book will be a notable asset for researchers, teachers, scientists, practitioners, policy makers, entrepreneurs, and other stakeholders alike.

G.D.Sharma

Prof. G.D. Sharma
Vice-Chancellor, University of Science and Technology, Meghalaya, India
Founder Vice-Chancellor, Bilaspur University, Bilaspur, Chhattisgarh, India
Former Vice-Chancellor, Nagaland University, Kohima, Nagaland, India

Preface

Land degradation is a result of the over-exploitation of land by human beings to obtain their livelihoods. The main reasons for land deterioration are uncontrolled application of chemical fertilizers and pesticides, environmental pollution, and the changing environment. The major abiotic stresses are salinity and drought that limit sustainable agricultural production. Soil microflora is one of God's best gifts to human beings that have tremendous potential to cope with these stresses by strengthening antioxidant defence systems, nutrient uptake, and phytohormone regulation in crop plants. Soil microbial diversity plays an essential role in driving ecosystem functions. A number of native microorganisms under different environmental conditions have been identified for their important activities in agriculture and ecorestoration. In addition, several engineered microorganisms have been created to increase agricultural productivity and to remediate pollutants from contaminated soil systems. The microorganisms have a tendency to improve the soil-plant health and thus play a vital role in sustainable agriculture and ecorestoration through nitrogen fixation, nutrient uptake, mineral solubilization, siderophore production, and bioremediation of soil pollutants.

The present book consists of 13 chapters contributed by experts worldwide. Chapter 1 describes the role of soil microorganisms in sustainable crop production to achieve food security. Chapter 2 focuses on plant-microorganism interactions to explore microbial-mediated molecular mechanisms to cope up salinity and drought stress in plants, while Chapter 3 explores the role of arbuscular mycorrhizal fungi in horticultural crops. Chapter 4 deals with the role of endophytes in managing degraded land toward enhancing food, fodder, fuel, and fiber production. Chapter 5 focuses on microbial approaches to enhance the food crop productivity under salinity and drought stress conditions. Chapter 6 explores secondary metabolites and their role in the development of resilience in crops for agricultural sustainability. *Bacillus*-based biocontrol for sustainable agriculture is presented in Chapter 7. Chapter 8 focuses on the adaptation and potentialities of the phyllosphere microbiome under various distresses toward sustainable plant growth. Chapter 9 describes the significance of biofertilizers in *Curcuma longa*, while Chapter 10 explores the role of soil microorganisms in maintaining soil health. Chapter 11 deals with recent advances in the potentiality of microorganisms in promoting plant growth and managing degraded land. Chapter 12 focuses on productivity losses through bio-deterioration in water-deficit harvested sugarcane, whereas Chapter 13 explores bio-inoculants and the potential microorganisms for restoration of degraded soil.

Vimal Chandra Pandey
Umesh Pankaj

Acknowledgments

We wish to extend our sincere thanks to Irma Shagla Britton (Senior Editor), Rebecca Pringle (Editorial Assistant), and Shannon Welch (Editorial Assistant), CRC Press/Taylor & Francis for their whole-hearted support, guidance, and coordination to carry out this project. We would like to thank all the contributors for their excellent chapter contributions. We would like to thank all the reviewers for their time and expertise to review the chapters of this book. We are highly thankful to Prof. G.D. Sharma, Vice-Chancellor, University of Science and Technology Meghalaya, RiBhoi, Meghalaya, India, for writing the Foreword at short notice. Finally, diction is not enough to express our deepest sense of unbounded gratitude to our respected parents and family for their support, love, and encouragement for which we will remain indebted to them throughout our lives.

Acknowledgments

Editors

Dr. Vimal Chandra Pandey featured in the world's top 2% scientists curated by Stanford University, United States. Dr. Pandey is a leading researcher in the field of environmental engineering, particularly phytomanagement of polluted sites. His research focuses mainly on the remediation and management of degraded lands, including heavy metal-polluted lands and post-industrial lands polluted with fly ash, red mud, mine spoil, and others, to regain ecosystem services and support a bio-based economy with phytoproducts through affordable green technology such as phytoremediation. Dr. Pandey's research interests also lie in exploring industrial crop-based phytoremediation to attain bioeconomy security and restoration, adaptive phytoremediation practices, phytoremediation based biofortification, carbon sequestration in waste dumpsites, climate resilient phytoremediation, fostering bioremediation for utilizing polluted lands, and attaining the United Nations Sustainable Development Goals. His phytoremediation work has led to the extension of phytoremediation beyond its traditional application. He is now engaged to explore profitable phytoremediation with least risk, low input and minimum care. Dr. Pandey worked as a CSIR-Pool Scientist and DS Kothari Postdoctoral Fellow at Babasaheb Bhimrao Ambedkar University, Lucknow; Consultant at the Council of Science and Technology, Uttar Pradesh and DST-Young Scientist at CSIR-National Botanical Research Institute, Lucknow. He is the recipient of a number of awards / honors / fellowships and is a member of the National Academy of Sciences India. Dr. Pandey serves as a subject expert and panel member for the evaluation of research and professional activities in India and abroad for fostering nature sustainability. He has published over 100 scientific articles/book chapters in peer-reviewed journals/books. Dr. Pandey is also the author and editor of several books published by Elsevier, with several more forthcoming. He is associate editor of *Land Degradation and Development* (Wiley); editor of *Restoration Ecology* (Wiley); associate editor of *Environment, Development and Sustainability* (Springer); associate editor of *Ecological Processes* (Springer Nature); academic editor of *PLoS ONE* (PLoS); advisory board member of *Ambio* (Springer); editorial board member of *Environmental Management* (Springer); *Discover Sustainability* (Springer Nature) and *Bulletin of Environmental Contamination and Toxicology* (Springer). He also works/ worked as guest editor for many reputed journals.

Dr. Umesh Pankaj is presently working as a Teaching-cum-Research Associate in Microbiology at Rani Lakhsmi Bai Central Agricultural University, Jhansi, Uttar Pradesh, India. His doctoral degree was awarded in field of Agricultural Microbiology from Jawaharlal Nehru University, New Delhi, and research work carried out from CSIR-Central Institute of Medicinal and Aromatic Plants, Lucknow, India. His area of research is mainly focused on plant-microorganism interaction with special emphasis on the phytoremediation and restoration of salt-affected soil using halotolerant bio-inoculants like plant growth-promoting microorganisms and arbuscular mycorrhiza fungi. Currently, he is working on biofertilizers development and their use for sustainable development. He is also engaged in teaching for the courses at post-graduate and undergraduate levels. He has published 12 research articles in peer-reviewed international journals and 5 book chapters to his credit. He has received Young Microbiologist Award from Agro-Environmental Development Society, Uttar Pradesh. He is a life member of various scientific societies.

Contributors

Kanchan Yadav
CSIR–Institute of Himalayan Bioresource Technology
Palampur, India
and
Academy of Scientific and Innovative Research (AcSIR)
Ghaziabad, India

Kunal Singh
CSIR–Institute of Himalayan Bioresource Technology
Palampur, India
and
Academy of Scientific and Innovative Research (AcSIR)
Ghaziabad, India

Amrinder Kaur
Department of Plant Pathology
Punjab Agricultural University
Ludhiana, India

Ritu Rani
Department of Plant Pathology
Punjab Agricultural University
Ludhiana, India

Hari Kesh
Department of Genetics and Plant Breeding
CCS Haryana Agricultural University
Hisar, India

Savitha T.
Department of Microbiology
Tiruppur Kumaran College for Women
Tiruppur, India

Pooja Yadav
CSIR–Institute of Himalayan Bioresource Technology
Palampur, India
and
Academy of Scientific and Innovative Research (AcSIR)
Ghaziabad, India

Ashutosh Srivastava
Department of Basic Sciences
College of Agriculture
Rani Lakshmi Bai Central Agricultural University
Jhansi, India

Prashant Kaushik
Kikugawa Research Station
Shizuoka, Japan
and
Instituto de Conservación y Mejora de la Agrodiversidad
València, Spain

Deepamala Maji
Distant Research Centre
CSIR–National Botanical Research Institute
Lucknow, India

Suman Singh
Department of Botany
University of Lucknow
Lucknow, India

Ashutosh Awasthi
College of Agriculture Sciences
Teerthanker Mahaveer University
Moradabad, India

Abhishek Sharma
C.G. Bhakta Institute of Biotechnology
Uka Tarsadia University
Surat, India

Priyanka Sati
Graphic Era (Deemed to Be) University
Dehradun, India

Eshita Sharma
CSIR–Institute of Himalayan Bioresource Technology
Palampur, India

Ruchi Soni
Regional Centre of Organic Farming
Ghaziabad, India

Shweta Singh
Biosystems and Integrative Sciences Institute (BioISI)
Plant Functional Biology Center
University of Minho, Campus de Gualtar
Braga, Portugal

Ashraf Y. Z. Khalifa
Department of Biological Sciences
College of Science
King Faisal University
Riaydh, Kingdom of Saudi Arabia
and
Department of Botany and Microbiology Faculty of Science
University of Beni-Suef
Beni-Suef, Egypt

A. Sankaranarayanan
C.G. Bhakta Institute of Biotechnology
Uka Tarsadia University
Surat, India

Deepa Minakshi
Graphic Era (Deemed to Be) University
Dehradun, India

Praveen Dhyani
Department of Biotechnology
Kumaun University
Bhimtal, India

Lav Sharma
Centre for the Research and Technology of Agro-Environment and Biological Sciences
Universidade de Trás-os-Montes e Alto Douro
Vila Real, Portugal
and
Battelle UK Ltd.
Crop Protection Division
Langstone Technology Park
Havant, United Kingdom

Rupesh Kumar Singh
Department of Protection of Specific Crops
InnovPlant Protect Collaborative Laboratory
Estrada de Gil Vaz
Elvas, Portugal
and
Department of Biology
Centre of Molecular and Environmental Biology
University of Minho
Braga, Portugal

Krishan K. Verma
Key Laboratory of Sugarcane Biotechnology and Genetic Improvement
Guangxi Academy of Agricultural Sciences
Guangxi, China

Rakesh Yonzone
College of Agriculture (Extended Campus)
Uttar Banga Krishi Viswavidyalaya
Majhian, India

M. Soniya Devi
College of Agriculture
Rani Lakshmi Bai Central Agricultural University
Jhansi, India

Bandi Arpitha Shankar
Sardar Vallabhai Patel University of Agriculture and Technology
Meerut, India

Yumnam Bijilaxmi Devi
College of Horticulture and Forestry
Rani Lakshmi Bai Central Agricultural University
Jhansi, India

Vishnu D. Rajput
Academy of Biology and Biotechnology
Southern Federal University
Rostov-on-Don, Russia

Svetlana Sushkova
Academy of Biology and Biotechnology
Southern Federal University
Rostov-on-Don, Russia

Tatiana Minkina
Academy of Biology and Biotechnology
Southern Federal University
Rostov-on-Don, Russia

Pulak Bhaumik
Uttar Banga Krishi Viswavidyalaya
Majhian, India

Thounaojam Thomas Meetei
School of Agriculture
Lovely Professional University
Phagwara, India

Thorny Chanu Thounaojam
Department of Botany
Cotton University
Guwahati, India

Varucha Misra
ICAR–Indian Institute of Sugarcane Research
Lucknow, India

Santeshwari
ICAR–Indian Institute of Sugarcane Research
Lucknow, India

Sonika Pandey
ICAR–Indian Institute of Pulses Research
Kanpur, India

A.K. Mall
ICAR–Indian Institute of Sugarcane Research
Lucknow, India

Shubha Trivedi
College of Agriculture
Rani Lakshmi Bai Central Agricultural University
Jhansi, India

Mukesh Srivastava
Rani Lakshmi Bai Central Agricultural University
Jhansi, India

1 Role of Soil Microorganisms in Sustainable Crop Production

Amrinder Kaur and Ritu Rani
Punjab Agricultural University, Ludhiana, India

CONTENTS

DOI: 10.1201/9781003147077-1

1.1 INTRODUCTION

The microorganisms and their biosynthetic activities under different environmental conditions can be exploited in agriculture and other allied fields. They play an important role in agriculture by improving the soil quality, plant health and nutritional status (Barea et al. 2013a; Lugtenberg 2015) and have also been used in many other ways to enhance food processing, safety and quality, human and animal health, environmental protection, treatment of municipal wastes, etc. However, the efficacy of these microorganisms relies upon appropriate environmental conditions such as availability of nutrients, temperature, pH, oxygen, etc., for metabolising the substrates. Considering their usefulness in solving problems linked to the application of chemical pesticides and fertilisers, they are nowadays widely exploited in natural and organic farming (Higa 1994, 1995).

Soil consists of an indifferent kind of microorganisms that are responsible for the healthy growth and functioning of the plants. These communities are known as "soil microbial biomass", which are mainly involved in nutrient uptake and soil respiration (Buee et al. 2009; Raaijmakers et al. 2009). Growth and development of the crops are closely related to the type of the soil microflora, particularly those in close association to the rhizosphere. The beneficial rhizospheric microbial communities, such as plant growth-promoting fungi (PGPF), plant growth-promoting rhizobacteria (PGPR), cyanobacteria, etc., provide enormous benefits to agriculturally important crops through increased nutrient availability, increased disease suppression and tolerance to biotic and abiotic stresses, thus enhancing their productivity (Mishra et al. 2016). Nowadays, excessive use of pesticides and synthetic fertilisers, soil erosion, mixing of effluents into river waters, etc., has led to the degradation of soil health and contamination of surface and groundwater, which thus ultimately disturbs the ecological balance. Several researchers are currently trying to analyse the role of different root-associated microorganisms which are required to solve the issue related to both ecological and economic sustainability (Nautiyal et al. 2010; Chaudhary et al. 2012). Application of different types of biopesticides being developed due to comprehensive research related to these beneficial soil microorganisms (BSMs) will further help in fulfilling different nutritional requirements of the crops (Bashan et al. 2014; Malusa and Vassilev 2014; Owen et al. 2015).

In recent years, microbial ecologists and soil microbiologists have tried to differentiate the various soil microorganisms into two categories, namely, harmful or beneficial, on the basis of their functionality and interaction with the plants. However, due to the massive complication of these microbial entities in the soil on both micro- and macrolevels, it is a big task to exploit them for farming. Thus, detailed knowledge is needed for characterising various soil microbial diversities for better functioning and their ultimate impact on soil health management and ecological sustainability. Thus, in this chapter, we discuss the different types of BSMs, how they tend to improve soil and plant health, how they support the plants to overcome different biotic and abiotic stresses and what further future challenges need to be addressed.

1.2 TYPES OF SOIL MICROORGANISMS FOR SUSTAINABLE DEVELOPMENT AND ENVIRONMENT MANAGEMENT

Among the different types of living entities present in the world, active microbial populations existing in the soil are found to be most significant as they are directly enrolled in improving soil fertility and promote growth of plants by decreasing various abiotic and biotic stresses (Glick 2010). Blackwell (2011) stated that 1 gram of soil contains millions of fungal species and thousands of bacterial species, among which major proportions are considered to be advantageous for both soil and plants.

Complex soil matrices being formed by the BSMs tend to enhance plant growth, improve nutrient uptake and help in defence response by plants through exudations of various active metabolites. Additionally, they also improve tolerance to unfavourable environmental conditions like nutrient deficiency, drought, salt stress and heavy metal (HM) toxicity. Recently, scientists have found that these BSMs play an important role in the soil ecosystem as well for their roles in organic waste decomposition, in the alleviation of different soil stressors and in detoxifying toxic substances such as pesticides (Ma et al. 2016). Direct interaction of these microorganisms with the root system results in increase of mineral and nutrient uptake from the decayed soil organic matter, thus increasing the plant's growth (Nihorimbere et al. 2011). Some other groups of microorganisms that have harmful effects to plants by causing various diseases tend to reduce their productivity. On the other side, the presence of these beneficial microorganisms makes soil healthier, leading to the suppression of the growth of unhealthy soil microflora. Various kinds of beneficial microorganisms present in the soil are discussed as follows.

1.2.1 Cyanobacteria

Cyanobacteria, considered to be an important component of soil, are ubiquitous photosynthetic prokaryotes frequently present in wetlands, streams, ponds, lakes and rivers (Singh et al. 2016a, b). They were first observed in the rice fields by Fritsch (1907) and are considered to enrich the soil fertility of rice fields by fixing nitrogen and can act as a suitable natural source for enhancing soil fertility (Song et al. 2005). Symbiotic or free-living blue-green algae have a good historical record in sustainable agriculture. As per the report by Smil (1999), symbiotic cyanobacteria (water fern, *Azolla*) or free-living cyanobacteria fix 4–6 billion kg of nitrogen annually. Effective nitrogen-fixing cyanobacteria such as *Anabaena*, *Nostoc*, *Calothrix*, *Aulosira* spp. have been identified in diverse agro-ecological regions and are effectively applied in rice fields to enhance their production (Prasad and Prasad 2001), and apart from rice, their use is also being exploited in other cropping systems.

Despite their association with global nitrogen fixation, studies also suggest that microorganisms are also involved in phytohormone production in symbiotic and free-living associations. Different species of *Calothrix*, *Anabaena*, *Nostoc*, *Chlorogloeopsis*, *Plectonema*, *Cylindrospermum* and *Anabaenopsis* have been reported to be involved in indole acetic acid (IAA) production in the rhizospheric region of the soil in paddy and wheat fields (Karthikeyan et al. 2007; Manjunath et al. 2011) and also enhance the integrity of the soil through secretion of various types of biomolecules (Kaushik 2014; Rosa and Philippis 2015). Due to their ability to grow in saline soils, cyanobacteria tend to increase the fertility of such soils where most of the other plant species are unable to grow (Singh and Dhar 2010).

1.2.2 Plant Growth-Promoting Rhizobacteria

Plant growth can be promoted by different types of beneficial microorganisms (Mishra and Arora 2012; Ahemad and Kibret 2014; Goswami et al. 2016). PGPRs such as *Azotobacter, Pseudomonas, Bacillus, Azospirillum, Rhizobium, Arthrobacter*, are some of the bacterial entities that have the capability of enhancing plant growth and soil fertility (Cheng 2009; Sharan and Nehra 2011; Mishra and Arora 2012; Arora 2015). Growth promotion by these PGPRs involves siderophore production, nitrogen fixation, phytohormone production and solubilisation of inorganic substances such as Zn, P, K (Beneduzi et al. 2012), making them easily availability to plants and indirectly benefiting the plants by inhibiting the growth of various plant pathogens by one of numerous mechanisms, such as production of anti-fungal metabolites or antibiotics, depletion of iron from the rhizospheric soil through siderophore production, induced systemic resistance (ISR) and production of chitinase responsible for cell wall degradation (Vejan et al. 2016).

In addition, PGPRs are also considered to act as potential microorganisms by providing protection to the plants from various environmental stresses (Khare and Arora 2011; Kang et al. 2014). Earlier, the role of PGPRs was discovered for the enhancement of crop yield, but recent studies have also proved them to play a major role in the proper functioning of the agro-ecological systems (Ahemad and Kibret 2014; Arora and Mishra 2016; Choudhary and Agrawal 2017). Apart from this, PGPRs are also reported to be helpful in the restoration of degraded land, improving the soil quality, reducing environmental contaminants and participating in climate change mitigation.

1.2.3 Plant Growth-Promoting Fungi

A wide variety of pathogenic and beneficial fungi are present in the soil, which directly or indirectly affect the plant growth at different stages. It is estimated that out of 1.5 million fungal species which are known to be present in nature, only 5–10% have been formally described (Laughlin and Stevens 2012). According to Robertson and Groffman (2007), majority of the soil microbial communities are constituted by fungi and are considered to be the major contributors to soil nutrient cycling processes including nitrogen mineralisation, immobilisation and transformation (Selim et al. 2012). Plant-fungi interaction starts when roots release some specific biomolecules or chemical compounds that act as nutrient sources for PGPF and thus provide defence against pathogenic microorganisms. For example, exudations from the root cortical and epidermal cells stimulate the proliferation of fungi internally and externally of the root, whereas growth of entomopathogenic fungi is inhibited by the release of some phenolic compounds from these roots (Selim et al. 2012). Different fungal species such as *Aspergillus* spp., *Candida* spp., *Paecilomyces* spp.,

Alternaria spp., *Aureobasidium* spp., *Cladosporium* spp., *Phoma* spp., *Penicillium* spp. and *Sporobolomyces* spp. are present in soil, which are considered to be agriculturally important and the main focus of considerable research over many years.

Another beneficial fungi found in soil is mycorrhizal fungi, well known to establish a symbiotic relationship with the roots of more than 80% of terrestrial plants (Gaur and Kaushik, 2011; Marcel et al. 2015; Pankaj et al. 2017). Advantageous properties of mycorrhizae are amenable to natural ecosystem where intervention activities are very less (Lehman et al. 2015). Two types of mycorrhizae, ectomycorrhizae and endomycorrhizae, are important for sustainable agriculture (Smith and Read 2008). Arbuscular mycorrhiza (AM), one of the endomycorrhiza formerly known as vesicular arbuscular mycorrhiza (VAM), is also known to exhibit an obligate interaction with plant's root system (Presto et al. 2013). According to Parniske (2008), plants can enhance 4–20% of their photosynthate to support AM fungi, which is equivalent to the utilisation of around 5 billion tonnes of carbon/year by AM fungi.

The symbiotic association by AM fungi with their host plants involves spore formation in the root zone vicinity that is capable of germination under harsh environmental conditions and also helps in alleviating abiotic stresses in the plant (Becerra et al. 2014; Pankaj et al. 2021). They are also known to play a significant role in soil structuring by formation of a network of extraradical hyphae which holds soil particles intact with the host roots. Enhanced phosphorus uptake by the host plants is also considered to be the most significant effect of AM fungi (Aggarwal et al. 2011). They also interrelate with phosphate-solubilising bacteria (PSB) and nitrogen-fixing bacteria which synergistically assist in the growth and development of the plants. The dual inoculation treatment (bacteria and arbuscular mycorrhizal fungi [AMF]) has significantly improved the plant weight and nitrogen as well as phosphorus uptake in plant tissues (Puppi et al. 1994; Bianciotto et al. 2000; Pankaj et al. 2019a). Ample literature is also available on the tripartite symbiosis of mycorrhiza-legume-*Rhizobium* (Barea 2000) and their role in the improvement of nodulation and N_2 fixation (Nair et al. 1991; Nadeem et al. 2014).

1.3 MECHANISMS OF PLANT GROWTH PROMOTION BY BENEFICIAL SOIL MICROORGANISMS

Soil beneficial microorganisms' role is to promote plant growth through various methods such as siderophore production, mineral solubilisations, bioremediation of HM salts, etc. Among BSM, PGPRs directly promote plant growth through nitrogen fixation, phosphorous and potassium solubilisation as well as the production of siderophores and phytohormones (Castro et al. 2009; Pankaj et al. 2020; Figure 1.1). All of these BSMs help the plants to grow and maintain soil fertility. Indirect methods are the supplementary provision provided by these BSMs that help maintain a healthy plant rhizosphere but do not directly affect soil fertility. This includes the production of antibiotics, polysaccharides, various hydrolytic enzymes and cyanide compounds that lead to ISR (Figure 1.1).

Plant growth can be directly encouraged by PGPR by making easy accessibility of N, P, Fe, etc., through nutrient mineralisation and solubilisation along with greater levels of phytohormones like indol-3-acetic acid (IAA) (Barazani and

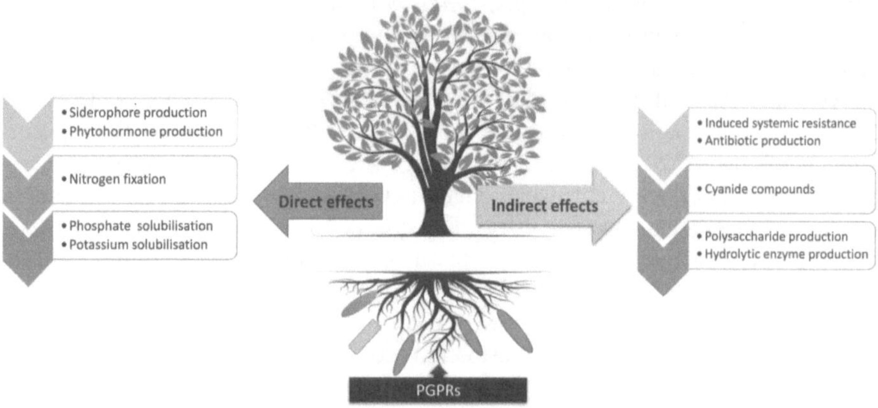

FIGURE 1.1 Mechanisms of growth promotion by PGPRs.

Friedman 1999). Considering the environmental safety, PGPRs are also substantial disease suppressors through the production of chemicals such as pyoluteorinare and 2,4-diacetylphloroglucinol (2,4-DAPG) (Mazurier et al. 2009; Beneduzi et al. 2012), and offer resistance to plants against various plant pathogens and insect pests, through initiation of systemic acquired resistance or ISR (Pieterse et al. 2009) by activation of certain defence-related enzymes such as lipoxygenase (LOX), peroxidase (PO), phenylalanine ammonia lyase (PAL), chitinase, catalase (CAT), β-1, 3-glucanase, superoxide dismutase, polyphenol oxidase and ascorbate peroxidase (Saha et al. 2016; Sharma et al. 2016; Shrivastava et al. 2016; Sindhu et al. 2016).

1.3.1 Mineral Solubilisation by Soil Microorganisms

Various soil microorganisms are known to mobilise the inorganic forms of minerals and make them accessible to plants. PSB solubilise inorganic phosphates of Ca, Fe and Al in the soil by producing siderophores, organic acids, hydroxyl and carboxyl groups, chelating them to the bound phosphates and the existing calcium (Sharma et al. 2013). Other isolates of *Enterobacter* spp., *Pantoea* spp. and *Klebsiella* spp. are known to dissolve more calcium phosphate than iron and aluminium phosphates (Chung et al. 2005). Therefore, the role of microorganisms is important in the circulation of soil phosphorus by transporting to the plant (Richardson and Simpson 2011). The accessibility of potassium in the soil, though adequate in India, still needs much in the form of K fertiliser. It is worth mentioning here that the use of available natural resources such as K-feldspar, muscovite, biotite, phlogopite and sand and waste mica can be very helpful (Nishanth and Biswas 2008) and bio-interventions help increase K lake. But at the same time, partial access to potassium-containing minerals interrupts their function as K fertilisers in agriculture. Potassium-solubilising bacteria, potassium-dissolving bacteria, and potassium-solubilising rhizobacteria have important K solubilisation potential. This soil type of bacteria mainly includes *Bacillus circulans*, *Pseudomonas* sp., *Acidithiobacillus ferroxidans*, *Bacillus mucilaginosus*,

Burkholderia spp. and *Paenibacillus* spp. (Saiyad et al. 2015; Verma et al. 2015). It is also conveyed by several workers that potassium-solubilising microorganisms as bio-inoculants play valuable roles in agricultural sustainability (Meena et al. 2013a, b, 2015; Zhang and Kong 2014; Singh et al. 2015, 2016a, b).

1.3.2 Production of Siderophores

Iron is one of the most important minerals needed for the preservation of all life. In its oxidised state, Fe_{31} is a highly ionic form, capable of forming insoluble oxyhydroxides and hydroxides, resulting in its inaccessibility to plants and microorganisms while the low-pH Fe_{21} ionic form is freely available and plants can easily uptake it (Rajkumar et al. 2009). The absorption of iron by bacteria and fungi is due to the occurrence of siderophores, which have certain characteristics and the ability to chew on iron. Siderophores are amino acids (non-proteins) less than 1000 Da and are associated with rhizospheric microorganisms (Krewulak and Vogel 2008; Lemanceau et al. 2009). Similarly, plants contain phytosiderophores, which are essential for finding micronutrients in the rhizosphere. Bacteria such as *Burkholderia, Grimontella, Pseudomonas* and *Enterobacter* have high siderophore production (Dimkpa et al. 2009).

1.3.3 Bioremediation for Plant Growth Promotion

Bioremediation refers to the use of biological agents to remove impurities, contaminants and toxins from soil, water and other areas. The success of phytoremediation of any soil depends largely on the level of pollution and the contamination factors. In the surrounding environment, plants with a high concentration of iron usually grow at the lowest rate in the presence of rhizoshperic metal concentration.

The method of extracting HMs from contaminated soil is still being questioned, although extensive research work has been done. The essential conditions required for such hyperaccumulator plants are high levels of biomass growth that can withstand high-stress conditions (Nie et al. 2002). The microbial communities of the soil are affected by the contamination of HMs in many ways: First, the reduction of microbial biomass (Brookes and McGrath 1984; Fliessbach et al. 1994); second, by reducing the microflora living in certain soils (Chaudri et al., 1993); and lastly, with the prevention of the entire microbial community, which may result in a negative soil profile (Sandaa et al. 1999; Gray and Smith 2005). Organic remedies are considered the most suitable for the removal of toxic metals, due to their low cost and eco-friendly environment (Ali et al. 2013; Ullah et al. 2015). The use of BSM, especially PGPR, for soil preparation under climate change and overuse of fertilisers is beneficial (Nautiyal et al. 2013; Tiwari et al. 2016). Plant growth and high survival rates under stressful conditions are common in BSMs, demonstrating their ability to adapt by converting complex substances into simple and non-toxic compounds.

The release of the industrial polluted waste/wastewater into water sources has led to HM contamination of the soil and is today the nastiest environmental threat with thoughtful impressions on human health and agriculture. Excessive use of nitrogen, phosphorus and potassium (NPK) fertilisers as well as unnecessary use of several agrochemicals (herbicides, pesticides, fungicides, etc.) has exacerbated the situation (Saberi and Hassan 2014), so treatment of such metal-pressed soils has become a major concern for agriculture. Among HM contaminants, lead and cadmium are the most abundant and have the highest potential for toxicity. The resistance to toxic HMs makes PGPR an appropriate bioremediating agent (Mustapha and Holimoon 2015; Tiwari and Lata 2018) and promotes metal chelation by lowering soil pH due to its ability to produce non-toxic organic acids (Mishra et al. 2017).

1.3.3.1 Bioremediation by Bacteria

Among the various rhizosphere microorganisms, important microflora-forming bacteria are responsible for reducing HM toxicity (Hassan et al. 2017) and promoting plant growth and productivity through the secretion of growth regulators that help plants absorb nutrients(Nadeem et al. 2014). The formation of metal complexes from siderophore complexes, metabolites and transporter proteins from bacteria limits metal contamination (Ahemad 2012). These BSMs have special metabolic functions to reduce metal stress primarily through cation of metal-ion attachments with food transport, cell wall adhesion and accretion across the cytoplasmic membrane and other organs, thus making them of major importance in the field of agricultural science (Ullah et al. 2015). Such evidence has already been observed, in which the removal of toxic metals results in the formation of a completely new species through the newest genetic modulation technique (Ullah et al. 2015). Some methods of gene insertion to obtain over-expression or mutation of the target gene can serve as a new alternative to obtain the desired microorganisms with high throughput efficiency of metal sequestration, translocation, etc. (Singh et al. 2011). Some of the important examples of bio-remediation of HMs by PGPR are listed in Table 1.1.

1.3.3.2 Bioremediation by Fungi

Mycorrhizal fungi are considered important migrants of various soil microorganisms. Root-colonising mycorrhizae have the potential to improve the absorption of various important elements along with HM ions (Kumar et al. 2015; Shukla et al. 2015; Singh et al. 2019). Soil metal bioavailability can be rapidly altered by bacteria. The accumulation of HMs at different environmental trophic levels can be reduced by developing mycorrhiza strategies in the presence of mixed systems of microbial culture (Kramer 2005).

The association of AMF (AM or VAM fungi) in plants has been found to promote their growth – especially due to their extensive mycelium mats that phosphorus not only receives nutrients and provides them to plants but also binds them to the soil. Various studies have shown that VAM fungi can effectively prevent HM stress (Meherg 2003; Kumar et al. 2015). Therefore, VAM-mediated nutrient effects can be attributed to HM displacement as well as HM removal from the nutrient pool available at the soil and plant interface.

Filamentous fungi such as *Trichoderma*, *Penicillium* and *Aspergillus* can tolerate metallic stress in soil (Ezzouhri et al. 2009; Oladipo et al. 2017). The fungal cell wall plays a significant role in metal chelation that has been attributed to negatively charged functional groups such as carboxylic, sulfhydryl and phosphate (Ong et al. 2017). Effective reduction of toxic metal concentrations in soil with the help of

TABLE 1.1
Bioremediation of Heavy Metals by PGPR

Metal	PGPR	Host Plant	Results Obtained	References
Rhizobium sp.	Chromium VI	*Pisum sativum*	Decreased the chromium toxicity and improved the nitrogen concentration in plants under glasshouse conditions	Soni et al. (2014)
Brevundimonas diminuta, Alcaligenes faecalis	Mercury	*Scripus mucronatus*	Decreased toxicity in the soil and increased phytoremediation under greenhouse conditions	
Bradyrhizobium japonicum CB1809	Arsenic	*Helianthus annuus, Tritucuma estivum*	Decreases the arsenic toxicity among the plants under pot house studies	Yavar et al. (2014)
Bacillus megatarium	Lead	*Brassica napus*	Decreased the soil contamination by this metal and improves the yield among plants under field conditions	Reichman (2014)
Bacillus, Staphylococcus, Aerococcus	Chromium, cadmium, copper, zinc and lead	*Prosopis juliflora, Lolium multiforum*	Improves phytoremediation under greenhouse conditions	Wani and Khan (2012)

filamentous spore-forming fungus *Aspergillus niger* has been reported by Corenõ-Alonso et al. (2014) and *Trichoderma* sp. and in chickpea plants by Tripathi et al. (2013, 2017). On the other hand, AMFs also form important microflora in soils and connect soil and plant roots, providing nutrients by increasing root surface area as well as increasing soil binding (Saxena et al. 2017).

1.4 ABILITY OF SOIL MICROORGANISMS FOR STRESS ALLEVIATION IN AGRICULTURALLY IMPORTANT CROPS

A variety of stress factors, such as salinity, drought, nutrient deficiencies, pollution, diseases and pests, can alter plant-microbial interactions in the rhizosphere. Recent research shows that plant perception of environmental stress signals stimulates the activation of signalling molecules, phytohormones, which play an important role. Since plants are subjected to multiple stresses simultaneously, an appropriate meta-analysis reveals a complex regulation of plant growth and immunity. Understanding how phytohormones interact in signalling networks is fundamental to unravelling how plant-microbiome systems grow and survive in stressful environments (Pozo et al. 2015). The mechanisms involved in plant-microbial interactions under stress conditions are not properly understood. However, ongoing research is providing evidence that various changes in plant morphology, physiology and root exudation profiles may induce the plant to recruit stress-reducing microorganisms responsible for the crop and also help to increase productivity (Zolla et al. 2013; Damodaran et al. 2019). Since these various stress factors can have detrimental effects on the productivity of agricultural systems, the role of rhizosphere microorganisms in plant growth under adverse conditions has been discussed by Barea et al. (2013b) and Pankaj et al. (2020).

1.4.1 Improving the Ability of Soil Microorganisms for Biological Control of Plant Pathogens

Plant root endophytes and rhizospheric microorganisms such as PGPM, *Trichoderma* spp. and AMF protect the plants from pathogens by inducing space and nutrients, antibiotics (for PGPR and Trichoderma), mycoparasitism (for *Trichoderma*) and plant defence mechanisms (Barea et al. 2013a). Defence priming is a precondition of immunity induced by the migration of various microorganisms, which acts as an effective defence against various plant pathogens. It also works systematically on the distal parts of shoots and roots to produce systemic resistance (ISR) against many leaf and root pathogens in plants (Selosse et al. 2014).

The symbiotic relationship of AM fungi with plants is a well-known example of the protection of host plants from harmful organisms, including microbial pathogens, herbivorous insects and other parasitic plants. Jasmonic acid (JA) plays an important role where AM colonisation prioritises plant immunity by enhancing plant ability to respond to pathogenic attack. The ISR in AM symbiosis is also known as mycorrhiza-induced resistance (MIR). The identification of defence control precursors for AM development and MIR integration is a major challenge for research that facilitates the development of biotechnological strategies to improve the use of AM fungi in integrated insect pest and disease management programmes (Jung et al. 2012; Pozo et al. 2013).

In addition to AM fungi, PGPR, *Trichoderma* spp. and species of non-pathogenic *Fusarium* spp., local resistance and ISR are also prominent through the production of microbial-associated molecular models (MAMPs) that activate immune responses. The priming effects of MAMPs depend on the activation of the JA signalling pathway, which regulates the protection of inactivated plants (Pozo et al. 2015) and is thought to be the key hormone to switch from plant growth to protective responses (Pangesti et al. 2013). Since the impact of these biological control methods is influenced by harsh environmental conditions, research on biocontrol needs to focus on the challenge of finding appropriate screening methods for the selection of microorganisms living in changing climatic conditions. Understanding the impact of the environment on the performance of biocontrol agents (BCAs) can help evaluate the resulting product and develop effective combinations of anti-microbial. A major challenge in rhizosphere biotechnology is to discover the increased potential for biological safety through a combination of mechanisms that each organism uses individually.

Unlike the various pests and diseases that attack farm crops, the damage caused to plants by various parasitic weeds cannot be avoided. Parasitic weeds are difficult to control because most of their life cycle takes place underground. A new class of plant hormones called strigolactones stimulates root germination and plant seed germination. Furthermore, these hormones have also been shown to be involved in root colonisation by AM fungi. After AM colonisation, plants reduce the production of strigolactones, which reduces germination and infection by various parasite plant seeds, and consequently, these weeds have detrimental effects on plant fitness and yield (López-Ráez et al. 2012). Several workers have discussed the potential for AM symbiosis in control based on the production of strigolactones (Jung et al. 2012; Pozo et al. 2013).

1.4.2 Improving the Ability of Soil Microorganisms to Overcome the Deleterious Effect of Different Osmotic Stressors

The increased aridity level in many parts of the world has led to drought and salinity problems. To cope with such osmotic pressure, plants need to develop a number of adaptive mechanisms, such as water efficiency and fine control of the transpiration rate, to overcome the high production of reactive oxygen species (ROS) caused by stress conditions. Maintaining these two simple mechanisms means that water and ROS balance can be achieved by establishing AM symbiosis and immunisation with PGPR, both of which are as diverse as cell osmoregulation, antioxidant defence, ionic homeostasis, and control and redistribution of root water. Such microbial activity contributes to better regulation of plant water quality and contributes to better plant resistance to osmotic pressure conditions (Bal et al. 2013; Calvo-Polanco et al. 2013; Barzana et al. 2014; Pankaj et al. 2019a, 2020). Finally, the improved water-holding capacity of plants planted with these BSMs allows them to increase transpiration rates and have higher photosynthesis rates in water-deficient conditions.

Nowadays, the focus is mainly on the role played by AMF and other rhizosphere microorganisms in improving plant water conditions based on root hydraulic conduction improvement, which ultimately depends on the performance of aquaporin (Aroca et al., 2012; Calvo-Polanco et al. 2013; Barzana et al. 2014). Aquaporins are membrane endogenous proteins that allow water and other small neutral molecules to pass through biofilms following the osmotic gradient (Chaumont and Tyerman, 2014; Li et al. 2014). BSM vaccination improves plant nutritional status with the release of volatile organic compounds (VOCs) by certain microorganisms and modulates the activity of root aquaporins and thereby root hydraulic circulation (Groppa et al. 2012; Barzana et al. 2014). Since it is clear that AM fungi and other rhizosphere microorganisms can exhibit resistance/tolerance to osmotic pressures, further studies are needed to transfer this knowledge to natural ecology. This is fundamental because soil and rhizosphere microorganisms are important factors for plant survival in changing climates, where plants are likely to be exposed to adverse conditions in the coming years.

1.4.3 Phytoremediation of Contaminated Soils Using Soil Microorganisms

Plant-related microorganisms, such as AMF and bacteria, are capable of treating HMs, organic xenobiotics including VOCs, oil-derived alkanes or polycyclic aromatic hydrocarbon-contaminated (phytoremediation) environments. Presently, microorganisms with suitable host plants are extensively used for the efficient phytoremediation of toxic chemical compounds by removing, stabilising, volatilisation, reducing and/or decomposing pollutants in the environment (Azcón et al. 2013;

Singh et al. 2019, 2020). For successful phytoremediation, plants and their associated soil microorganisms must be able to thrive in polluted environments (Pongrac et al. 2013; Singh et al. 2020).

Much research is being conducted on this topic for HMs scavenging or biological xenobiotic degradation (Azcón et al. 2013). Mechanisms associated with the role of plant-associated bacteria in the phytoremediation of environments contaminated with HM or organic xenobiotics, in general and in alkanes, have been studied specifically by Germaine et al. (2013) and Afzal et al. (2013), which mainly includes improved plant growth, nutrient (P and N) supply, plant hormone production and Fe-binding siderophore production, increased ACC-deaminase activity (ethylene depletion), and biological xenobiotic depletion. Another important mechanism for improving phytoremediation is the bio-enrichment of plant-related microbial communities based on horizontal gene transfer, which will be a challenge for future research (Germaine et al. 2013). This mechanism is mainly dependent on the immune genes involved in HM bioremediation processes and is present in more transliterated plasmids in bacterial communities (Afzal et al. 2013; Germaine et al. 2013).

AM fungi control the entry of toxic metal species and cause metal ingestion and/or increased flux, metal stabilisation, etc. (Ferrol et al. 2009; González-Guerrero et al. 2009, 2010). It has also been reported that plants with mycorrhizal associations have the ability to fight the oxidative stress produced by HMs or to correct oxidative damage, which may be related to extensive changes in genetic expression and symbiotic-induced protein synthesis. Glomalin-related soil proteins produced by AM fungi can sequester HM (Cornejo et al. 2008; Singh et al. 2019) irreversibly, thereby contributing to metal stabilisation in the soil. Interactions between HM-adapted rhizobacteria and AM fungi have been investigated in various experiments by Medina and Azcon (2010). They mainly observed that the target bacteria accumulated large amounts of HM; co-vaccination enhances plant establishment and growth and reduces HM concentrations in plants and increases the enzymatic activity of HM-friendly bacteria and plant hormone production in the mycorrhizosphere. Automated AM fungicides and PGPR vaccination, combined with improved phytoextraction to purify agro-waste residues, altered bacterial community composition, and HM contaminated soils, but these results are interchangeable under field conditions.

1.5 APPLICATION OF RHIZOSPHERIC MICROORGANISMS FOR ENHANCING FOOD GRAIN PRODUCTION WITH INCREASED CROP YIELDS

The applications of different soil microorganisms for increasing production of crops yields through several ways are discussed below.

1.5.1 Biological Control Agents

Different soil microorganisms such as PGPF and PGPR present in rhizosphere can act as potential biocontrol agents against different pests and osmotic stresses. With an advantage of having the same ecological niche as of the pathogen, PGPRs can promote growth of plants directly or indirectly through many ways in different crops (Kloepper et al. 1980; Hallman et al. 1997). Several mechanisms of biological disease control are exhibited by these PGPR, among which the important ones are production of metabolites such as cell wall degrading enzymes, antibiotics, siderophores and hydrogen cyanide (HCN) (Kloepper et al. 1980; Weller 1988) and competition which directly affects the pathogen.

Various PGPRs such as *Rhizobium leguminosarum*, *Pseudomonas monteilii*, *Bacillus subtilis*, *Azototobater*, *Pseudomonas fluorescens*, *Bradyrrhizobium japonicum*, *Bacillus coagulans* and *Paenibacillus brasilensis* are used as potential BCAs in different crops, and the most important point is to focus on the commercialisation of these biocontrol agents. For this, the important points to be taken care are that the formulation should be economical with good shelf life and must be in a suitable form for storage, shipping and application. Risk assessment to the environment and human health are prior areas of concern before the release of new formulations, and in the early screening process, the microorganisms having good antagonistic potential but are able to grow at human body temperature should be discarded (National Academy of Sciences 1989).

1.5.2 Biofertilisers

Biofertilisers are those substances containing living microorganisms that are capable of colonising the root zone area of plants and thus tends to increase the growth and yield of such plants through increased nutrient uptake called biofertiliser (Bhattacharjee and Dey 2014; Pankaj 2020). Today, they are being applied to the agricultural fields as a replacement to conventional fertilisers. Different beneficial microorganisms have been used as biofertilisers in various crops (Table 1.2).

In agriculture, biofertilisers are gaining importance due to their ability to minimise environmental pollution and maintaining soil health by lowering the use of chemicals pesticides (Saeed et al. 2015). PGPRs, known for their quality of N_2 fixing, potassium and phosphate solubilisers, are recommended as a sustainable solution to improve plant-nutrient uptake and production of the crop (Bhattacharjee and Dey 2014; Pankaj et al. 2019b). According to an estimate, cyanobacteria in symbiotic association contribute 7–85 kg N_2/ha/year, whereas its free-living form can fix 15 kg N_2/ha/year (Elkan 1992). It has been observed that cereal crops can obtain up to 30% of their required nitrogen dose from biological nitrogen fixation when fertilised with higher amounts of potassium and phosphorus as well as with other microelements (Pedraza 2008; Mmbaga et al. 2014). Rhizosphere-associated *Rhizobacteria* can fix nitrogen as well as solubilised phosphorus has been used as inoculum in legume crops (Marks et al. 2007) and non-leguminous crops such as wheat, rice, maize and sugarcane (Dobereiner 1997). Many countries are promoting the use of rhizobial inoculants for nitrogen uptake, for example, in Brazil, application of rhizobial inoculants in soybean crop provides N_2 with a value equivalent to US$7 billion for fertilisers (Hungria et al. 2007). Rennie and Hynes (1997) stated that Canada has a substantial market of *Rhizobium* inoculant products.

Azospirillum is the most extensively studied genus among non-symbiotic N_2-fixing bacteria. It not only increases the nitrogen content of the plants but also

TABLE 1.2
List of Important Beneficial Soil Microorganisms Being Used as Biofertilisers

Type of Biofertiliser	BSM	Characteristics	References
Phosphate solubilisers	*Pseudomonas* spp., *Burkholderia* spp., *Rhizobium* spp. *Bacillus* spp. and mycorrhiza	• They help in uptake of the insoluble phosphate to the plants in different ways	Duarah et al. (2011)
Nitrogen fixers	*Rhizobium* spp.	• It is root-nodule-forming bacteria which fix atmospheric nitrogen in symbiotic association with roots of leguminaceous plants	Baset Mia and Shamsuddin (2010)
	Bradyrhizobium spp.	• Its application increases the P, S and N content and improves seed and straw yield of soybean under field conditions	Chibeba et al. (2015), Raja and Takankhar (2018)
	Azotobacter spp.	• It is a fast-growing bacteria and helps in nitrogen fixation	Siddiqui et al. (2014)
	Cyanobacteria	• The presence of this bacteria in rice fields helps in fast uptake of nitrogen	Kaushik (2014)
	Azolla spp.	• This water fern has asymbiotic relation with blue and green algae and can help in cultivation of rice or other crops	Carrapico et al.(2000)
Zinc solubilisers	*Bacillus* and *Pseudomonas* spp.	• Solubilises various zinc compounds such as zinc carbonate ($ZnCO_3$), zinc oxide (ZnO) and zinc sulphide (ZnS)	Ahemad and Kibret (2014)
	Funneliformis mosseae and *Rhizophagus intraradices*	• Increases Zn uptake both under pot and field conditions and thus leads to enhanced growth of the plants in lettuce	Konieczny and Kowalaska (2016)

(Continued)

TABLE 1.2 *(Continued)*
List of Important Beneficial Soil Microorganisms Being Used as Biofertilisers

Type of Biofertiliser	BSM	Characteristics	References
Potassium solubilisers	*Bacillus* spp., *Pseudomonas* spp. and *Burkholderia* spp.	• Solubilise potassium, present in the insoluble form of rocks and silicate minerals	Shanware et al. (2014)
Phytohormone producers	*Bacillus* spp., *Pseudomonas* spp., *Rhizobium* spp. *Burkholderia* spp. and mycorrhiza	• Phytohormones such as gibberellins, auxins and cytokinins produced by these microorganisms enhance proliferation of the cells in the root area of the plants through overproduction of lateral roots and root hairs and thus subsequently increase water and nutrient uptake	Ahemad and Kibret (2014)
	Rhizobium daejeonense, *Acinetobacter calcoaceticus* and *Pseudomonas mosselii*	• Application of these microorganisms significantly increased plant growth through nutrient solubilisation and phytohormones production under pot/field experiments	Torre-Ruiz et al. (2016)

improves its growth by phytohormones production such as gibberellins, cytokinins and auxins (Steenhoudt and Vanderleyden 2000). Since long AM fungi and PSB are reported to solubilise insoluble phosphates and thus help in promoting yield in various crops (Fernandez et al. 2007; Shahab et al. 2009), various studies indicate that application of rock phosphate in conjunction with PSB could lower the cost of applying chemical fertilisers by 50% (Yazdani et al. 2009). Many research papers have also demonstrated that application of PSB-based biofertilisers increases P content in rice (Sarkar et al. 2012), wheat (Sarkar et al. 2014), sugarcane (Sundara et al. 2002), mung bean (Vikram and Hamzehzarghani 2008) and maize (Oliveira et al. 2009). PGPRs are also reported to have good iron uptake capabilities under iron stress conditions through siderophore production and enhance plant productivity. Siderophores are small, iron-chelating compounds produced by different microorganisms (Miller 2008;Saha et al. 2015). Microorganisms release siderophores to absorb iron from these mineral phases by the formation of soluble Fe^{3+} complexes that can be taken up by active transport mechanisms (Raymond et al. 2003). Biofertilisers for solubilising "nonexchangeable" K are also in practice in some agro-ecosystems (Sangeeth and Suseela 2015).

Zinc (Zn) is also regarded as an essential micronutrient in plants (Sauchelli 1969) but only small amount of it is available to plant, as most of it is converted to different chemical forms. Various studies have indicated that the release of fixed and insoluble forms of Zn by zinc solubilising bacteria is an important aspect of enhancing Zn availability in the soil through production and excretion of organic acids (Bhupinder et al. 2005). PGPF have been extensively studied for solubilisation of insoluble zinc compounds both under *in vitro* and *in vivo* conditions. However, some genera of PGPR such as *Pseudomonas*, *Acinetobacter*, *Bacillus* and *Gluconacetobacter* have also been reported to solubilise insoluble zinc.

1.5.3 Biopesticides

Different pests (invertebrates, plant pathogens and weeds) are responsible for lowering about 50% of the potential global crop yield before it is harvested and another 20% is destroyed in post-harvest handling (Bailey et al. 2010). However, for sustainable agriculture, use of chemical-based pesticides should be relinquished because of their hazardous impact on the soil health and environment. Workers all around the world are now more focussed on the replacement of the excessive use of chemical pesticides in agriculture with the eco-friendly biopesticides. Biopesticides act like other chemical pesticides but originally derived from natural resources such as animals, plants, bacteria and minerals (USEPA 2016). In India, a key technological revolution in the field of biocontrol occurred when synthetic insecticides failed to control *Helicoverpa armigera*, *Spodoptera litura* and other pests of cotton (Kranthi et al. 2002). Biocontrol is the process that is safe, cost-effective and eco-friendly to control the widespread resistance of chemical insecticides towards pest insects. Later, biopesticides became a part of integrated pest management which was earlier completely based on the use of chemical pesticides (Mishra et al. 2020). Microbial pesticides are also known as BCAs. Microorganisms as biopesticides offer the advantage of higher selectivity and lower or no toxicity in comparison to the conventional chemical pesticides (MacGregor 2006). BCAs can be grouped into three broad categories, namely bacterial, fungal and viral. Strains of *Bacillus thuringiensis* (Bt) are the most widely used bacterial pesticides accounting for 90% of the biopesticide market in the United States (Chattopadhyay et al. 2004). Several species of *Pseudomonas* are also reported to exhibit biocontrol potential against various phytopathogens (Arora et al. 2008; Kang et al. 2014). For example, biopesticides containing *Pseudomonas fluorescens* and *Pseudomonas syringae* have been used at large scale now (Arora 2015).

Among the fungal biopesticides, *Trichoderma harzianum* exhibit strong antagonist behaviour against several soilborne fungi such as *Rhizoctonia*, *Pythium*, *Fusarium* and other phytopathogens (Hartmann et al. 2008). *Baevuria bassiana* and *Metarhizium anisopliae* are naturally occurring entomopathogenic fungi considered as good BCA and infect-sucking pests including *Nezara viridula* L. (green vegetable bug) and *Creontiades* sp. (green and brown mirids) (Sosa-Gomez and Moscardi, 1998). Baculovirus is the main viral BCA that is commercially used for designing phage pesticides. Some major biopesticides used in agro-ecosystems are mentioned in Table 1.3.

1.5.4 Bioherbicides

Bioherbicides are used to control different herbs/weeds which are harmful to crops, in an eco-friendly manner (Harding and Raizada 2015). Mainly fungi-based bioherbicides have been explored more and the potential of bacteria is has not been explored

TABLE 1.3
List of Some Important Biopesticides Being Used in Agroecosystems

Target Organism	BSM	Commercial Name
Plant growth-promoting rhizobacteria (PGPR)		
Lepidopteran pests	*B. thuringiensis* subsp. *aizawai*	Turex
	B. thuringiensis subsp. *israelensis*	VectoBac
	B. thuringiensis subsp. *kurstaki*	Dipel WP, Batik
Fusarium, *Pythium* and *Rhizoctonia*	*Bacillus subtilis*	Kodiak
Post-harvest diseases	*Pseudomonas syringae*	Bio-Save 10LP
Turf fungal diseases	*Pseudomonas aureofaciens*	Spot-Less
Crown gall disease	*Agrobacterium radiobacter* strain K-84	NoGall
Japanese beetle grubs	*Bacillus popilliae*	Milky spore powder
Downy mildew, *Fusarium* wilt	*Bacillus licheniformis*	EcoGuard
Plant growth-promoting fungi (PGPF)		
Thrips, whitefly and mites	*Beauveria bassiana*	*Naturalis* L., Botanigard
Soilborne pathogens	*Trichoderma harzianum*	Biozim, Phalada 105
	Trichoderma viride	Monitor, Trichoguard
Root rots	*Pythium oligandrum*	Polyversum
Soil and foliar diseases	*Trichoderma polysporum*	Binab T
Post-harvest disease	*Candida oleophila*	NEXY

Source: Modified from Mishra et al. (2015).

much in weed control, and on the other hand, viruses prove difficult to handle on the ground because of their host specificity and dependence on vectors (Charudattan 2005). Different trade names have commercialised as bioherbicides against the rust fungus, *Puccinia canaliculata* and *Chondrostereum purpureum*. Plant pathogenic bacteria *Xanthomonas campestris* pv. Poannua (Xcp) and *P. syringae* pv. *tagetis* (Pst) have also been used in the formulations of bioherbicides (Johnson et al. 1996).

1.6 CONCLUSION AND FUTURE PERSPECTIVES

In conclusion, we can say that many achievements have been made in the field of agriculture through the use of microbial biotechnology, but there are still many challenges and opportunities to be explored to achieve environmentally safe and sustainable agriculture. The use of BSM in sustainable agriculture and environmental management offers innumerable benefits. Their use as biofertilisers and biopesticides has gained importance by providing significant support to the agro-ecosystem. Their ability to compete with pathogenic microorganisms and survive in harsh environmental conditions makes them effective candidates in a wide variety of stress management. Our awareness of the response of BSMs in agro-ecosystems is growing and their potential significant impact on environmental recovery is also strengthening and helping to achieve the collective sustainable development goal collectively. The dangerous effects of climate change on reducing agricultural production can also be mitigated through the use of BSM. However, with regard to the perspective of climate change, the effective use of BSMs requires further research and the elaboration of mechanisms involved in their interactions with plants under extreme conditions, which have been found to have a greater potential for their broad application in environmental sustainability.

Researchers have led new and environmental research into the daily growth of the human population, climate change, increased per capita food and energy demand, declining soil health, and declining ecosystem efficiency due to increased reliance on the use of chemical pesticides and fertilisers. To meet this need in a sustainable manner, there is a need to adopt an interdisciplinary approach that combines plant and microbiological ecology by mitigating climate change problems and maintaining a diverse microbial population in the rhizosphere. Nature has provided us with a large amount of soil microflora, which helps in soil formation and has a variety of beneficial properties that help maintain soil health and environmental balance. The use of interactions between plants and soil microorganisms is essential to meet the growing food production demand of the world population at low environmental cost. This requires extensive knowledge of interdisciplinary strategies to manage soil microorganisms, mainly in relation to different microbial communities.

Despite the beneficial role of these microorganisms, our knowledge of their diversity and the important role that these organisms play in today's global living systems is still in its infancy. Furthermore, many of the tests previously performed by various workers with beneficial microorganisms were mainly under controlled conditions, where the bio-formulation of BSM provided very good results, but there was no need to develop bio-formulation with increased shelf life. Necessary to control various pests and diseases in agriculturally important crops, thrifty and can withstand harsh weather conditions in field tests and their large-scale application should be effective. Undoubtedly, the achievement of a biased rhizosphere opens up new possibilities for future agricultural development plans based on reducing the inputs of agrochemicals and thereby utilising the services provided by beneficial microorganisms to achieve sustainable environment and economic goals.

REFERENCES

Afzal, M., Yousaf, S., Reichenauer, T.G., Sessitsch, A. 2013. Ecology of alkane-degrading bacteria and their interaction with the plant. In: F.J. de Bruijn (ed.), *Molecular Microbial Ecology of the Rhizosphere*, vol. 2. Wiley Blackwell, Hoboken, New Jersey, pp. 975–989.

Aggarwal, A., Kadian, N., Tanwar, A., Yadav, A., Gupta, K.K. 2011. Role of arbuscular mycorrhizal fungi (AMF) in global sustainable development. *Journal of Applied and Natural Science* 3: 340–351.

Ahemad, M. 2012. Implication of bacterial resistance against heavy metals in bioremediation: A review. *Journal of Institute of Integrative Omics and Applied Biotechnology* 3: 39–46.

Ahemad, M., Kibret, M. 2014. Mechanisms and applications of plant growth promoting rhizobacteria: Current perspective. *Journal of King Saud University Science* 26: 1–20.

Ali, H., Khan, E., Sajad, M.A. 2013. Phytoremediation of heavy metals – Concepts and applications. *Chemosphere* 91: 869–881.

Aroca, R., Porcel, R., Ruíz-Lozano, J.M. 2012. Regulation of root water uptake under abiotic stress conditions. *Journal of Experimental Botany* 63: 43–57.

Arora, N.K. ed. 2015. *Plant Microbes Symbiosis: Applied Facets*, Springer, Dordrecht, The Netherlands.

Arora, N.K., Mishra, J. 2016. Prospecting the roles of metabolites and additives in future bioformulations for sustainable agriculture. *Applied Soil Ecology*. https://doi.org/10.1016/j.apsoil.2016.05.020.

Arora, N.K., Khare, E., Verma, A., Sahu, R.K. 2008. In-vivo control of *Macrophomina phaseolina* by a chitinase and β-1,3-glucanaseproducing *Pseudomonas* NDN1. *Symbiosis* 46: 129–135.

Azcon, R., Medina, A., Aroca, R., Ruiz-Lozano, J.M. 2013. Abiotic stress remediation by the arbuscular mycorrhizal symbiosis and rhizosphere bacteria/yeast interactions. In: F.J. de Bruijn (ed.), *Molecular Microbial Ecology of the Rhizosphere*, Wiley Blackwell, Hoboken, New Jersey, pp. 991–1002.

Bailey, A., Chandler, D., Wyn, P., Greaves, G.J., Prince, G., Tatchell, M. 2010. *Biopesticides: Pest Management and Regulation*, CAB International, Wallingford, UK, p. 240.

Bal, H.B., Nayak, L., Das, S., Adhya, T.K. 2013. Isolation of ACC deaminase producing PGPR from rice rhizosphere and evaluating their plant growth promoting activity under salt stress. *Plant Soil* 366: 93–105.

Barazani, O., Friedman, J. 1999. Is IAA the major root growth factor secreted from plant growth-mediating bacteria? *Journal of Chemical Ecology* 25: 2397–2406.

Barea, J.M. 2000. Rhizosphere and mycorrhiza of field crops. In: J.P. Toutant, E. Balazs, E. Galante, J.M. Lynch, J.S. Schepers, D. Werner and P.A. Werry (eds.), *Connecting Science and Policy (OECD), INRA Biological Resource Management*, Springer-Verlag, Berlin, pp. 110–125.

Barea, J.M., Pozo, M.J., Azcón, R., Azcón- Aguilar, C. 2013a. Microbial interactions in the rhizosphere. In: F.J. de Bruijn (ed.), *Molecular Microbial Ecology of the Rhizosphere*, vol. 1, Wiley Blackwell, Hoboken, New Jersey, pp. 29–44.

Barea, J.M., Pozo, M.J., Lopez-Ráez, J.A., Aroca, R., Ruiz-Lozano, J.M., Ferrol, N., Azcon, R., Azcón-Aguilar, C. 2013b. Arbuscular mycorrhizas and their significance in promoting soil-plant systems sustainability against environmental stresses In: B. Rodelas and J. González-López (eds.), *Beneficial Plant-Microbial Interactions: Ecology and Applications*, CRC Press, Boca Raton, Florida, pp. 353–387.

Barzana, G., Aroca, R., Bienert, G.P., Chaumont, F., Manuel Ruiz-Lozano, J. 2014. New insights into the regulation of aquaporins by the arbuscular mycorrhizal symbiosis in maize plants under drought stress and possible implications for plant performance. *Molecular Plant-Microbe Interaction* 27: 349–363.

Baset Mia, M.A., Shamsuddin, Z.H. 2010. *Rhizobium* as a crop enhancer and biofertilizer for increased cereal production. *African Journal of Biotechnology* 9: 6001–6009.

Bashan, Y., de Bashan, L.E., Prabhu, S.R., Hernandez, J.P. 2014. Advances in plant growth promoting bacterial inoculant technology: Formulations and practical perspectives (1998–2003). *Plant Soil* 378: 1–33.

Becerra, A., Bartoloni, N., Cofré, N., Soteras, F., Cabello, M. 2014. Arbuscular mycorrhizal fungi in saline soils: Vertical distribution at different soil depth. *Brazilian Journal of Microbiology* 45: 585–594.

Beneduzi, A., Ambrosini, A., Luciane, M.P., Passaglia, L.M.P. 2012. Plant growth promoting rhizobacteria (PGPR): Their potential as antagonists and biocontrol agents. *Genetics and Molecular Biology* 35:1044–1051.

Bhattacharjee, R., Dey, U. 2014. Biofertilizer, a way towards organic agriculture: A review. *African Journal of Microbiology Research* 8:2332–2343.

Bhupinder, S., Senthil, A.N, Singh, B.K., Usha, K. 2005. Improving zinc efficiency of cereals under zinc deficiency. *Current Science* 88: 36–44.

Bianciotto, V., Lumini, E., Lanfranco, L., Minerdi, O., Bonfante, P., Perotto, S. 2000. Detection and identification of bacterial endosymbionts in arbuscular mycorrhizal fungi belonging to the family Gigasporaceae. *Applied Environmental Microbiology* 46: 4503–4509.

Blackwell, M. 2011. The fungi: 1, 2, 3 … 5.1 million species? *American Journal of Botany* 3: 426–438.

Brookes, P.C., McGrath, S.P. 1984. Effects of metal toxicity on the size of the soil microbial biomass. *Journal of Soil Science* 35: 341–346.

Buee, M., De Boer, W., Martin, F., van Overbeek, L., Jurkevitch, E. 2009. The rhizosphere zoo: An overview of plant associated communities of microorganisms, including phages, bacteria, archaea and fungi and some of their structuring factors. *Plant Soil* 321: 189–212.

Calvo-Polanco, M., Sanchez-Romera, B., Aroca, R. 2013. Arbuscular mycorrhizal fungi and the tolerance of plants to drought and salinity. In: R. Aroca (ed.), *Symbiotic Endophytes*, Springer-Verlag, Berlin, Heidelberg, pp. 271–288.

Carrapico, F., Teixeira, G., Diniz, M.A. 2000. *Azolla* as a biofertilizer in Africa: A challenge for the future. *Revista de Ciências Agrárias* 23: 120–138.

Castro, R.O., Cornejo, H.A.C., Rodriguez, L.M., Bucio, J.L. 2009. The role of microbial signals in plant growth and development. *Plant Signaling and Behaviour* 4: 701–712.

Charudattan, R. 2005. Ecological, practical, and political inputs into selection of weed targets: What makes a good biological control target? *Biological Control* 35: 183–196.

Chattopadhyay, A., Bhatnagar, N.B., Bhatnagar, R. 2004. Bacterial insecticidal toxins. *Critical Reviews in Microbiology* 30: 33–54.

Chaudhary, V., Rehman, A., Mishra, A., Chauhan, P.S., Nautiyal, C.S. 2012. Changes in bacterial community structure of agricultural land due to long term organic and chemical amendments. *Microbial Ecology* 64: 450–461.

Chaudri, A.M., McGrath, S.P., Giller, K.E., Rietz, E., Sauerbeck, D.R. 1993. Enumeration of indigenous *Rhizobium leguminosarum* biovar *trifolii* in soils previously treated with metal-contaminated sewage sludge. *Soil Biology and Biochemistry* 253: 301–309.

Chaumont, F., Tyerman, S.D. 2014. Aquaporins: Highly regulated channels controlling plant water relations. *Plant Physiology* 164: 1600–1618.

Cheng, W. 2009. Rhizosphere priming effect: Its functional relationships with microbial turnover, evapotranspiration, and C-N budgets. *Soil Biology and Biochemistry* 41: 1795–1801.

Chibeba, A.M., de Fatima Guimaraes, M., Brito, O.R. et al. 2015. Co-inoculation of soybean with *Bradyrhizobium* and *Azospirillum* promotes early nodulation. *American Journal of Plant Sciences* 6: 1641–1649. https://doi.org/10.4236/ajps.2015.610164.

Choudhary, K.K., Agrawal, S.B. 2017. Effect of UV-B radiation on leguminous plants. In: E. Lichtfouse (ed.), *Sustainable Agricultural Reviews*, Springer International Publishing, Cham, Switzerland, pp. 115–162.

Chung, H., Park, M., Madhaiyana, M., Seshadri, S., Song, J., Cho, H., et al. 2005. Isolation and characterization of phosphate solubilizing bacteria from the rhizosphere of crop plants of Korea. *Soil Biology and Biochemistry* 3:1970–1974.

Corenõ-Alonso, A., Solé, A., et al. 2014. Mechanisms of interaction of chromium with *Aspergillus niger* var. *tubingensis* strain Ed8. *Bioresource Technology* 158: 188–192.

Cornejo, P., Meier, S., Borie, G., Rillig, M., Borie, F. 2008. Glomalin-related soil protein in a Mediterranean ecosystem affected by a copper smelter and its contribution to Cu and Zn sequestration. *Science of Total Environment* 406: 154–160.

Damodaran, T., Rajan, S., Gopal, R., Yadav, A. et al. 2019. Successful community based management of banana wilt caused by *Fusarium oxysporum* f.sp. *cubense* tropical race-4 through ICAR-FUSICONT. *Journal of Applied Horticulture* 21: 37–41.

Dimkpa, C.O., Merten, D., Svatos, A., Büchel, G., Kothe, E. 2009. Siderophores mediate reduced and increased uptake of cadmium by *Streptomyces tendae* F4 and sunflower (*Helianthus annuus*), respectively. *Journal of Applied Microbiology* 5: 687–1696.

Dobereiner, J. 1997. Biological nitrogen fixation in the tropics: Social and economic contributions. *Soil Biology and Biochemistry* 29: 771–774.

Duarah, M., Saikia, N.D., Boruah, H.P.D. 2011. Phosphate solubilizers enhance NPK fertilizer use efficiency in rice and legume cultivation. *Biotechnology* 4: 227–238.

Elkan, G.H. 1992. Biological nitrogen fixation systems in tropical ecosystems: An overview. In: K. Mulongoy, M.E. Gueye and D.S.C. Spencer (eds.), *Biological Nitrogen Fixation and Sustainability of Tropical Agriculture*, John Willey Sons, Chichester, pp. 27–40.

Ezzouhri, L., Castro, E., Moya, M., Espinola, F., Lairini, K. 2009. Heavy metal tolerance of filamentous fungi isolated from polluted sites in Tangier, Morocco. *African Journal of Microbiology Research* 3(2): 35–48.

Fernandez, L.A., Zalba, P., Gomez, M.A., Sagardoy, M.A. 2007. Phosphate solubilization activity of bacterial strains in soil and their effect on soybean growth under greenhouse conditions. *Biology and Fertility of Soils* 43: 805–809.

Ferrol, N., González-Guerrero, M., Valderas, A., Benabdellah, K., Azcón-Aguilar, C. 2009. Survival strategies of arbuscular mycorrhizal fungi in Cu-polluted environments. *Phytochemistry Reviews* 8: 551–559.

Fliessbach, A., Martens, R., Reber, H.H. 1994. Soil microbial biomass and microbial activity in soils treated with heavy metal contaminated sewage sludge. *Soil Biology and Biochemistry* 26: 1201–1205.

Fritsch, F.E. 1907. A general consideration of the subaerial and fresh water algal flora of Ceylon. A contribution to the study of tropical algal ecology: I. Subaerial algae and algae of inland fresh water. *Proceedings of Royal Society of London*, Series B, 79: 197–254.

Gaur, S., Kaushik, P. 2011. Analysis of vesicular arbuscular mycorrhiza associated with medicinal plants in Uttarakhand state of India. *World Applied Science Journal* 14: 645–653.

Germaine, K.J., McGuinness, M., Dowling, D.N. 2013. Improving phytoremediation through plant-associated bacteria. In: F.J. de Bruijn (ed.), *Molecular Microbial Ecology of the Rhizosphere*, vol. 2. Wiley Blackwell, Hoboken, New Jersey, pp. 963–973.

Glick, B.R. 2010. Using soil bacteria to facilitate phytoremediation. *Biotechnology Advances* 28: 367–374.

González-Guerrero, M., Benabdellah, K., Ferrol, N., Azcón-Aguilar, C. 2009. Mechanisms underlying heavy metal tolerance in arbuscular mycorrhizas. In: C. Azcón-Aguilar, J.M. Barea, S. Gianinazzi and V. Gianinazzi-Pearson (eds.), *Mycorrhizas Functional Processes and Ecological Impact*, Springer-Verlag, Berlin-Heidelberg, pp. 107–122.

González-Guerrero, M., Benabdellah, K., Valderas, A., Azcon-Aguilar, C., Ferrol, N. 2010. GintABC1 encodes a putative ABC transporter of the MRP subfamily induced by Cu, Cd, and oxidative stress in *Glomus intraradices*. *Mycorrhiza* 20: 137–146.

Goswami, D., Thakker, J.N., Dhandhukia, P.C. 2016. Portraying mechanics of plant growth promoting rhizobacteria (PGPR): A review. *Cogent Food and Agriculture* 2:1127500.

Gray, E.J., Smith, D.L. 2005. Intracellular and extracellular PGPR: Commonalities and distinctions in the plant-bacterium signalling processes. *Soil Biology and Biochemistry* 37: 395–412.

Groppa, M.D., Benavides, M.P., Zawoznik, M.S. 2012. Root hydraulic conductance, aquaporins and plant growth promoting microorganisms: A revision. *Applied Soil Ecology* 61: 247–254.

Hallman, J., Quadt-Hallman, A., Mahafee, W.F., Kloepper, J.W. 1997. Bacterial endophytes in agricultural crops. *Canadian Journal of Microbiology* 43: 895–914.

Harding, D.P., Raizada, M.N. 2015. Controlling weeds with fungi, bacteria and viruses: A review. *Frontiers in Plant Science* 6: 659.

Hartmann, A., Rothballer, M., Schmid, M., Hiltner, L. 2008. A pioneer in rhizosphere microbial ecology and soil bacteriology research. *Plant Soil* 312: 7–14.

Hassan, T.U., Bano, A., Naz, I. 2017. Alleviation of heavy metals toxicity by the application of plant growth promoting rhizobacteria and effects on wheat grown in saline sodic field. *International Journal of Phytoremediation* 19: 522–529.

Higa, T. 1991. Effective microorganisms: A biotechnology for mankind. In: J.F. Parr, S.B. Hornick and C.E. Whitman (eds.), *Proceedings of the First International Conference on Kyusei Nature Farming*, U.S. Department of Agriculture, Washington, DC, pp. 8–14.

Higa, T. 1994. Effective microorganisms: A new dimension for nature farming. In: J.F. Parr, S.B. Hornick and M.E. Simpson (eds.), *Proceedings of the Second International Conference on Kyusei Nature Farming*, U.S. Department of Agriculture, Washington, DC, pp. 20–22.

Higa, T. 1995. Effective microorganisms: Their role in Kyusei Nature Farming and sustainable agriculture. In: J.F. Parr, S.B. Hornick and M.E. Simpson (eds.), *Proceedings of the Third International Conference on Kyusei Nature Farming*, U.S. Department of Agriculture, Washington, DC.

Hungria, M., Campo, R.J., Mendes, I.C. 2007. *Aimportância do processo de fixaçãobiológica do nitrogênio para a cultura de soja: Componente essencial para a competitividade do produtobrasileiro*, Embrapa Soja, Londrina, Brazil, p. 80.

Johnson, D.R., Wyse, D.L., Jones, K.J. 1996. Controlling weeds with phytopathogenic bacteria. *Weed Technology* 10: 621–624.

Jung, S., Martínez-Medina, A., López-Ráez, J.A., Pozo, M.J. 2012. Mycorrhiza-induced resistance and priming of plant defences. *Journal of Chemical Ecology* 38: 651–664.

Kang, S.M., Khan, A.L., Waqas, M., You, Y., Kim, J., Hamayun, M., et al. 2014. Plant growth promoting rhizobacteria. *Annual Review of Microbiology* 63: 541–556.

Karthikeyan, N., Prasanna, R.L., Nain, L., Kaushik, B.D. 2007. Evaluating the potential of plant growth promoting cyanobacteria as inoculants for wheat. *European Journal of Soil Biology* 43: 23–30.

Kaushik, B.D. 2014. Developments in cyanobacterial biofertilizer. *Proceeding of the Indian National Science Academy* 80: 379–388.

Khare, E., Arora, N.K. 2011. Physiologically stressed cells of fluorescent *Pseudomonas* EKi as better option for bioformulation development for management of charcoal rot caused by *Macrophomina phaseolina* in field conditions. *Current Microbiology* 62: 1789–1793.

Kloepper, J.W., Schroth, M.N., Miller, T.D. 1980. Effects of rhizosphere colonization by plant growth promoting rhizobacteria on potato plant development and yield. *Phytopathology* 70: 1078–1082.

Konieczny, A., Kowalska, I. 2016.The role of arbuscular mycorrhiza in zinc uptake by lettuce grown at two phosphorus levels in the substrate. *Agricultural and Food Science* 25(2). doi: 10.23986/afsci.55534.

Kramer, U. 2005. Phytoremediation: Novel approaches to cleaning up polluted soils. *Current Opinion in Biotechnology* 16: 133–141.

Kranthi, K.R., Russell, D., Wanjari, R., Kherde, M., Munje, S., Lavhe, N., Armes, N. 2002. In-season changes in resistance to insecticides in *Helicoverpa armigera* (Lepidoptera: Noctuidae) in India. *Journal of Economic Entomology* 95:134–142. https://doi.org/10.1603/0022-0493-95.1.134.

Krewulak, H.D., Vogel, H.J. 2008. Structural biology of bacterial iron uptake. *Biochimica et Biophysica Acta* 1778: 1781–1804.

Kumar, R., Mishra, R.K., Qidwai, A., Shukla, S.K., Mishra, V., Pandey, A., et al. 2015. Detoxification and tolerance of heavy metals in plants. *Plant Metal Interaction: Emerging Remediation Techniques*, Chapter 13. Elsevier Inc., Amsterdam, pp. 335–360, ISBN: 978-0-12-803158-2.

Laughlin, R.J., Stevens, R.J. 2012. Evidence for fungal dominance of denitrification and co-denitrification in a grassland soil. *Soil Science Society of America Journal* 66: 1540–1548.

Lehman, R.M., Cambardella, C.A., Stott, D.E., et al. 2015. Understanding and enhancing soil biological health: The solution for reversing soil degradation. *Sustainability* 7: 988–1027.

Lemanceau, P., Bauer, P., Kraemer, S., Briat, J.F. 2009. Iron dynamics in the rhizosphere as a case study for analysing interactions between soils, plants and microbes. *Plant Soil* 321: 513–535.

Li, G., Santoni, V., Maurel, C. 2014. Plant aquaporins: Roles in plant physiology. *Biochimica et Biophysica Acta – General Subjects* 1840: 1574–1582.

López-Ráez, J.A., Bouwmeester, H., Pozo, M.J. 2012. Communication in the rhizosphere, a target for pest management In: E. Lichtfouse (ed.), *Sustainable Agriculture Reviews,* Vol. 8. *Agroecology and Strategies for Climate Change,* vol. 8. *Sustainable Agriculture Reviews*, Springer, Dordrecht, The Netherlands, pp. 109–133.

Lugtenberg, B. 2015. Life of microbes in the rhizosphere. In: B. Lugtenberg (ed.), *Principles of Plant Microbe Interactions*, Springer, Berlin-Heidelberg, pp. 7–15.

Ma, Y., Oliveira, R.S., Freitas, H., Zhang, C. 2016. Biochemical and molecular mechanisms of plant-microbe-metal interactions: Relevance for phytoremediation. *Frontiers in Plant Science* 10.3389, 00918.

MacGregor, J.T. 2006. *Genetic toxicity assessment of microbial pesticides: Needs and recommended approaches*. International Association of Environmental Mutagen Societies, Arnold, MD, pp. 1–17.

Malusa, E., Vassilev, N. 2014. A contribution to set a legal framework for biofertilizers. *Applied Microbiology and Biotechnology* 98: 6599–6607.

Manjunath, M., Prasanna, R., Shama, P., Nain, L., Singh, R. 2011. Developing PGPR consortia using novel genera *Providencia* and *Alcaligenes* along with cyanobacteria for wheat. *Archives of Agronomy and Soil Science* 57: 873–887.

Marcel, G.A., Heijden, V., Martin, F.M., Selosse, M.A., Sanders, I.R. 2015. Mycorrhizal ecology and evolution: The past, the present, and the future. *New Phytologist* 205: 1406–1423.

Marks, B.B., Megías, M., Nogueira, M.A., Hungria, M. 2007. Biotechnological potential of rhizobial metabolites to enhance the performance of *Bradyrhizobium* spp. And *Azospirillum brasilense* inoculants with soybean and maize. *AMB Express* 20: 3–21.

Mazurier, S., Corberand, T., Lemanceau, P., Raaijmakers, J.M. 2009. Phenazine antibiotics produced by fluorescent pseudomonads contribute to natural soil suppressiveness to *Fusarium* wilt. *ISME Journal* 3: 977–991.

Medina, A., Azcon, R. 2010. Effectiveness of the application of arbuscular mycorrhiza fungi and organic amendments to improve soil quality and plant performance under stress conditions. *Journal of Soil Science and Plant Nutrition* 10: 354–372.

Meena, V.S., Maurya, B.R., Bohra, J.S., Verma, R., Meena, M.D. 2013a. Effect of concentrate manure and nutrient levels on enzymatic activities and microbial population under submerged rice in alluvium soil of Varanasi. *Crop Research* 45 (1–3):6–12.

Meena, V.S., Maurya, B.R., Verma, R., Meena, R.S., Jatav, G.K., Meena, S.K. 2013b. Soil microbial population and selected enzyme activities as influenced by concentrate manure and inorganic fertilizer in alluvium soil of Varanasi. *Bioscan* 8 (3): 931–935.

Meena, V.S., Verma, J.P., Meena, S.K. 2015. Towards the current scenario of nutrient use efficiency in crop species. *Journal of Cleaner Production* 102: 556–557.

Meharg, A.A. 2003. The mechanistic basis of interactions between mycorrhizal associations and toxic metal cations. *Mycological Research* 107: 1253–1265.

Miller, M.J. 2008. *Siderophores (microbial iron chelators) and siderophore-drug conjugates (new methods for microbially selective drug delivery)*, University of Notre Dame, Louisiana.

Mishra, S., Arora, N.K. 2012. Evaluation of rhizospheric *Pseudomonas* and *Bacillus* as biocontrol tool for *Xanthomonas campestris* pv. *campestris*. *World Journal of Microbiology and Biotechnology* 28: 693–702.

Mishra, J., Tewari, S., Singh, S., Arora, N.K. 2015. Biopesticides: Where we stand? In: N.K. Arora (ed.), *Plant Microbes Symbiosis: Applied Facets*, Springer-India, New Delhi, pp. 37–76.

Mishra, J., Prakash, J., Arora, N.K. 2016. Role of beneficial soil microbes in sustainable agriculture and environment management. *Climate Change and Environ Sustainability* 4(2):137–149.

Mishra, J., Singh, R., Arora, N.K. 2017. Alleviation of heavy metal stress in plants and remediation of soil by rhizosphere microorganisms. *Frontiers in Microbiology*. https://doi.org/10.3389/fmicb.2017.01706.

Mishra, J., Dutta, V., Arora, N.K. 2020. Biopesticides in India: Technology and sustainability linkages. *3 Biotechnology* 10: 210. https://doi.org/10.1007/s13205-020-02192-7.

Mmbaga, G.W., Mtei, K.M., Patrick, A., Ndakidemi, P.A. 2014. Extrapolations on the use of rhizobium inoculants supplemented with phosphorus (P) and potassium (K) on growth and nutrition of legumes. *Journal of Agricultural Sciences* 5:1207–1226.

Mustapha, M.U., Halimoon, N. 2015. Screening and isolation of heavy metal tolerant bacteria in industrial effluent. *Procedia Environmental Sciences* 30: 33–37.

Nadeem, S.M., Ahmad, M., Zahir, Z.A., Javaid, A., Ashraf, M.2014. The role of mycorrhizae and plant growth promoting rhizobacteria (PGPR) in improving crop productivity under stressful environments. *Biotechnology Advances* 32:429–448.

Nair, M.G., Safir, G.R., Siqueira, J.O. 1991. Isolation and identification of vesicular-arbuscular mycorrhiza-stimulatory compounds from clover (*Trifolium repens*) roots. *Applied Environmental Microbiololgy* 57: 434–439.

National Academy of Sciences. 1989. *Alternative Agriculture*. Committee on the Role of Alternative Agriculture Farming Methods and Modern Production Agriculture, National Research Council, Board on Agriculture, National Academy Press, Washington, DC, p. 448.

Nautiyal, C.S., Chauhan, P.S., Bhatia, C.R. 2010. Changes in soil physiochemical properties and microbial functional diversity due to 14 years of conversion of grassland to organic agriculture in semi-arid agro-ecosystems. *Soil Tillage Research* 109: 55–60.

Nautiyal, C.S., Srivastava, S., Chauhan, P.S., Seem, K., Mishra, A., Sopory, S.K. 2013. Plant growth-promoting bacteria *Bacillus amyloliquefaciens* NBRISN13 modulates gene expression profile of leaf and rhizosphere community in rice during salt stress. *Plant Physiology and Biochemistry* 66: 1–9.

Nie, L., Shah, S., Burd, G.I., Dixon, D.G., Glick, B.R. 2002. Phytoremediation of arsenate contaminated soil by transgenic canola and the plant growth-promoting bacterium *Enterobacter cloacae* CAL2. *Plant Physiology and Biochemistry* 40: 355–361.

Nihorimbere, V., Ongena, M., Smargiassi, M., Thonart, P. 2011. Beneficial effect of the rhizosphere microbial community for plant growth and health. *Biotechnology Agronomy and Society and Environment* 2: 327–337.

Nishanth, D., Biswas, D.R. 2008. Kinetics of phosphorus and potassium release from rock phosphate and waste mica enriched compost and their effect on yield and nutrient uptake by wheat (*Triticum aestivum* L.). *Bioresource Technology* 99: 3342–3354.

Oladipo, O.G., Awotoye, O.O., Olayinka, A., Bezuidenhout, C.C., Maboeta, M.S. 2017. Heavy metal tolerance traits of filamentous fungi isolated from gold and gemstone mining sites. *Brazilian Journal of Microbiology* 49: 29–37.

Oliveira, C.A., Sa, N.M.H., Gomes, E.A., Marriel, I.E., Scotti, M.R., Guimaraes, C.T., et al. 2009. Assessment of the mycorrhizal community in the rhizosphere of maize. *Zea mays* L. genotypes contrasting for phosphorus efficiency in the acid savannas of Brazil using denaturing gradient gel electrophoresis DGGE. *Applied Soil Ecology* 41: 249–258.

Ong, G.H., Ho, X.H., Shamkeeva, S., Fernando, M.S., Shimen, A., Wong, L.S. 2017. Biosorption study of potential fungi for copper remediation from Peninsular Malaysia. *Remediation Journal* 27: 59–63.

Owen, D., Williams, A.P., Griffith, G.W., Withers, P.J.A. 2015. Use of commercial bio-inoculants to increase agricultural production through improved phosphorus acquisition. *Applied Soil Ecology* 86: 41–54.

Pangesti, N., Pineda, A., Pieterse, C.M.J., Dicke, M., van Loon, J.J.A. 2013. Two-way plant-mediated interactions between root-associated microbes and insects: From ecology to mechanisms. *Frontiers in Plant Science* 4. https://doi.org/10.3389/fpls.2013.00414.

Pankaj, U. 2020. Bio-fertilizers for management of soil, crop and human health. In: C. Singh, S. Tiwari, J.S. Singh and A.N. Yadav (eds.), *Microbes as Biofertilizers in Agricultural Fields*, CRC Press/Taylor & Francis, Boca Raton, Florida, pp. 71–85.

Pankaj, U., Verma, S.K., Semwal, S., Verma, R.K. 2017. Assessment of natural mycorrhizal colonization and soil fertility status of lemongrass [(*Cymbopogon flexuosus*, Nees ex Steud) W. Watson] crop in subtropical India. *Journal of Applied Research on Medicinal and Aromatic Plants* 5: 41–46.

Pankaj, U., Singh, D.N., Singh, G., Verma, R.K. 2019a. Microbial inoculants assisted growth of *Chrysopogon zizanioides* promotes phytoremediation of salt affected soil. *Indian Journal of Microbiology* 59(2): 137–146.

Pankaj, U., Singh, G., Verma, R.K. 2019b. Microbial approaches in management and restoration of marginal lands. In: J.S. Singh (ed.), *New and Future Developments in Microbial Biotechnology and Bioengineering: Microbes in Soil, Crop and Environmental Sustainability*, Elsevier, Cambridge, MA, pp. 295–305.

Pankaj, U., Singh, D.N., Mishra, P., Gaur, P., Vivekbabu, C.S., Shanker, K., Verma, R.K. 2020. Autochthonous halotolerant plant growth promoting rhizobacteria promote bacoside A yield of *Bacopa monnieri* (L) Nash and phytoextraction of salt-affected soil. *Pedosphere* 30(5): 671–683.

Pankaj, U., Kurmi, A., Lothe, N.B., Verma, R.K. 2021. Influence of the seedlings emergence and initial growth of palmarosa (*Cymbopogon martinii* (Roxb.) Wats. Var. Motia Burk) by arbuscular mycorrhizal fungi in soil salinity conditions. *Journal of Applied Research on Medicinal and Aromatic Plants* 24: 100317.

Parniske, M. 2008. Arbuscular mycorrhiza: The mother of plant root endosymbiosis. *Nature Reviews Microbiology* 6: 763–775.

Pedraza, R. 2008. Recent advances in nitrogen-fixing acetic acid bacteria. *International Journal Food Microbiology* 125: 25–35.

Pieterse, C.M., Leon-Reyes, A., Van der Ent, S., Van Wees, S.C. 2009. Networking by small-molecule hormones in plant immunity. *Nature Chemical Biology* 5: 308–316.

Pongrac, P., Vogel-Mikus, K., Poschenrieder, C., Barceló, J., Tolrà, R., Regvar, M. 2013a. Arbuscular mycorrhiza in glucosinolate-containing plants: The story of the metal hyperaccumulator *Noccaea* (Thlaspi) praecox (Brassicaceae). In: F.J. de Bruijn (ed.), *Molecular Microbial Ecology of the Rhizosphere*, vol. 2. Wiley Blackwell, Hoboken, New Jersey, pp. 1023–1032.

Pozo, M.J., Jung, S.C., Martínez-Medina, A., López- Ráez, J.A., Azcón-Aguilar, C., Barea, J.M. 2013. Root allies: Arbuscular mycorrhizal fungi help plants to cope with biotic stresses. In: R. Aroca (ed.), *Symbiotic Endophytes*, Springer-Verlag, Berlin-Heidelberg, pp. 289–307.

Pozo, M., López-Ráez, J., Azcón-Aguilar, C., García- Garrido, J. 2015. Phytohormones as integrators of environmental signals in the regulation of mycorrhizal symbioses. *New Phytology* 205:1431–1436.

Prasad, R.C., Prasad, B.N. 2001. Cyanobacteria as a source biofertilizer for sustainable agriculture in Nepal. *Journal of Plant Science Botanica Orientalis* 8: 127–133.

Presto, S., Angelinic, P., Bianciottob, V., Bonfanteab, P., Girlandaab, M., Kulld, T., et al. 2013. Interactions of fungi with other organisms. *Plant Biosystems* 147: 208–218.

Puppi, G., Azcon, R., Hoflich, G. 1994. Management of positive interactions of arbuscular mycorrhizal fungi with essential groups of soil microorganisms. In: S. Gianinazzi and H. Schüepp (eds.), *Impact of Arbuscular Mycorrhizas on Sustainable Agriculture and Natural Ecosystems*, Birkhäuser Verlag, Basel, Switzerland, pp. 201–215.

Raaijmakers, J.M., Paulitz, T.C., Steinberg, C., Alabouvette, C., Moenne-Loccoz, Y. 2009. The rhizosphere: A playground and battlefield for soil borne pathogens and beneficial microorganisms. *Plant Soil* 321: 341–336.

Raja, D., Takankhar, V.J. 2018. Response of liquid biofertilizers (*Bradyrhizobium* and PSB) on nutrient content in soybean. *International Journal of Current Microbiology and Applied Sciences* 7: 3701–3706. doi:10.20546/ijcmas.2018.705.428.

Rajkumar, M., Vara Prasad, M.N., Freitas, H., Ae, N. 2009. Biotechnological applications of serpentine soil bacteria for phytoremediation of trace metals. *Critical Reviews in Biotechnology* 29: 120–130.

Raymond, K.N., Dertz, E.A., Kim, S.S. 2003. *Enterobactin*: An archetype for microbial iron transport. *Proceedings of the National Academy of Sciences*, 100: 3584–3588.

Reichman, S.M. 2014. Probing the plant growth-promoting and heavy metal tolerance characteristics of *Bradyrhizobium japonicum* CB1809. *European Journal of Soil Biology* 63: 7–13.

Rennie, R.J., Hynes, R.K. 1997. Scientific and legislative quality control of legume inoculants for lentil and field pea. *Journal of Production Agriculture* 6: 569–574.

Richardson, A.E., Simpson, R.J. 2011. Soil microorganisms mediating phosphorus availability update on microbial phosphorus. *Plant Physiology* 156: 989–996.

Robertson, G.P., Groffman, P.M. 2007. Nitrogen transformation. In: E.A. Paul (ed.), *Soil Microbiology, Ecology and Biochemistry*, Springer, New York, pp. 341–364.

Rosa, F., Philippis, R.D. 2015. Role of cyanobacterial exopolysaccharides in phototrophic biofilms and in complex microbial mats. *Life* 5: 1218–1238.

Saberi, A.R., Hassan, S.A. 2014. The effects of nitrogen fertilizer and plant density on mustard (*Brassica juncea*): An overview. *Global Advance Research Journal of Agricultural Sciences* 3: 205–210.

Saeed, K.S., Ahmed, S.A., Hassan, I.A., Ahmed, P.H. 2015. Effect of bio-fertilizer and chemical fertilizer on growth and yield in cucumber *Cucumissativus* in green house condition. *Pakistan Journal of Biological Sciences*18: 129–134.

Saha, M., Sarkar, S., Sarkar, B., Sharma, B.K., Bhattacharjee, S., Tribedi, P. 2015. Microbial siderophores and their potential applications: A review. *Environmental Science and Pollution Research* 10: 1–16.

Saha, M., Maurya, B.R., Meena, V.S., Bahadur, I., Kumar, A. 2016. Identification and characterization of potassium solubilizing bacteria (KSB) from Indo-Gangetic Plains of India. *Biocatalalysis and Agricultural Biotechnology* 7: 202–209.

Saiyad, S.A., Jhala, Y.K., Vyas, R.V. 2015. Comparative efficiency of five potash and phosphate solubilizing bacteria and their key enzymes useful for enhancing and improvement of soil fertility. *International Journal of Scientific Research and Publication* 5: 1–6.

Sandaa, R.A., Torsvik, V., Enger, O. 1999. Analysis of bacterial communities in heavy metal-contaminated soils at different levels of resolution. *FEMS Microbiology and Ecology* 30: 237–251.

Sangeeth, K.P., Suseela, B.R. 2015. Integrated plant nutrient system with special emphasis on mineral nutrition and biofertilizers for black pepper and cardamom. *Critical Reviews in Microbiology*. doi:10.3109/1040841X.2014.958433.

Sarkar, A., Islam, T., Biswas, G.C., Alam, S., Hossain, M., Talukder, N.M. 2012. Screening for phosphate solubilizing bacteria inhabiting the rhizoplane of rice grown in acidic soil in Bangladesh. *Acta Microbiologica Immunologica Hungarica* 59: 199–213.

Sarkar, A., Mohammad, N.T., Islam, M.T. 2014. Phosphate solubilizing bacteria promote growth and enhance nutrient uptake by wheat. *Plant Science Today* 1: 86–93.

Sauchelli, V. 1969. *Trace Elements in Agriculture*, D Van Nostrand Reinhold Co., New York, p. 248.

Saxena, B., Shukla, K., Giri, B. 2017. Arbuscular mycorrhizal fungi and tolerance of salt stress in plants. In: Q.S. Wu (ed.), *Arbuscular Mycorrhizas and Stress Tolerance of Plants*, Springer, Singapore, pp. 67–97.

Selim, K.A., El-Beih, A.A., Abdel-Rahman, T.M., El-Diwany, A.I. 2012. Biology of endophytic fungi. *Current Research in Environmental and Applied Mycology* 1: 31–82.

Selosse, M., Bessis, A., Pozo, M. 2014. Microbial priming of plant and animal immunity: Symbionts as developmental signals. *Trends in Microbiology* 22: 607–613.

Shahab, S., Ahmed, N., Khan, N.S. 2009. Indole acetic acid production and enhanced plant growth promotion by indigenous PSBs. *African Journal of Agricultural Research* 4: 1312–1316.

Shanware, A.S., Kalkar, S.A., Trivedi, M.M. 2014. Potassium solubilisers: Occurrence, mechanism and their role as competent biofertilizers. *International Journal Current Microbiology and Applied Sciences* 3: 622–629.

Sharan, V.S., Nehra, V. 2011. Plant growth promoting rhizobacteria: A critical review. *Life Sciences and Medicine Research* 11: 1–30.

Sharma, S.B., Sayyed, R.Z., Trivedi, M.H., Gobi, T.A. 2013. Phosphate solubilizing microbes: Sustainable approach for managing phosphorus deficiency in agricultural soils. *Spring* 2: 1–14.

Sharma, A., Shankhdhar, D., Shankhdhar, S.C. 2016. Potassium-solubilizing microorganisms: Mechanism and their role in potassium solubilization and uptake. In: V.S. Meena, B.R. Maurya, J.P. Verma and R.S. Meena (eds.), *Potassium Solubilizing Microorganisms for Sustainable Agriculture*, Springer, New Delhi, India, pp. 203–219.

Shrivastava, M., Srivastava, P.C., D'Souza, S.F. 2016. KSM soil diversity and mineral solubilization, in relation to crop production and molecular mechanism. In: V.S. Meena, B.R. Maurya, J.P. Verma and R.S. Meena (eds.), *Potassium Solubilizing Microorganisms for Sustainable Agriculture*, Springer, New Delhi, India, pp. 221–234.

Shukla, S.K., Mishra, R.K., Pandey, M., Mishra, V., Pathak, A., Pandey, A., et al. 2015. Land reformation using plant growth promoting rhizobacteria with context to heavy metal contamination, *Plant Metal Interaction: Emerging Remediation Techniques*, Chapter, 21. Elsevier Inc., Amsterdam, pp. 499–530, ISBN: 978-0-12-803158-2.

Siddiqui, A., Shivle, R., Magodiya, N., Tewari, K. 2014. Mixed effect of *Rhizobium* and *Azotobacter* as biofertilizer on nodulation and production of Chick pea, *Cicer arietinum*. *Bioscience Biotechnology Research Communications* 1: 46–49.

Sindhu, S.S., Parmar, P., Phour, M., Sehrawat, A. 2016. Potassium-solubilizing microorganisms (KSMs) and its effect on plant growth improvement. In: V.S. Meena, B.R. Maurya, J.P. Verma and R.S. Meena (eds.), *Potassium Solubilizing Microorganisms for Sustainable Agriculture*, Springer, New Delhi, India, pp. 171–185.

Singh, N.K., Dhar, D.W. 2010. Cyanobacterial reclamation of salt affected soil. In: E. Lichtfouse (eds.), *Genetic Engineering, Biofertilisation, Soil Quality and Organic Farming*, Springer, Dordrecht, The Netherlands, pp. 245–24.

Singh, J.S., Abhilash, P.C., Singh, H.B., Singh, R.P., Singh, D.P. 2011. Genetically engineered bacteria: An emerging tool for environmental remediation and future research perspectives. *Gene* 480: 1–9.

Singh, N.P., Singh, R.K., Meena, V.S., Meena, R.K. 2015. Can we use maize (*Zea mays*) rhizobacteria as plant growth promoter? *Vegetos* 28: 86–99.

Singh, J.S., Kumar, A., Rai, A.N., Singh, D.P. 2016a. Cyanobacteria: A precious bioresource in agriculture, ecosystem, and environmental sustainability. *Frontiers in Microbiology* 7: 529.

Singh, M., Dotaniya, M.L., Mishra, A., Dotaniya, C.K., Regar, K.L., Lata, M. 2016b. Role of biofertilizers in conservation agriculture. In: J.K. Bisht, V.S. Meena, P.K. Mishra and A. Pattanayak (eds.), *Conservation Agriculture: An Approach to Combat Climate Change in Indian Himalaya*, Springer, Singapore, pp. 113–134.

Singh, G., Pankaj, U., Chand, C., Verma, R.K. 2019. Arbuscular mycorrhizal fungi-assisted phytoextraction of toxic metals by *Zea mays* L. from tannery sludge. *Soil and Sediment Contamination: An International Journal* 28(8): 729–746.

Singh, G., Pankaj, U., Ajayakumar, P.V., Verma, R.K. 2020. Phytoremediation of sewage sludge by *Cymbopogon martinii* (Roxb.) Wats. Var. motia Burk. Grown under soil amended with varying levels of sewage sludge. *International Journal of Phytoremediation* 22 (5): 540–550.

Smil, V. 1999. Nitrogen in crop production: An account of global flows. *Global Biogeochemical Cycles* 13: 647–662.

Smith, S.E., Read, D.J. 2008. *Mycorrhizal Symbiosis*, Academic Press, London, p. 787.

Song, T., Martensson, L., Eriksson, T., Zheng, W., Rasmussen, U. 2005. Biodiversity and seasonal variation of the cyanobacterial assemblage in a rice paddy field in Fujian, China. *FEMS Microbiology Ecology* 123: 54131–1400.

Soni, S.K., Singh, R., Singh, M. et al. 2014. Pretreatment of Cr(VI)-amended soil with chromate-reducing rhizobacteria decreases plant toxicity and increases the yield of *Pisum sativum*. *Archives of Environmental Contamination and Toxicology* 66: 616–627. https://doi.org/10.1007/s00244-014-0003-0.

Sosa-Gomez, D.R., Moscardi, F. 1998. Laboratory and field studies on the infection of stink bugs, *Nezara viridula*, *Piezodorus guildinii*, and *Euschistus heros* (Hemiptera: Pentatomidae) with *Metarhizium anisopliae* and *Beauveria bassiana* in Brazil. *Journal of Invertebrate Pathology* 2:115–120.

Steenhoudt, O., Vanderleyden, J. 2000. *Azospirillum*, a free-living nitrogen-fixing bacterium closely associated with grasses: Genetic, biochemical and ecological aspect. *FEMS Microbiology Reviews* 24: 487–506.

Sundara, B., Natarajan, V., Hari, K. 2002. Influence of phosphorus solubilizing bacteria on the changes in soil available phosphorus and sugarcane and sugar yields. *Field Crop Research* 77: 43–49.

Tiwari, S., Lata, C. 2018. Heavy metal stress, signalling, and tolerance due to plant-associated microbes: An overview. *Frontiers in Plant Science*. https://doi.org/10.3389/fpls.2018.00452.

Tiwari, S., Lata, C., Chauhan, P.S., Nautiyal, C.S. 2016. *Pseudomonas putida* attunes morphophysiological, biochemical and molecular responses in *Cicer arietinum* L. during drought stress and recovery. *Plant Physiology and Biochemistry* 99: 108–117.

Torre-Ruiz, N.D.L., Ruiz-Valdiviezo, V.M., Rincon-Molina, C.I., Rodríguez- Mendiola, M., Arias-Castroa, C., Gutiérrez-Miceli, F.A., et al. 2016. Effect of plant growth-promoting bacteria on the growth and fructan production of *Agave americana* L. *Brazilian Journal of Microbiology* 47: 587–596. doi:10.1016/j.bjm.2016.04.010.

Tripathi, P., Singh, P.C., Mishra, A., Chaudhry, V., Mishra, S., Tripathi, R.D., et al. 2013. Trichoderma inoculation ameliorates arsenic induced phytotoxic changes in gene expression and stem anatomy of chickpea (*Cicer arietinum*). *Ecotoxicology and Environmental Safety* 89: 8–14.

Tripathi, P., Singh, P.C., Mishra, A., Srivastava, S., Chauhan, R., Awasthi, S., et al. 2017. Arsenic tolerant *Trichoderma* sp. Reduces arsenic induced stress in chickpea (*Cicer arietinum*). *Environmental Pollution* 223: 137–145.

Ullah, A., Heng, S., Munis, M.F.H., Fahad, S., Yang, X. 2015. Phytoremediation of heavy metals assisted by plant growth promoting (PGP) bacteria: A review. *Environmental and Experimental Botany* 117:28–40.

United States Environmental Protection Agency (USEPA). 2016. https://www.epa.gov/ingredients-used-pesticide-products/what-are-biopesticides.

Vejan, P., Abdullah, R., Khadiran, T., Ismail, S., Nasrulhaq Boyce, A. 2016. Role of plant growth promoting rhizobacteria in agricultural sustainability –A review. *Molecules* 21 (5): 573.

Verma, P., Yadav, A.N., Khannam, K.S., Panjiar, N., Kumar, S., Saxena, A.K., et al., 2015. Assessment of genetic diversity and plant growth promoting attributes of psychotolerant bacteria allied with wheat (*Triticum aestivum* L.) from the northern hills zone of India. *Annals of Microbiology*. https://doi.org/10.1007/s13213-014-1027-4.

Vikram, A., Hamzehzarghani, H. 2008. Effect of phosphate solubilizing bacteria on nodulation and growth parameters of green gram *Vigna radiata* L. Wilczek. *Research Journal of Microbiology* 3: 62–72.

Wani, P.A., Khan, M.S. 2012. Bioremediation of lead by a plant growth promoting rhizobium species RL9. *Bacteriology Journal* 2: 66–78.

Weller, D.M. 1988. Biological control of soil borne plant pathogens in the rhizosphere with bacteria. *Annual Review of Phytopathology* 26: 379–407.

Yavar, A., Sarmani, S., Hamzah, A., Khoo, K.S. 2014. Phytoremediation of mercury contaminated soil using *Scripusmucronatus* exposed by bacteria. *International Conference on Agricultural, Ecological and Medical Sciences* 2: 6–7.

Yazdani, M., Bahmanyar, M.A., Pirdashti, H., Esmaili, M.A. 2009. Effect of phosphate solubilization microorganisms and plant growth promoting rhizobacteria on yield and yield components of corn. *International Journal of Biological and Life Sciences* 1: 2–6.

Zhang, C., Kong, F. 2014. Isolation and identification of potassium-solubilizing bacteria from tobacco rhizospheric soil and their effect on tobacco plants. *Applied Soil Ecology* 82: 18–25.

Zolla, G., Bakker, M.G., Badri, D.V., Chaparro, J.M., Sheflin, A.M., Manter, D.K., Vivanco, J. 2013. Understanding root-microbiome interactions. In: F.J. de Bruijn (ed.), *Molecular Microbial Ecology of the Rhizosphere*, vol. 2. Wiley Blackwell, pp. 745–754.

2 Microbial-Mediated Molecular Mechanisms to Cope Up Salinity and Drought Stress in Plants

Kanchan Yadav, Pooja Yadav, and Kunal Singh
CSIR–Institute of Himalayan Bioresource Technology, Palampur, India
and
AcSIR, Ghaziabad, India

CONTENTS

DOI: 10.1201/9781003147077-2

2.1 INTRODUCTION

A key concern is the food supply and agricultural productivity for an ever-increasing population. The use of chemical fertilisers to increase food production is not environmentally friendly and is also health hazardous. Furthermore, environmental stress and climate change have a negative impact on plant development, lowering plant yield (Kaushal and Wani 2016). Harsh climatic conditions such as drought, salinity, freezing conditions, or soil pollution severely impact plant growth and development. According to a few estimates, as much as 50% of agricultural productivity gets lost due to several biotic and abiotic stresses. For example, salinity leads to water-deficit conditions, as more concentration of sodium ions changes the soil texture and alters soil porosity by decreasing water uptake. Therefore, the plants' growth and productivity is affected by high salt stress and considered an important limiting factor (Singh et al. 2018). Similarly, when the plants suffer from the drought during their initial growth phase, nutrient uptake, germination, and seedling setup are deeply affected due to low water availability and impaired enzyme functions (Taiz and Zeiger 2010). The severe drought condition decreases water potential of soil, leading to osmotic stress and inhibiting plant growth, ultimately causing death. Rhizospheric microorganisms provide a strong alternative and sustainable solution to resolve the negative effects of such abiotic stresses on plants and to increase crop production in an environmentally friendly manner.

The application of plant growth-promoting microorganisms (PGPM) inhabiting the rhizospheric soil increases the tolerance against several abiotic stress through direct and indirect mechanisms such as nutrient mobilisation, phytohormone production (auxin, gibberellin, and abscisic acid), and atmospheric nitrogen fixation (Meena et al. 2017). Many bacterial species have been quoted in this context for helping to overcome abiotic stress, e.g., *Pseudomonas*, *Microbacterium*, *Enterobacterium*, *Bacillus*, *Azospirillum*, and *Rhizobium* (Vurukonda et al. 2016; Pankaj et al. 2020). In addition, many genes providing tolerance against stress have been reported to get positively regulated by these PGPM inoculations and facilitate plant in tolerating stress conditions (Kaushal and Wani 2016).

2.2 PHYSIOLOGICAL AND BIOCHEMICAL CHANGES IN PLANT TISSUES UNDER THE EFFECT OF SALINE SOIL

High concentration of salt increases the turgor pressure in plant tissues by decreasing water and osmotic potential (Chakdar et al. 2019). As a result, water movement to leaf tissue from root decreases due to decrease in water potential. In response to saline environment, various anatomical changes occur in leaf such as reduced leaf area, photosynthetic activity, chlorophyll content and quick leaf defoliation, thicker epidermis, and larger palisade tissue (Parida et al. 2004; Zhang et al. 2016; Pankaj et al. 2020). For example, reduction in relative water content in cucumber leaves after salt exposure has been reported by Stepien and Klobus (2006). Photosystem II is affected by saline conditions, which inhibit photosynthesis (Tavakkoli et al. 2011). Various cellular enzymes, e.g., DNase, RNase, proteases, and nitrogen metabolism-involved enzymes, are also affected by salt stress conditions (Nathawat et al. 2005; Siddiqui et al. 2008; Kumar et al. 2020).

2.2.1 Microorganisms to Cope Up Salinity Stress

The salts that are soluble in water accumulate in the upper layer of the soil to a high extent, which cause hindrance in agricultural productivity. The total amount of dissolved mineral salts present in soil and water is known as salinity (Rao et al. 2016). Higher salinity results in lower water potential that decreases the water and nutrient uptake from soil, thus causing osmotic stress. Henceforth, salt stress is the excessive salt accumulation in soil that inhibits plant growth and may lead to even plant death (Rahman et al. 2019). Various physiological and metabolic changes in plants are adversely affected due to salt stress (Gupta and Huang 2014; Pankaj et al. 2019a). The untoward effect of salt stress varies with plant species, plant growth stages, climatic conditions, irrigation management practices, and soil conditions (Rao et al. 2019). At present according to an estimate in the world, approximately 50% of irrigated land is facing the challenge of salt stress with continuous increase everyday due to drastic environmental changes happening around, caused by global warming with human intervention as one of the major factors (Kumar et al. 2017). In this aspect, the solution for keeping the arable land quality high lies in the application of natural resources with soilborne beneficial microorganisms with potential to keep sustainable agriculture intact. There are millions of microorganisms inhabiting plant root systems, which enhance the growth and productivity of plants and keep the rhizospheric environment healthy. Among them, many microorganisms with potential of ACC deaminase (ACCd) activity, indole acetic acid (IAA) production, and other plant growth-promoting attributes can alleviate salt stress by various mechanisms such as increasing the concentration of compatible solutes, altering photosynthetic and antioxidant pathway, etc. (Rao et al. 2019; Pankaj et al. 2020). Yet, a long path has to be walked down to fully understand the various mechanisms deployed by microorganisms in helping the plant to tolerate the adversity. In this chapter, a detailed discussion is applied to bring all the major work related to microbial-assisted salt stress alleviation focussing on understanding the mechanisms with respect to molecular, genetic, physiological, and biochemical changes. This will lead to better utilisation of management practices under adverse conditions.

2.2.2 Microorganism-Mediated Enhancement of Salinity Tolerance in Plants

Microorganisms can survive in adverse conditions such as high saline environment through their metabolism (Meena et al. 2017). The microorganisms combat the salinity effect by reducing the ROS accumulation, maintaining the osmotic balance, and increasing the nutrient mobilisation to plants. The microorganisms such as halotolerant bacteria are capable of promoting plant growth by their PGP activities such as phytohormones production and solubilisation of nutrients. Table 2.1 summarises such halotolerant bacteria involved in plant salinity stress. Numerous traits, i.e., cytokinin, IAA, abscisic acid, ACCd, volatile organic compounds, and exopolysaccharides production, provide salinity tolerance to treated plants (Abbas et al. 2019). Several PGP bacterial strains even regulate gene transcription in plants for improvement of salt tolerance (Nautiyal et al. 2013). The different microorganisms improve

TABLE 2.1
Microorganisms Involved in Salinity Amelioration and Their Tolerance in Plants

Bacteria	Plant	Mechanism	References
Azotobacter sp.	*Zea mays*	• Improved nutrition	Rojas-Tapias (2012)
Arthrobacter sp.	*Triticum aestivum, Pisum sativum*	• Nutrient uptake increased	Barnawal et al. (2014)
Achromobacter sp.	*Lycopersicon esculentum*	• Ethylene level reduced	Mayak et al. (2004b)
Azospirillum sp.	*Helianthus annuus, Z. mays*	• Improved chlorophyll content	Naz and Bano (2015)
Acinetobacter sp.	*Cucumis sativus*	• Ethylene content reduced	Kang et al. (2014)
Aeromonas sp.	*T. aestivum*	• EPS production	Ashraf et al. (2004)
Burkholderia sp.	*C. sativus*	• Increased content of water and chlorophyll	Kang et al. (2014)
Brachybacterium sp.	*Arachis hypogaea*	• K^+ content higher	Shukla et al. (2012)
Bacillus sp.	*C. sativus, Codonopsis pilosula, Oryza sativa, Z. mays, Arabidopsis thaliana, Glycine max, Solanum lycopersicum, Puccinellia tenuiflora, T. aestivum, Vigna radiata* L.	• ACC deaminase activity improved plant growth	Upadhyay et al. (2011)
Enterococcus sp.	*Vigna radiate*	• Uptake of less sodium	Panwar et al. (2016)
Curtobacterium sp.	*Hordeum vulgare*	• Proline production	Cardinale et al. (2015)
Bacillus amyloliquefaciens	*O. sativa*	• Modifies microbial community and gene expression in rhizosphere	Nautiyal et al. (2013)
Pseudomonas pseudoalcaligenes and *Pseudomonas alcaligenes*		• Alleviates salinity stresses adverse effects and maintains bacterial diversity	Rangarajan et al. (2002)
B. amyloliquefaciens		• Increases plant biomass, proline, and water content and decreases reactive oxygen species	Chauhan et al. (2019)
P. pseudoalcaligenes and *Bacillus pumilus*		• Reduces superoxide dismutase activity, lipid peroxidation, and plant growth and development increased	Jha and Subramanian (2014)

(Continued)

TABLE 2.1 ***(Continued)***
Microorganisms Involved in Salinity Amelioration and Their Tolerance in Plants

Bacteria	Plant	Mechanism	References
B. amyloliquefaciens	*T. aestivum*	• Modulates the leaf gene expression profile	Nautiyal et al. (2013)
B. pumilus		• Increased antioxidant production and limited nutrient uptake	Khan et al. (2016)
Burkholderia sp.		• ACC deaminsae production	Sarkar et al. (2018)
Azotobacter vinellandii		• Higher GAs, IAA, proline, and malondialdehyde	Sahoo et al. (2014)
P. pseudoalcaligenes		• Glycine betaine-like quaternary • Compound concentration	
Sphingomonas pokkalii sp.		• Regulates salinity negative effect	Palaniyandi et al. (2014)
Curtobacterium albidum		• EPS production, IAA, and ACC deaminase activity under saline environment	Vimal et al. (2018)
Klebsiella sp.	*Avena sativa, T. aestivum*	• Increase proline and K^+ level	Sapre et al. (2018)
Micrococcus sp.	*A. thaliana* and *O. sativa*	• Siderophore and IAA production	Sukweenadhi et al. (2015)
Ochrobactrum sp.	*A. hypogaea*	• ACC deaminase and production	Paulucci et al. (2015)
Pseudomonas sp.	*A. hypogaea, Brassica napus, V. radiata, C. sativus Gossypium hirsutum, H. annuus, Glycyrrhiza uralensis, G. max, Cappisum annuum, T. aestivum, S. lycopersicum, Z. mays, Silybum marianum*	• Improves plant growth by IAA, EPS, and proline production	Abbas et al. (2019)
Rhizobium sp.	*V. radiata, P. sativum, Z. mays*	• Increases photosynthetic rate	Barnawal et al. (2014); Cardinale et al. (2015)

(Continued)

TABLE 2.1 ***(Continued)***
Microorganisms Involved in Salinity Amelioration and Their Tolerance in Plants

Bacteria	Plant	Mechanism	References
Serratia sp.	*T. aestivum*	• Exopolysaccharide production	Singh and Jha (2016)
Endophytic Fungi			
Piriformospora indica	*S. lycopersicum* and *Medicago truncatula*	• Increased proline and glycine betaine content	Abdelaziz et al. (2019)
Yarrowia lipolytica	*Z. mays*	• Modulates the proline production	Jan et al. (2019)
Epichloë bromicola	*Hordeum vulgare*	• Proline increased	Chen et al. (2019)
Trichoderma harzianum	*Brassica juncea*	• Modulates the antioxidant and osmolytes and improves the uptake of essential elements	Ahmad et al. (2015)
Trichoderma longibrachiatum	*T. aestivum*	• Increases chlorophyll and proline content	Zhang et al. (2016)
P. indica	*T. aestivum*	• Controls photosynthetic and compatible solute level	Zarea et al. (2012)

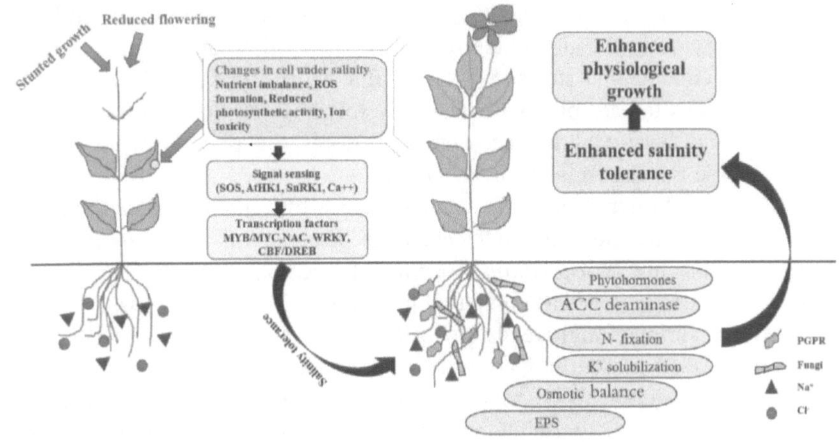

FIGURE 2.1 Morphological and physiological changes in plants under salinity stress ameliorated by microorganism-mediated enhancement of plant growth.

the health of plant under salinity stress by selectivity of Na^+, K^+, and Ca^{2+} and maintain the high ratio of K^+/Na^+ on roots of plants (Timmusk et al. 2017). Some microorganisms are known for production of phytohormones like abscisic acid (ABA), IAA, cytokinin, gibberellin that alleviate the physiological processes of plants and help to combat saline conditions as illustrated in Figure 2.1. For example, species belonging to genera *Stenotrophomonas*, *Rhizobium*, and *Pseudomonas* increase plant growth under saline conditions (100 mM NaCl) by producing IAA phytohormone. The reductive effect of saline environment on seed germination and plant root growth is alleviated by IAA production (Egamberdieva et al. 2011). *Azospirillum* sp. and *Bacillus amyloliquefaciens* increase the tolerance against salinity conditions by producing ABA phytohormone (Shahzad et al. 2017). Khan et al. (2011) showed that gibberellin-producing *Aspergillus fumigatus* increases the growth and photosynthetic rate of inoculated plants as compared to uninoculated plants.

Microorganisms increase the salinity stress tolerance by raising the osmolytes such as proline, sugar content, and malondialdehyde. Their accumulation increases by inoculation of microorganisms such as *Bacillus*, *Arthrobacter*, *Azospirillum*, and *Serratia marcescens* (Zarea et al. 2012; Singh and Jha 2016). For instance, *Bacillus pumilus* provides salinity tolerance by biofilm formation, maintaining the osmotic balance and extracellular polymeric substance (EPS) and ACCd production. The PGP activities of bacteria diminish the Na^+ ion supply to plant and improve the physiological traits in inoculated plants as compared to non-inoculated. According to Khan et al. (2019), the various plant growth-promoting rhizobacteria (PGPR) strains – *Arthrobacter woluwensis*, *Arthrobacter aurescens*, *Bacillus megaterium*, *Bacillus aryabhattai*, and *M. oxydans* – facilitate the salinity tolerance by activating the antioxidants and modulating phytohormones level and osmotic balance. In addition to osmolyte accumulation, microorganisms are also reported to mitigate the salt stress by activating the antioxidant enzymes as ROS formation is the most common phenomena during salinity stress, which can damage cell structures. *Pseudomonas mendocina*, *Bacillus* sp., and *Curtobacterium albidum* are reported to induce the antioxidant enzymes, i.e., superoxide dismutase, peroxidase and catalase, and are implicated in reducing salt stress (Singh and Jha 2016). *B. megaterium* strain A12 (BMA12) provides tomato plants with salinity tolerance by maintaining the redox homeostasis and hormonal balance, which restores the photosynthetic efficacy through multiple mechanisms for plant survival.

A few beneficial microorganisms under salinity stress conditions lead to biofilm formation, which alleviates the adverse effects. *Pseudomonas stutzeri*-inoculated chili pepper plants and *Azospirillum*-inoculated lettuce plants showed less negative effects against salinity stress as compared to controls (Bacilio et al. 2016). In addition to bacteria, various fungi are also reported to provide resistance against salinity stress. Various fungi present in plant roots help the plant in adapting to its environments, including saline soils (Parniske 2008). Under salt stress, fungal endophytes such as *Fusarium culmorum* isolate provide resistance against salinity stress in rice as this endophytic fungi have been isolated from monocot (*Leymus mollis*) (Redman et al. 2011). *Trichoderma* mediates the tolerance by alleviating the ABA, proline, and ascorbic acid level in treated plants of *Arabidopsis* with *Trichoderma atroviride* and *Trichoderma virens*. The IAA-producing fungi through endogenous signalling mediate root hairs and lateral root development, which helps plant in salt adaptation (Contreras-Cornejo et al. 2014). The arbuscular mycorrhizal fungi (AMF) counteracts salt stress, by maintaining osmotic balance inside the plant cells via osmolyte concentration increment, and also increases carbohydrates and antioxidant systems (Evelin et al. 2013; Pankaj et al. 2019b). Under saline conditions, the *Jatropha curcas* inoculated with AM fungi shows increased levels of soluble sugars and proline in the leaves (Dodd and Perez-Alfocea 2012).

2.2.3 Molecular Mechanisms of Salt Tolerance

Plants have evolved genetic and epigenetic systems in a response to abiotic stresses, including salt. Salinity tolerance is regulated by interaction of many genes. It includes a number of physiological and biochemical processes interacting with one other to provide resistance at cellular and molecular level (Shahbaz and Ashraf 2013). The traditional method of plant breeding and molecular techniques are used to increase resistance to salt stress.

2.2.4 Genes Involved in Salinity Stress Mitigation

A large number of genes and transcription factors are upregulated in different plant species under stress. Transcription factors (TFs) play important role in tolerance as they modulate the stress-responsive genes. These transcription factors comprise many gene families such as bZIP, MYB, NAC, AP2/ERF, C2H2, DREB, and WRKY that are involved in salinity stress response and their modulation or induction can lead to better adaptation to such environment (Gupta and Huang 2014). The TFs and their role in tolerance against salt stress have been studied abundantly. Overexpression of gene *CkdREB* in *Caragana korshinski* (Wang et al. 2011) and gene *bZIP* in *Tamarix hispida* (Wang et al. 2010) provide resistance to salinity stress.

PGPR also increases the expression of various stress-responsive genes, e.g., under *Enterobacter* spp. (Kim et al. 2014) inoculation, the salt-responsive genes like *RAB18, DREB2b, RD29B*, and *RD29A* in *Arabidopsis* increased in expression under salt-stress conditions. *Dietzia natronolimnaea* (*STR1*) and *Arthrobacter protophormiae* (SA3) provide tolerance against salinity stress in wheat plants by regulating the *CTR1* (Constitutive Triple Response1) component expression of the ethylene signalling pathway (Kumar et al. 2020). Bharti et al. (2016) show that the PGP *D. natronolimnaea* induce the *TaWRKY* and *TaMYB* gene expressions in wheat. *Klebsiella* sp. modulates *rbcL* and *WRKY1* (Sapre et al. 2018) gene expressions in *Arabidopsis*. Over-expression of NAC genes confers salt resistance in rice and wheat, like transcription factors OsNAC5 and ZFP179 (Song et al. 2011) that are upregulated under salinity stress in rice, which regulates the synthesis and accumulation of sugar, proline, and LEA proteins. In *Arabidopsis*, AtWRKY8 upregulates under salt stress, which binds to the RD29A promoter (Hu et al. 2013).

It has been reported that *PsbA* gene transcription and translation lowers under excessive salt accumulation conditions. *PsbA* plays a key role in repairing the damaged photosystem (Krishna et al. 2013). It has been reported that BMA12 increases the expression of *PsbA* gene under stress. It has also been observed that *PBGD* gene expression is increased by symbiosis of BMA12, which led to higher chlorophyll biosynthesis in plants (Akram et al. 2019). *Pseudomonas* PS01-inoculated *Arabidopsis* shows improvement in the germination rate under salt conditions. The *GLYI7* and *APX2* transcriptional levels increase in seedlings inoculated with PS01 was low in comparison to untreated plants. This reduced level of *GLYI7* and *APX2* shows that *Pseudomonas* may reduce the stress in plants (Chu et al. 2019). *A. woluwensis, A. aurescens, B. aryabhattai, B. megaterium*, and *M. oxydans* regulate the salt tolerance gene (*GmST1*) and IAA-mediating (*GmLAX3*) gene (Khan et al. 2019).

2.2.4.1 Salt Overly Sensitive Stress Genes

The salt overly sensitive (SOS) stress signalling pathway involves the three major genes *SOS1, SOS2*, and *SOS3* (Hasegawa et al. 2000). The above genes act in cascade manner where Na^+/H^+ antiporter is encoded by *SOS1*; *SOS2* is Ser/Thr kinase and *SOS3* encodes a protein which induces the Ca^+ binding and Na^+ efflux from cell. The *SOS1* gene expression increases during salt stress in soybean and durum wheat. Through transgenic approach, this gene is transformed to *Arabidopsis* and tolerance to stresses has been achieved (Zhao et al. 2017). Microorganisms are reported to increase the SOS gene expression and *D. natronolimnaea*-inoculated wheat plants show modulation of genes – *SOS4* and *SOS1*, and *NHX* and *HKT* genes encoding ion transporters are also modulated (Chakdar et al. 2019). Nautiyal et al. (2013) showed that *B. amyloliquefaciens* rice plants increase the salt tolerance in plants by upregulation of *EREBP, SOS1, NADP-Me2*, and *SERK1* genes.

2.2.4.2 Abscisic Acid Responsive Genes

ABA is important hormone in salt stress as its level increases in the water-deficit soil and its application improves the stress effect. ABA is crucial cellular signal that regulates the expression of various water-deficit and salt-responsive genes. ABA treatment instigates the MAPK4-like *GLP1* and *TIP1* gene expressions in wheat under salt stress (Keskin et al. 2010).

2.2.5 Osmolyte Accumulation

Water homeostasis and accumulation of osmolytes are helpful in improving the impact of salinity stress. The PGPR *B. megaterium*-inoculated maize plants show increased *ZmPIP* gene expression (Marulanda et al. 2010). The *B. aryabhattai* (H19-1)- and *B. mesonae* (H20-5)-inoculated plants show abscisic acid (ABA), proline, and antioxidant enzyme activities significantly high as compared to uninoculated and H19-1-inoculated plants under saline conditions, and ABA-response element-binding proteins 1 (*AREB1*) and 9-cisepoxycarotenoid dioxygenase 1 (NCED1) genes were also upregulated (Yoo et al. 2019).

2.2.6 Extracellular Polymeric Substance Production

A number of multifunctional polysaccharides including structural, intracellular, and extracellular or exo-polysaccharides are synthesised by microorganisms under natural conditions. EPS plays an important role in salinity stress alleviation (Upadhyay et al. 2011). EPS binds to cations like Na^+, reducing their availability for absorption by plants. Recent research showed that microorganisms' physical and chemical attributes are modulated by EPS under saline conditions. It has also been reported that structural stability of the biofilms is maintained through EPS (Zheng et al. 2016). The secreted substances determine biofilm strength against salinity stress. A high saline environment disrupts biofilms produced by affecting the physiological processes and microbial metabolism (Bassin et al. 2011). The *C. albidum* strain SRV4 shows alleviation against stress by EPS production as well as IAA production, ACCd activity, and nitrogen fixation (Vimal et al. 2018).

2.2.7 Essential Nutrient Uptake

The essential elements such as phosphorus, nitrogen, and potassium accumulation are reduced under saline conditions due to ion toxicity and high osmotic pressure (Kumar et al. 2020). The microorganisms' association (symbiotic and non-symbiotic) with plants alleviates salt stress through increasing the nutrient uptake. The plant growth-promoting bacteria (PGPB) convert the unavailable phosphorus form to available form by chelation. The phosphate-solubilising bacteria make PO^{3-} available to plants under stress conditions, e.g., *Burkholderia* solubilises the PO^{2-} from minerals and makes it available to plants (Kang et al. 2014). In another study, it has been shown that K and P uptake increases in tomato and wheat plants when treated with *Achromobacter piechaudi*, and *Bacillus aquimaris*, respectively (Mayak et al. 2004b).

2.2.8 miRNAs in Salinity Stress

The short non-coding RNAs having a length of 19–23 nucleotides, which play a regulatory role in many biological processes, are known as miRNAs (Budak et al. 2015). According to rindade et al. (2010), different miRNAs such as miR159, miR393, miR402, miR408, and miR397b in *Phaseolus*, *Medicago*, *Arabidopsis*, rice, and other plants play regulatory roles for various abiotic stresses such as cold, salinity, and drought. The miR393 has a regulatory role for salinity tolerance in *Arabidopsis*; the over-expression of *OsMIR393* provides tolerance to excess salt (Gao et al. 2011). The miR169 alleviates salinity stress through a nuclear transcription factor expression in rice (Zhao et al. 2009).

2.2.9 Fungi-Mediated Salinity Stress Tolerance

The endophytic and mycorrhizal fungi are known to induce systemic resistance and mitigate the stress conditions through alleviating the osmo-protectant level, activating the antioxidant system and modulating the phytohormone level (Li et al. 2017). *Trichoderma asperellum*-inoculated cucumber plants mitigate the salt stress effect by altering the phytohormone level and its phosphate solubilisation activity (Lei and Zhang 2015). It has been reported that *Trichoderma harzianum*-inoculated salt-stressed mustard plants showed enhanced oil content, enhanced antioxidants, and osmolyte accumulation (Ahmad et al. 2015). Various *Trichoderma* strains are reported to enhance the salinity tolerance through ACCd production. Root colonising fungal endophyte *Piriformospora indica* induces tolerance against drought in Chinese cabbage and tolerance to salinity stress in barley by increasing the antioxidant level and other growth processes (Sun et al. 2010).

AM fungi allow the high water flow rate by upregulation of aquaporin genes. It has been observed that PIP gene (*PvPIP1-1*, *PvPIP1-3*, and *PvPIP2-1*) expressions increased in AM-inoculated roots (Aroca et al. 2007). In one more study, the tonoplast aquaporin (*LeTIP*) and plasma membrane aquaporin (*LePIP2* and *LePIP1*) transcript levels were found to be lower in AM-inoculated roots than in non-AM-inoculated roots, while in leaves, all three aquaporin gene expressions were significantly enhanced by AMF colonisation (Gupta et al. 2020).

2.3 MICROORGANISMS TO COPE UP DROUGHT STRESS

Among many abiotic factors, water scarcity is a key environmental factor for reduced production under agriculture (Langeroodi et al. 2020). Prolonged water scarcity is recognised as drought conditions that pose a threat to plant growth, productivity, and even survival (Naveed et al. 2014). The major features of plant-water interactions, viz., transpiration rate, relative water content, osmoticum, water, and pressure potential, become crucially impacted leading to decreased crop productivity (Kirkham 2005). To endure drought stress, plants have developed mechanisms such as morphological and osmotic adaptations, water supply optimisation, antioxidant systems, and stress-responsive proteins that help in reducing the harmful effects of drought (Farooq et al. 2009). The knowledge of such adaptive responses by plants to drought stress is important in understanding their behaviour during stress. Likewise, it has been reported and observed for decades that drought can cause decreased nutrient uptake. There are many reasons for such nutrient unavailability during water-deficient conditions such as decreased nutrient flow and diffusion in soil, nutrient uptake proteins in roots are downregulated (Robredo et al. 2011), and decreased activity of microorganisms (Sanaullah et al. 2012). This is also evident that amount of nutrients is more in upper layer of soil as compared to deeper soil layer, thus affecting nutrient availability to plants during drought stress (Bradford and Hsiao, 1982). Beneficial microorganisms aid the plants in overcoming nutrient deficiencies in drought-affected nutrient-poor soil. Such microorganisms help in the uptake of nutrients, thereby enabling their solubility and absorption by the plant (Glick 2012). Figure 2.2 illustrates the complex interaction of these microorganisms (fungi, bacteria, viruses, archaea, protozoa, etc.) with plants having the potential to improve yield under stress conditions (Dastogeer and Wylie 2017). Fungal endophytes are one such microorganism that has the ability for plant tissue colonisation asymptomatically and have the potential to confer stress tolerance to plants under natural ecosystem (Saikia et al. 2018). Endophytic fungi such as *F. culmorum* and *Curvularia protuberate* have been found to increase the crop production of water-deficient rice plants (Redman et al. 2011). Endophytes also increase the development of osmolytes affecting plant-water equilibrium and photosynthesis rate, antioxidant generation, and other structural parameters, thus improving the ability of plants to combat stress (Dastogeer and Wylie 2017). Another set of beneficial microorganisms called PGPRs also carry the ability to improve growth of plants under drought stress through a broad range of mechanisms like siderophore production, nitrogen fixation, phosphate solubilisation, and phytohormone signalling (Bhattacharyya and Jha 2012).

2.3.1 Microorganisms' Role in Drought Stress through Modulation of Plant Hormone Levels

Microorganisms providing drought stress tolerance improve plant growth through enhancing nutrient absorption and soil fertility, thus reducing water scarcity and salinity (Kaushal and Wani 2016; Vurukonda et al. 2016). Many known microorganisms mitigating drought stress are catalogued in Table 2.2, along with their functions. Tolerance against water scarcity may be enhanced by the utilisation of these stress-tolerant microorganisms (Nadeem et al. 2014). These microorganisms impart some shielding effect to plants against the negative effect of drought even in very low water availability by providing nutrients and preferable environmental conditions to the rhizosphere. Broadly, plant growth in drought stress can be enhanced through microorganisms by both direct and indirect mechanisms. The direct mechanism is provided by enhancing phytohormone production and reduction of various abiotic factors such as regulation of ethylene through production of ACCd, and an indirect mechanism improves nutrient availability to plants through IAA production, biological nitrogen fixation, phosphate solubilisation, and siderophore production (Farooq et al. 2009). Consequently, under stress conditions, PGPR can stimulate production of plant hormones like IAA, ABA, ACC, gibberellic acid, and cytokinin.

Auxins are known to increase tolerance to water scarcity of which IAA is the most prominent member that influences the differentiation and shoot growth and cell division during drought conditions (Ullah et al. 2019). There are two major pathways for IAA production, viz., tryptophan dependent and independent with PGP microorganisms preferring the first one. Tryptophan-dependent pathways are used

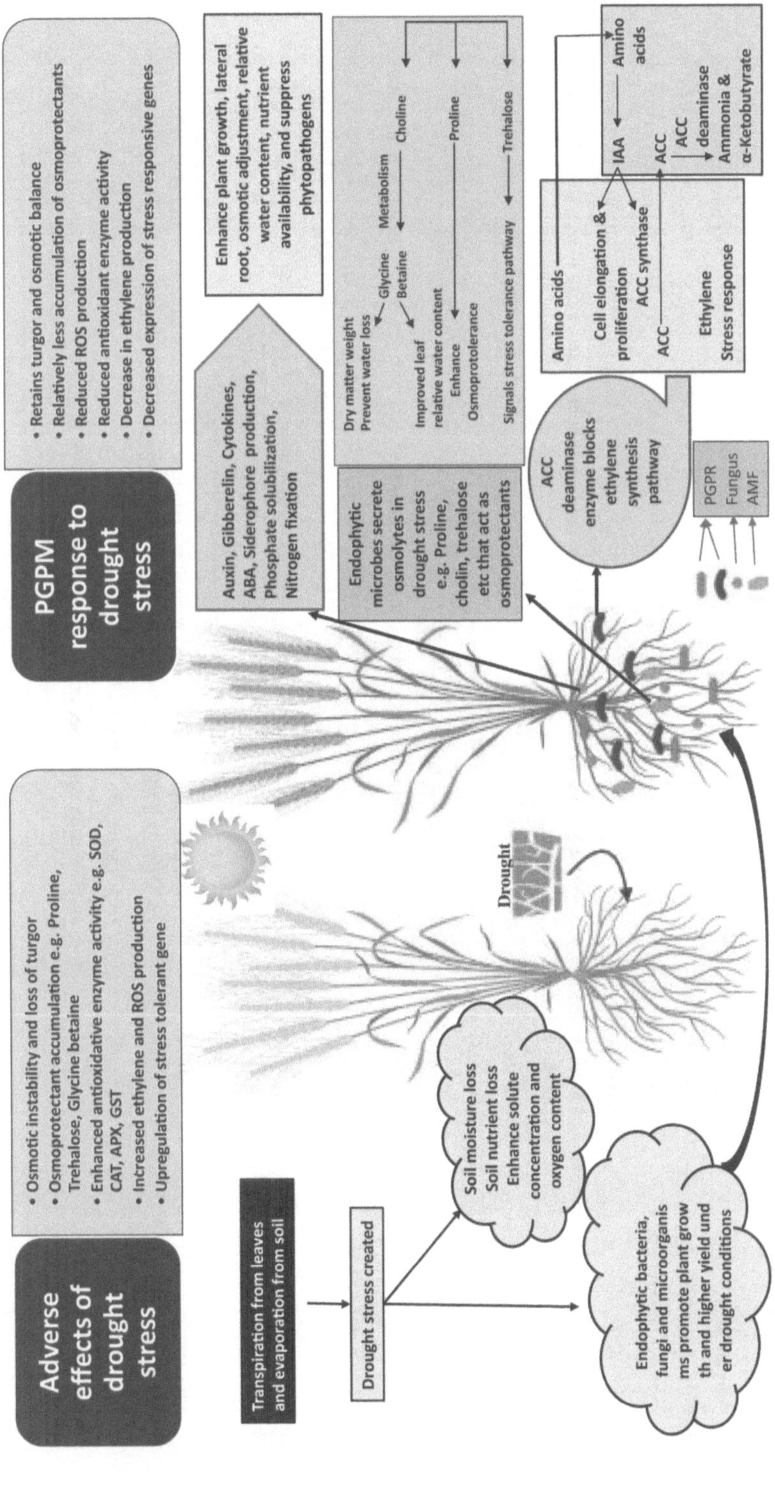

FIGURE 2.2 A summarised overview of plant growth-promoting microorganism-mediated drought tolerance in plants.

TABLE 2.2

Cataloguing of Prominent Bacterial Examples Assisting Drought Stress Tolerance in Plants with Their Putative Function during Interaction

Microorganism	Crop	Function	References
Pantoea agglomerans	*Triticum aestivum*	EPS	Amellal et al. (1998)
Paenibacillus polymyxa B2	*Arabidopsis thaliana*	Early dehydration response induction 15 (ERD15)	Timmusk et al. (2014, 2017)
Rhizobium spp.	*Helianthus annuus*	EPS	Alami et al. (2000)
Achromobacter piechaudii ARV8	*Solanum lycopersicum* and *Piper nigrum*	Ethylene production is decreasing while fresh and dry weight of plant are increasing	Mayak et al. (2004a)
Bacillus	*Lactuca sativa*	Cytokinin	Arkhipova et al. (2007)
Rhizobium tropici	*Phaseolus vulgaris*	Level of nitrogen and nodulation enhanced	Figueiredo et al. (2008)
P. polymyxa	*P. vulgaris*	Enhance root nodulation	Figueiredo et al. (2008)
Variovorax paradoxus	*Pisum sativum*	Ethylene production is reduced due to ACC deaminase	Belimov et al. (2009)
Pseudomonas putida	*H. annuus*	Production of EPS is responsible for soil health maintenance	Vardharajula et al. (2009)
Bacillus spp.	*Zea mays*	Increased accumulation of proline, decreased APX, GPX, CAT, and EPS	Vardharajula et al. (2011)
Bacillus subtilis, *Bacillus cereus*, *Serratia* spp.	*Cucumis sativus*	cAPX, rbcL, rbcS gene, and chlorophyll content increase	Wang et al. (2012)
V. paradoxus	*P. sativum*	Enhance nutrient availability, ACC deaminase-reduced root ABA	Jiang et al. (2012)
Enterobacter cloacae, *Enterobacter cancerogenus*	*Jatropha curcas*	Low amount of ACC triggers less endogenous ethylene, which eliminates potentially inhibitory effects of stress-induced more ethylene	Jha et al. (2012)
Bacillus licheniformis	*P. nigrum*	Gene expression of VA, Cadhn, sHSP	Lim and Kim (2013)
Azospirillum lipoferum	*Z. mays*	Increase accretion of free amino acids, soluble sugar, and proline	Bano et al. (2013)

(Continued)

TABLE 2.2 *(Continued)*
Cataloguing of Prominent Bacterial Examples Assisting Drought Stress Tolerance in Plants with Their Putative Function during Interaction

Microorganism	Crop	Function	References
Bacillus amyloliquefaciens (5113) and *Azospirillum brasilense* (NO40)	*T. aestivum*	Stress-responsive genes APX1, SAMS1, and HSP are all upregulated	Kasim et al. (2013)
Phyllobacterium brassicacearum strain (STM196)	*Arabidopsis thaliana*	Reduction of leaf transpiration rate so that increased ABA content improves biomass and water use efficiency and delay in reproductive development	Bresson et al. (2013)
Bacillus thuringiensis	*T. aestivum*	GR, SOD, CAT, IAA, and alginate activity all increased Stress volatile emissions are reduced	Timmusk et al. (2014)
Pseudomonas aeruginosa	*Vigna radiate*	Upregulation of genes, such as DREB2A, DHN, and CAT1 Improve SOD, CAT, and POX activity	Sarma and Saikia (2014)
Acinetobacter, *Pseudomonas*	*Vitis vinifera*	EPS and IAA	Rolli et al. (2014)
Pseudomonas sp.	*Z. mays*	EPS increased soil aggregation and water content	Naseem and Bano (2014)
Burkholderias, *Enterobacter* sp., *Phytofirman*	*Z. mays*	ACC deaminase activity enhances chlorophyll level	Naveed et al. (2014)
Glucon acetobacter diazotrophicus (PAL5)	*Saccharum officinarum*	Drought resistance was conferred by inoculation, which triggered ABA-dependent signalling genes	Vargas et al. (2014)
A. brasilense (NO40) *R. lleguminosarum* (LR-30) *Mesorhizobium ciceri* CR-30 and (CR-39)	*T. aestivum*	Catalase, EPS, and Rhizobacteria produced IAA, which helped in development	Hussain et al. (2014)

(Continued)

TABLE 2.2 ***(Continued)***
Cataloguing of Prominent Bacterial Examples Assisting Drought Stress Tolerance in Plants with Their Putative Function during Interaction

Microorganism	Crop	Function	References
B. thuringiensis	*Lavandula dentate*	Higher proline and K content were induced by IAA, which increased nutritional, physiological and metabolic activities while lowering GR and APX activity	Armada et al. (2014)
P. putida (H-2–3)	*Glycine max*	ABA, salicylic acid, and jasmonic acid levels are all low reduced SOD, flavonoids, and free radical scavenging activity, which modulated antioxidants	Kang et al. (2014)
B. thuringiensis	*Trifolium repens*	APX activity and proline decreased	Ortiz et al. (2015)
Azospirillum sp.	*L. sativa*	Increased levels of chlorophyll, aerial biomass, and ascorbic acid, improved all visual quality, chroma, and antioxidant ability	Fasciglione et al. (2015)
Azospirilum brasilense sp. (245)	*A. thaliana*	Plant seed yield, proline content, and relative leaf water all increased, while MDA, stomatal conductance, and also relative soil water level all decreased	Cohen et al. (2015)
P. putida	Chickpea *Cicer arietinum*	Stress-responsive gene expressions, osmolyte accumulation, and ROS scavenging capacity	Tiwari et al. (2016)
Pseudomonas libanensis (TR1) *Pseudomonas reactans* (Ph3R3)	Field mustard *Brassica oxyrrhina*	Improved growth of plant and relative water content as well as lower MDA and proline levels in leaves	Ma et al. (2016)
Sinorhizobium medicae	Burclover *Medicago truncatula*	During drought stress, root nodulation, and nutrient acquisition are essential	Staudinger et al. (2016)
Pantoea alhgai	*Alhagi sparsifolia*	Production of soluble sugars, EPS, siderophore, IAA, and protease increased, while proline decreased	Chen et al. (2017)
Bacillus megaterium	*T. aestivum*	Enhancing 59% relative water content, protein content, chlorophyll *a*, *b* and carotenoid, and proline content and reducing MDA content	Rashid et al. (2021)

by several microorganisms for IAA production such as *Agrobacterium tumefaciens*, *Azospirillum*, *Bacillus* sp., *Erwinia herbicola*, *Pseudomonas syringae*, *Pantoea alhagi*, and *Rhizobium* (Goswami et al. 2016; Chen et al. 2017). *P. alhagi*, an endophytic bacteria isolated from *Alhagi sparsifolia* leaves, produces IAA along with siderophore and ammonia, providing tolerance against drought by decreasing chlorophyll degradation and malondialdehyde (MDA) accumulation in wheat (Chen et al. 2017).

ABA is an essential component of plant photochemistry such as stomata closure, inhibition of plant growth, and breaking seed dormancy (Daszkowska-Golec 2016). During water-scarcity conditions, the ABA is responsible for root hydraulic conductivity and regulation of stress-responsive genes; thus, microorganisms such as *Azospirillum brasilense* provide tolerance for water deficits by increasing the ABA levels when inoculated to *Arabidopsis thaliana* (Cohen et al. 2015). Similarly, PGPM with ACCd enzymes can breakdown the ACC, i.e., ethylene precursor, thereby reducing the ethylene levels during drought stress (Glick 2005). ACC synthase converts SAM (*S*-adenosyl methionine) to ACC, which is then converted to ethylene by ACC oxidase (Bal et al. 2013). During water stress conditions, ethylene regulates the plant's homeostasis resulting in a decrease in growth. The presence of ACCd-producing microorganisms leads to the decay of ethylene in alpha-keto butyrate and thus reduces the harmful effects of ethylene and encourages plant growth (Glick 2005). The activity of microbial ACCd can further enhance the beneficial effects (i) defending the plant against the negative effects of different stresses (Glick 2014), (ii) enhancing the nodulation of legume (Belimov et al. 2009), (iii) causing late senescence of flower and leaves (Ali et al. 2012), and (iv) acting as a biocontrol agent for pathogens on plants (Hao et al. 2011). Well-known ACC-producing bacteria are *Alcaligenes* sp., *Achromobacter peichaudii* ARV8 (Mayak et al. 2004a), *Bacillus* sp., *Burkholderia*, *Pseudomonas* sp. (Noreen et al. 2012), *Varorax paradoxus* (5C-2) (Belimov et al. 2009), and *Ochrobactrum* sp. (RJ12) (Saikia et al. 2018). Opposed to ethylene, gibberellin and cytokinin are major group of phytohormones that are involved in many plant growth stages such as flowering, seed germination, fruit ageing, and stem elongation (Hedden and Phillips 2000). Beneficial bacteria such as *Bacillus aqimaris*, *Bacillus subtillus*, *Micrococcus luteus*, *Serratia marcecens*, and *P. putida* produce gibberelic acid in drought conditions (Kang et al. 2014). Cytokinins are the second group of phytohormones that stimulate the processes of plant cell division and cell differentiation with multiple effects on leaf senescence apical dominance and axillary bud development. Some of the cytokinin-producing microorganisms include *Proteus*, *Azospirullium*, *Bacillus*, *Klebsiella*, *Pseudomonas*, and *Xanthomonas* (Maheshwari et al. 2015).

Nitrogen is the most important macronutrient for growth and development of plant as it can be present in soluble form in very small amounts in certain soils. Therefore, to increase the yield in the plant, there is a demand for an excessive amount of nitrogenous fertiliser (Pathak et al. 2016). Some beneficial bacteria and endophytes help to complete nitrogen deficiency in the plant such as *A. brasilense sp*246, *Bradyrhizobium* sp., and *Rhizobium leguminosarum* E11. Further, some ACCd producing and stress tolerant bacteria play a role in atmospheric nitrogen fixation under water depletion conditions, for example, *Ocrobactrum pseudogrygnonens* (*RJ12*), *Pseudomonas* sp. (*RJ15*), and *Bacillus subtilis* (*RJ46*) (Saikia et al. 2018).

Phosphorous is another essential element that is required in a large amount not only for ATP and phospholipid production but also for photosynthesis and respiration as these activities required for plant growth and development. Drought tolerant microorganism dissolves inorganic phosphate (P) from water-scarce land and makes it available to plants even under very low moisture content such as *Bacillus halodenitrificans* (PU62), *Psychrobacter fozii* (IIWP-12), *Pseudomonas* sp., *Pseudomonas thivervalensis* (IHD-3), *B. subtilis* (IARI-IIWP-2), and *Pseudomonas monteilii* (IARI-IIWP-27) (Verma et al. 2014). Similarly, soil is rich in iron but this mineral occurs in the oxidised form in general and cannot be used by plants directly. In such conditions, PGPR produces siderophore that are Fe-chelating compounds responsible for acquiring of ferric ions, thus meeting the deficiency of iron (Whipps 2001). Diverse PGPRs isolated from wheat-growing sites like *Arthrobactor, Bacillus, Pseudomonas, Corynebacterium*, etc., were found to produce siderophores under water-scarce conditions (Whipps 2001).

The presence of AMF in the ecosystem is now generally recognised as shielding the host plants from drought stress (Wu and Xia 2006) through better uptake of water from soil (Duc et al. 2018), enhanced osmotic turgor adjustment, and enriched root hydraulic conductivity (Borowicz 2010). This symbiotic relationship enabled plants to grow despite various environmental stresses due to better nutrient acquisition (Brachmann and Parniske 2006). They produce high osmolytes during water-scarce conditions such as proline, glycine, and sugar, thus preserving tissues' water status in plants under stress. There are reports of many AMF imparting positive effects during drought stress such as *Funneliformis mosseae, Rhizophagus diaphagnum*, and *Glomus versiforme* that have been shown to increase the biomass of trifoliate orange along with total phosphate activity and phosphate content in water stress as provided in detail in Table 2.3 (Wu et al. 2011a). AMF have been shown to increase the trehalose and amino acid content of maize (Schellenbaum et al. 1998), increase nitrogen percentage in sunflowers despite the fact that in general nitrogen percentage is reduced under drought stress (Gholamhoseini et al. 2013), and mitigate the hydrogen peroxide-mediated oxidative damage with MDA content while increasing anitioxidants in *Arizona cypress* (Aalipour et al. 2020). In *G. versiforme*, AMF have been shown to increase the antioxidative enzyme activities, viz., APX, SOD, and CAT in citrus (Wu and Xia 2006).

2.3.2 Microbial-Enhanced Osmolytes and ROS Reduction Increases Plants' Water-Deficit Stress Tolerance Ability

During drought stress, plants undergo various biochemical and physiological changes in order to survive such as osmolyte regulation and ROS reduction. These changes positively regulated by many PGPM. Osmolytes are known to increase during water-deficient conditions imparting positive effects on membrane integrity. During drought stress, many plants tried to get protected via osmotic adjustment also termed as compatible solutes, i.e., accumulation of inorganic molecules and organic osmoticum in cells such as calcium, proline, malate, glycine beta, trehalose, and sucrose. When maize plants under drought stress were inoculated with *A. brasilense* for trehalose production, it was observed that root and leaf biomass

TABLE 2.3
Role of Mycorrhiza in Drought Stress Mitigation in Plants

Microorganism	Crop	Function	References
Funneliformis mosseae	*Zea mays*	Increase in trehalose content and higher trehalase activity accretion of amino acids and imino acids	Schellenbaum et al. (1998)
F. mosseae and *Claroideoglo musetunicatum*	*Triticum aestivum*	In mycorrhizal plants, higher Fe and P levels, also high biomass, and grain yield	Al-karaki et al. (2004)
Rhizophagus intraradices	*Glycine max*	Mycorrhizal protected plants against oxidative stress and greater leaf water potential in mycorrhizal plants	Porcel and Ruiz-Lozano (2004)
Glomus versiforme	*Citrus* sp.	Higher concentration of CAT, APX, and SOD	Wu and Xia (2006)
Rhizophagus intraradices	*Sorghum bicolor*	Mycorrhiza increased grain yield by 17.8% and minimised adverse effect of drought	Alizadeh et al. (2011)
R. diaphanum *F. mosseae* *G. versiforme*	*Poncirus trifoliate*	Drought-stressed mycorrhizal seedlings, especially *F. mosseae*, have higher biomass and plant growth, as well as activity of phosphatase, phosphorus contents in root and leaves	Wu et al. (2011b)
F. mosseae	*Helianthus annuus*	Inoculated plants had a higher nitrogen percentage than control plants with *F. mosseae*, also provided higher dry matter, seeds, and oil yields	Gholamhoseini et al. (2013)
F. mosseae	*P. trifoliate*	Higher root-hair growth and IAA synthesis	Liu et al. (2018)
F. mosseae	*Cupressus arizonica*	Improved growth water-deficit-induced H_2O_2 and MDA	Aalipour et al. (2020)

were also enhanced along with trehalose, indicating beneficial effect on both yield and survival (Rodriguez-Salazar et al. 2009). These compatible solutes have been reported to be increased by endophytic bacteria under drought stress, for instance, *Sphingomonas* sp. (LK11) isolated from *Tephrosia apollinea* leaves increased the production of amino acids (proline, glutamate, and glycine) and sugars in soybean (Asaf et al. 2017). Amino acids like proline, cysteine, aspartic acid, phenylalanine, and glutamic acid contents also get increased in rice plants after inoculation with *B. amyloliquefaciens*, an endophytic bacteria (Shahzad et al. 2017). Proline is one of the chief compatible solutes in response to water scarcity and plays an important role in osmotic adjustment, stabilising cellular structures, buffering of cellular redox, and scavenging free radicals (Huang et al. 2014). Inoculation with PGPM such as *Arthrobactor*, *Bacillus*, and *Burkholderia* (Sziderics et al. 2007) and endophyte *Chaetomium globosum* (Cong et al. 2015) are known to increase the proline content during drought stress, resulting in improvement in plants' stress tolerance capability. Glycine betain is an important osmolyte apart from proline that accumulates in plant cells during drought stress-mediating osmotic adjustments (Serraj and Sinclair 2002). PGP-containing microorganisms such as *Raoultella planticola* YL2 and *Klebsiella variicola* F2 increase the production of glycine betaine and choline (Gou et al. 2015).

Similar to PGPR, endophytic bacteria *Pseudomonas putida* can also enhance the osmolyte accumulation like proline, sugar, and MDA along with effectiveness in ROS reduction (Tiwari et al. 2016). Such a reduction in ROS levels is a major mechanism of many microorganisms for stress acclimatisation as ROS production is enhanced during water-deficient conditions leading to a hypersensitive response (HR) like situations causing oxidative damage to plants. In such conditions, superoxide dismutase (SOD) is the first enzyme to act against ROS production in plants (Das and Roychoudhury 2014). Microbial inoculation increases the expression of the SOD gene and its isoforms in drought stress, such as *P. indica* increases the production of SOD enzymes by over-expression of *CBL1*, *RD29A*, and *ANAC072* in *Brassica campestris* sp. chinensis roots (Sun et al. 2010). Table 2.4 catalogues many such AMF and endophytic bacteria that are helpful in decreasing ROS production under water-deficient conditions ultimately helping in drought stress alleviation. In *G. versiforme*, an AMF when inoculated in a citrus plant shows a high activity of antioxidants protects the plant from ROS (Wu and Xia 2006). When *Medicago sativa* was inoculated with *Sinorhizobium meliloti*, it increased the production of SOD by upregulating CU/ZnSOD in the root (Naya et al. 2007). Catalase, glutathione reductase, peroxidase, and ascorbate are also important for ROS scavenging (Gill and Tuteja 2010).

2.3.3 Genetic Modulation of Microorganism-Mediated Drought Stress Tolerance

When subjected to drought conditions, the plant responds by upregulating drought stress tolerant genes. Auxin-induced gene glucanase *At4g30280*, auxin-responsive gene *At1g56150*, and an extension carrier gene *At2g17500* were upregulated by *Pseudomonas chlororaphis* colonisation in *Arabidopsis* (Cho et al. 2013). An important class of genes in *Arabidopsis* that plays a role in detoxification and mitigating the damage caused to the cell is the one encoding the late embryogenesis abundant protein (At3g17520) known as LEA protein. This protein functions as a chaperone to protect molecules from cellular damage (Umezawa et al. 2004). Plants' resistance mechanisms and survival under dehydration are known to be enhanced by this class of proteins. Similarly, some other genes involved in drought stress tolerance are low-temperature-responsive (65 kDa) protein At3g15670, putative dehydrin At3g50980, and putative protein phosphatase 2C-At1g07430 (Cho et al. 2013). The rice OsbZIP46 gene also promotes drought tolerance by coding for the transcription factor of bZIP class (Lindemose et al. 2013). Under water-deficit conditions, two major pathways have been described for stress tolerance, one of which is ABA-dependent and

TABLE 2.4
Endophytic Fungi Assisting Drought Stress Tolerance in Plants with Their Major Function during Plant-Fungus Interaction

Microorganism	Crop	Function	References
Sinorhizobium meliloti	*Medicago sativa*	In roots, the SOD genes FeSOD and CU/ZnSOD upregulated	Naya et al. (2007)
Trichoderma hamatum (DIS 219b)	*Theobroma cacao*	In root, DREB2A, CBL1, ANAC072, and RD29A, ERD1 upregulated	Bae et al. (2009)
Piriformospora indica	*Brassica campestris* ssp. *Rosa chinensis*	Increase the levels of peroxidase, catalase, and superoxide dismutase (SOD), DREB2A, CBL1, ANAC072, and RD29A gene upregulate in root	Sun et al. (2010)
Trichoderma harzianum	*Oryza sativa*	Improve root development regardless of water availability and postpone drought stress	Shukla et al. (2012)
Gluconacetobacter diazotrophicus	*Saccharum officinarum*	IAA and proline production In shoots, the genes DREB1A/CBF3, ERD15, and DREB/CBF are all upregulated	Vargas et al. (2014)
T. harzianum (TH-56)	*O. sativa*	Aquaporin, dehydrin, and MDA genes are upregulated by DHN/AQU in the root	Pandey et al. (2016)

another is ABA independent (Abe et al. 2003). ABRE and DRE/CRT are essential cis-acting components of which ABRE is involved in the ABA-dependent pathway while DRE/CRT is involved in the ABA-independent pathway during drought stress. Stress-responsive genes are activated by suitable transcription factors, viz., ABF2/4/DREB2/AP2, and are responsible for the activation of COR78/RD29A/LTI78 having ABRE and DRE/CRT elements and shown in Figure 2.3 with their mechanisms (Shinozaki and Yamaguchi-Shinozaki 2007).

Sherameti et al. (2008) provided the molecular mechanism of water-deficiency tolerance by *P. indica* in *Arabidopsis* plants. In this study, genes associated with draught tolerance such as phospholipase D delta, Calcinurin B-like protein 1 (CBL1), CBL-interacting protein kinase 3, and histone acetyltransferase (HAT) were upregulated during water-deficit conditions. Similarly, the study by Kandasamy et al. (2009) explained the function of microbial-assisted stress mitigation during drought conditions by expression of drought stress-responsive genes, viz., *Hsp20, bZIP1*, and *AP2-EREBP* that are known to be implicated in ABA-dependent pathways. *Pseudomonas fluorescens* pf0-1

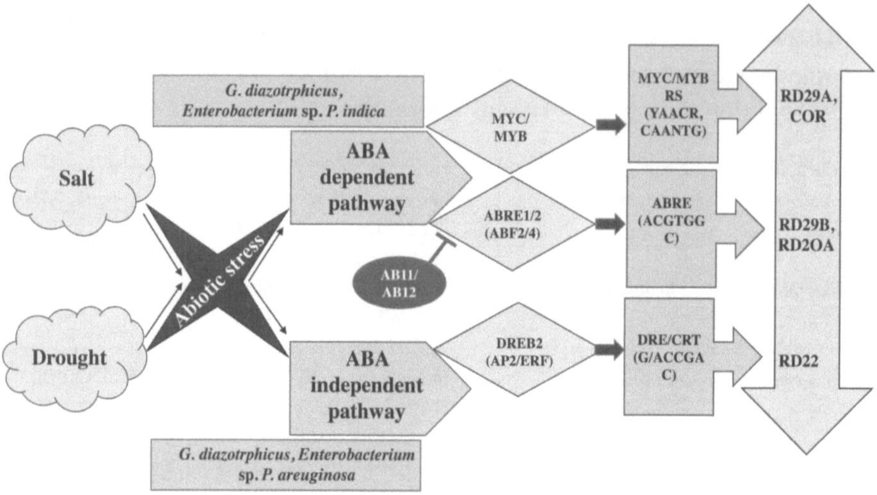

FIGURE 2.3 Microorganisms mediate abscisic acid (ABA)-dependent and -independent signalling for regulating genes involved in abiotic stress tolerance. The ABA-dependent pathway gets upregulated via MYC/MYB and ABF2/4 transcription factors while it is downregulated via AB11. In contrast, the ABA-independent pathway proceeds via signalling of DREB2 that modulates RD22 protein. Both the pathways ultimately lead to stress tolerance in plants.

promotes growth and nutrient uptake in plants, resulting in induced systemic tolerance to water stress. Similarly, proline was increased in plant shoots due to upregulation of DREB/CBF and ERD15, when treated with *Gluconacetobacter diazotrophicus* and *Saccharum officinarum* (Vargas et al. 2014). In another study, rhizobacteria *B. subtilis, Ocrobactrum pseudogrignonense* RJ12, and *Pseudomonas* RJ15 were found to be capable of reducing ethylene accumulation by downregulation of ACO mRNA, while under negative control, the same gene was being upregulated (Saikia et al. 2018). This interesting finding suggests a correlation of bacterial inoculation with lower ethylene levels during stress, resulting in better stress tolerance.

miRNA also have been implicated in having a role under abiotic stress such as cold, salinity, and drought. miR169 modulates the expression of nuclear transcription factor NF-YA alleviating drought along with salinity as discussed earlier (Zhao et al. 2009). In rice, *T. harzianum TH-56* DHN/AQU inoculation has been able to provide drought tolerance by upregulating dehydrin and aquaporin genes (Pandey et al. 2016), as over-expression of dehydrin gene helps in scavenging the reactive oxygen species, while aquaporin regulates the water homeostasis and xylem tube recovery (Holbrook and Zwieniecki 1999). In maize plants, trehalose biosyhthesis gene-fusion gene (*otsA/B*) was found to be effective to cope up drought conditions as showed by Rodriguez-Salazar et al. (2009). When they transformed *A. brasilense* with *otsA/B* gene and then inoculated to maize plants undergoing drought stress, 85% of the plants survived while only 55% of the plants survived after inoculation with wild-type strain. Henceforth, otsA/B may play a major role in *Azospirillum*-mediated stress mitigation. Proline content was also increased in plants under water stress conditions when treated with *B. subtilis* and was linked to the upregulation of *P5CS* gene responsible for proline biosynthesis (Yoshiba et al. 1997). *P. putida* strain GAP-P45 is shown to upregulate the proline metabolic genes in *Arabidopsis* during drought stress such as ornithine aminotransferase (*OAT*) and pyrroline-5-carboxylate synthetase (*P5CS1*), which are responsible for the proline production using P5C as an intermediate compound from ornithine and glutamate, respectively, while P5C was then converted to proline by pyrroline-5-carboxylate reductase-P5CR. The upregulation of these genes is important for the biosynthesis of proline under stress conditions. Proteins for proline catabolism are also upregulated like pyrroline-5-carboxylate dehydrogenase (P5CDH) and proline dehydrogenase (PDH1) whose function is to degrade proline. Thus, simultaneous metabolism and catabolism of proline is required for better plant growth and development during water-deficit conditions (Ghosh et al. 2017). Similarly, *Bacillus sublitis* gene *ProBA*, when introduced in *Arabidopsis*, showed an increased production of proline and enhanced the tolerance towards osmotic stress (Chen et al. 2007).

2.4 CONCLUSION AND FUTURE PERSPECTIVES

Various studies have shown that PGPM-inoculated plants provide drought and salinity stress tolerance under field conditions. Microbial consortia applications in field show promising results in field in alleviating stress conditions and sustainably enhancing agricultural productivity. Such PGPM enhances the growth under stress conditions by phytohormone regulation, antioxidant system enhancement, and nutrition improvement. Genes such as bZIP, MYB, NAC, DREB, DHN, and WRKY play a vital role in stress mitigation and these genes were modulated under effect of PGPM inoculation. Thus, these PGPMs can be used in the future for stress tolerance and would be good for sustainable agriculture.

ACKNOWLEDGEMENTS

Authors are thankful to the Director, CSIR–Institute of Himalayan Bioresource Technology, Palampur, for providing the necessary facility. We acknowledge the financial support from the Project OLP0042 and MLP0168, funded by CSIR, India. Kanchan Yadav and Pooja Yadav gratefully acknowledge UGC, the Government of India, for providing fellowships. The CSIR-IHBT publication number for this manuscript is 8471.

REFERENCES

Aalipour, H., Nikbakht, A., Etemadi, N., Rejali, F., Soleimani, M. 2020. Biochemical response and interactions between arbuscular mycorrhizal fungi and plant growth promoting rhizobacteria during establishment and stimulating growth of Arizona cypress (*Cupressus arizonica* G.) under drought stress. *Scientia Horticulturae* 261: 108923.

Abbas, R., Rasul, S., Aslam, K., Baber, M., Shahid, M., Mubeen, F. Naqqash, T. 2019. Halotolerant PGPR: A hope for cultivation of saline soils. *Journal of King Saud University – Science* 31(4): 1195–1201.

Abdelaziz, M.E., Abdelsattar, M., Abdeldaym, E.A., Atia, M.A., Mahmoud, A.W.M., Saad, M.M., Hirt, H. 2019. *Piriformospora indica* alters Na^+/K^+ homeostasis, antioxidant enzymes and LeNHX1 expression of greenhouse tomato grown under salt stress. *Scientia Horticulturae* 256: 108532.

Abe, H., Urao, T., Ito, T., Seki, M., Shinozaki, K., Yamaguchi-Shinozaki, K. 2003. Arabidopsis AtMYC2 (bHLH) and AtMYB2 (MYB) function as transcriptional activators in abscisic acid signaling. *Plant Cell* 15(1): 63–78.

Ahmad, P., Hashem, A., Abd-Allah, E.F., Alqarawi, A.A., John, R., Egamberdieva, D., Gucel, S. 2015. Role of *Trichoderma harzianum* in mitigating NaCl stress in Indian mustard (*Brassica juncea* L.) through antioxidative defense system. *Frontiers of Plant Science* 6: 868.

Akram, W., Aslam, H., Ahmad, S.R., Anjum, T., Yasin, N.A., Khan, W.U., Ahmad, A., Guo, J., Wu, T., Luo, W., Li, G. 2019. *Bacillus megaterium* strain A12 ameliorates salinity stress in tomato plants through multiple mechanisms. *Journal of Plant Interactions* 14(1): 506–518.

Alami, Y., Achouak, W., Marol, C., Heulin, T. 2000. Rhizosphere soil aggregation and plant growth promotion of sunflowers by an exopolysaccharide-producing *Rhizobium* sp. Strain isolated from sunflower roots. *Applied and Environmental Microbiology* 66(8): 3393–3398.

Ali, S., Charles, T.C., Glick, B.R. 2012. Delay of flower senescence by bacterial endophytes expressing 1-aminocyclopropane-1-carboxylate deaminase. *Journal of Applied Microbiology* 113(5): 1139–1144.

Alizadeh, O., Zare, M., Nasr, A.H. 2011. Evaluation effect of mycorrhiza inoculate under drought stress condition on grain yield of sorghum (*Sorghum bicolor*). *Advances in Environmental Biology* 5(8): 2361–2364.

Al-Karaki, G., McMichael, B.Z.A.K.J., Zak, J. 2004. Field response of wheat to arbuscular mycorrhizal fungi and drought stress. *Mycorrhiza* 14(4): 263–269.

Amellal, N., Burtin, G., Bartoli, F., Heulin, T. 1998. Colonization of wheat rhizosphere by EPS producing *Pantoea agglomerans* and its effect on soil aggregation. *Applied and Environmental Microbiology* 64(374): 3747.

Arkhipova, T.N., Prinsen, E., Veselov, S.U., Martinenko, E.V., Melentiev, A.I., Kudoyarova, G.R. 2007. Cytokinin producing bacteria enhance plant growth in drying soil. *Plant and Soil* 292(1): 305–315.

Aroca, R., Porcel, R., Ruiz-Lozano, J.M. 2007. How does arbuscular mycorrhizal symbiosis regulate root hydraulic properties and plasma membrane aquaporins in *Phaseolus vulgaris* under drought, cold or salinity stresses? *New Phytologist* 173: 808–816.

Asaf, S., Khan, M.A., Khan, A.L., Waqas, M., Shahzad, R., Kim, A.Y., Kang, S.M., Lee, I.J. 2017. Bacterial endophytes from arid land plants regulate endogenous hormone content and promote growth in crop plants: An example of *Sphingomonas* sp. and *Serratia marcescens*. *Journal of Plant Interactions* 12(1): 31–38.

Ashraf, M., Hasnain, S., Berge, O., Mahmood, T. 2004. Inoculating wheat seedlings with exopolysaccharide-producing bacteria restricts sodium uptake and stimulates plant growth under salt stress. *Biology and Fertility of Soils* 40(3): 157–162.

Bacilio, M., Moreno, M., Bashan, Y. 2016. Mitigation of negative effects of progressive soil salinity gradients by application of humic acids and inoculation with *Pseudomonas stutzeri* in a salt-tolerant and a salt-susceptible pepper. *Applied Soil Ecology* 107: 394–404.

Bae, H., Sicher, R.C., Kim, M.S., Kim, S.H., Strem, M.D., Melnick, R.L., Bailey, B.A. 2009. The beneficial endophyte *Trichoderma hamatum* isolate DIS 219b promotes growth and delays the onset of the drought response in *Theobroma cacao*. *Journal of Experimental Botany* 60(11): 3279–3295.

Bal, H.B., Nayak, L., Das, S., Adhya, T.K. 2013. Isolation of ACC deaminase producing PGPR from rice rhizosphere and evaluating their plant growth promoting activity under salt stress. *Plant and Soil* 366(1): 93–105.

Bano, Q.U.D.S.I.A., Ilyas, N., Bano, A., Zafar, N.A.D.I.A., Akram, A.B.I.D.A., Hassan, F. 2013. Effect of *Azospirillum* inoculation on maize (*Zea mays* L.) under drought stress. *Pakistan Journal of Botany* 45(S1): 13–20.

Barnawal, D., Bharti, N., Maji, D. 2014. ACC deaminase-containing *Arthrobacter protophormiae* induces NaCl stress tolerance through reduced ACC oxidase activity and ethylene production resulting in improved nodulation and mycorrhization in *Pisum sativum*. *Journal of Plant Physiology* 171: 884–894.

Bassin, J.P., Dezotti, M., Sant'Anna Jr, G.L. 2011. Nitrification of industrial and domestic saline wastewaters in moving bed biofilm reactor and sequencing batch reactor. *Journal of Hazardous Materials* 185(1): 242–248.

Belimov, A.A., Dodd, I.C., Hontzeas, N., Theobald, J.C., Safronova, V.I., Davies, W.J. 2009. Rhizosphere bacteria containing 1-aminocyclopropane-1-carboxylate deaminase increase yield of plants grown in drying soil via both local and systemic hormone signalling. *New Phytologist* 181(2): 413–423.

Bharti, N., Pandey, S.S., Barnawal, D., Patel, V.K., Kalra, A. 2016. Plant growth promoting rhizobacteria *Dietzia natronolimnaea* modulates the expression of stress responsive genes providing protection of wheat from salinity stress. *Scientific Reports* 6(6): 34768.

Bhattacharyya, P.N., Jha, D.K. 2012. Plant growth-promoting rhizobacteria (PGPR): Emergence in agriculture. *World Journal of Microbiology and Biotechnology* 28(4): 1327–1350.

Borowicz, V.A. 2010. The impact of arbuscular mycorrhizal fungi on strawberry tolerance to root damage and drought stress. *Pedobiologia* 53(4): 265–270.

Brachmann, A., Parniske, M. 2006. The most widespread symbiosis on earth. *PLoS Biology* 4(7): e239.

Bradford, K.J., Hsiao, T.C. 1982. Physiological responses to moderate water stress. In *Physiological Plant Ecology II*. Springer, Berlin-Heidelberg, 263–324.

Bresson, J., Varoquaux, F., Bontpart, T., Touraine, B., Vile, D. 2013. The PGPR strain *Phyllobacterium brassicacearum* STM 196 induces a reproductive delay and physiological changes that result in improved drought tolerance in *Arabidopsis*. *New Phytologist* 200(2): 558–569.

Budak, H., Kantar, M., Bulut, R., Akpinar, B.A. 2015. Stress responsive miRNAs and isomiRs in cereals. *Plant Science* 235: 1–3.

Cardinale, M., Ratering, S., Suarez, C., Montoya, A.M.Z., Geissler-Plaum, R., Schnell, S. 2015. Paradox of plant growth promotion potential of rhizobacteria and their actual promotion effect on growth of barley (*Hordeum vulgare* L.) under salt stress. *Microbiological Research* 181: 22–32.

Chakdar, H., Borse, D.N., Verma, S., Choudhary, P., Das, S. 2019. Microbial Management of Crop Salinity Stress: Mechanisms, Applications and Prospects. In *Salt Stress, Microbes, and Plant Interactions: Mechanisms and Molecular Approaches*, ed. Akhtar, M.S. Springer Nature, Singapore, 1–26.

Chauhan, P.S., Lata, C., Tiwari, S., Chauhan, A.S., Mishra, S.K., Agrawal, L., Chakrabarty, D., Nautiyal, C.S. 2019. Transcriptional alterations reveal *Bacillus amyloliquefaciens*-rice cooperation under salt stress. *Scientific Reports* 9(1): 1–13.

Chen, X.H., Koumoutsi, A., Scholz, R., Eisenreich, A., Schneider, K., Heinemeyer, I., Morgenstern, B., Voss, B., Hess, W.R., Reva, O., Junge, H. 2007. Comparative analysis of the complete genome sequence of the plant growth-promoting bacterium *Bacillus amyloliquefaciens* FZB42. *Nature Biotechnology* 25(9): 1007–1014.

Chen, C., Xin, K., Liu, H., Cheng, J., Shen, X., Wang, Y., Zhang, L. 2017. *Pantoea alhagi*, a novel endophytic bacterium with ability to improve growth and drought tolerance in wheat. *Scientific Reports* 7(1): 1–14.

Chen, T., Li, C., White, J.F., Nan, Z. 2019. Effect of the fungal endophyte *Epichloë bromicola* on polyamines in wild barley (*Hordeum brevisubulatum*) under salt stress. *Plant Soil* 436: 29–48.

Cho, S.M., Kang, B.R., Kim, Y.C. 2013. Transcriptome analysis of induced systemic drought tolerance elicited by *Pseudomonas chlororaphis* O6 in *Arabidopsis thaliana*. *Plant Pathology Journal* 29(2): 209.

Chu, T.N., Tran, B.T.H., Van Bui, L. et al. 2019. Plant growth-promoting rhizobacterium *Pseudomonas* PS01 induces salt tolerance in *Arabidopsis thaliana*. *BMC Research Notes* 12: 11.

Cohen, A.C., Bottini, R., Pontin, M., Berli, F.J., Moreno, D., Boccanlandro, H., Travaglia, C.N., Piccoli, P.N. 2015. *Azospirillum brasilense* ameliorates the response of *Arabidopsis thaliana* to drought mainly via enhancement of ABA levels. *Physiologia Plantarum* 153(1): 79–90.

Cong, G.Q., Yin, C.L., He, B.L., Li, L., Gao, K.X. 2015. Effect of the endophytic fungus chaetomium globosum ND35 on the growth and resistance to drought of winter wheat at the seedling stage under water stress. *Acta Ecologica Sinica* 35(18): 6120–6128.

Contreras-Cornejo, H.A., Macías-Rodríguez, L., Alfaro-Cuevas, R., López-Bucio, J. 2014. *Trichoderma* spp. improve growth of *Arabidopsis* seedlings under salt stress through enhanced root development, osmolite production, and Na^+ elimination through root exudates. *Molecular Plant-Microbe Interactions* 27: 503–514.

Das, K., Roychoudhury, A. 2014. Reactive oxygen species (ROS) and response of antioxidants as ROS-scavengers during environmental stress in plants. *Frontiers in Environmental Science* 2: 53.

Dastogeer, K.M., Wylie, S.J. 2017. Plant-fungi association: Role of fungal endophytes in improving plant tolerance to water stress. *Plant-Microbe Interactions in Agro-Ecological Perspectives*. Springer, Singapore, 143–159.

Daszkowska-Golec, A. 2016. The role of abscisic acid in drought stress: How ABA helps plants to cope with drought stress. *Drought Stress Tolerance in Plants*. Springer, Cham, 2, 123–151.

Dodd, I.C., Perez-Alfocea, F. 2012. Microbial amelioration of crop salinity stress. *Journal of Experimental Botany* 3415–3428.

Duc, N.H., Csintalan, Z., Posta, K. 2018. Arbuscular mycorrhizal fungi mitigate negative effects of combined drought and heat stress on tomato plants. *Plant Physiology and Biochemistry* 132: 297–307.

Egamberdieva, D., Kucharova, Z., Davranov, K., Berg, B., Makarova, N., Azarova, T., Chebotar, V., Tikhonovich, I., Kamilova, F., Validov, S.Z., Lugtenberg, B. 2011. Bacteria able to control foot and root rot and to promote growth of cucumber in salinated soils. *Biology and Fertility of Soils* 47: 197–205.

Evelin, H., Giri, B., Kapoor, R. 2013. Ultrastructural evidence for AMF mediated salt stress mitigation in *Trigonella foenum-graecum*. *Mycorrhiza* 23: 71–86.

Farooq, M., Wahid, A., Kobayashi, N., Fujita, D., Basra, S.M.A. 2009. Plant drought stress: Effects, mechanisms and management. *Agronomy for Sustainable Development* 29(1): 185–212.

Fasciglione, G., Casanovas, E.M., Quillehauquy, V., Yommi, A.K., Goñi, M.G., Roura, S.I., Barassi, C.A. 2015. *Azospirillum* inoculation effects on growth, product quality and storage life of lettuce plants grown under salt stress. *Scientia Horticulturae* 195: 154–162.

Figueiredo, M.V., Burity, H.A., Martinez, C.R., Chanway, C.P. 2008. Alleviation of drought stress in the common bean (*Phaseolus vulgaris* L.) by co-inoculation with *Paenibacillus polymyxa* and *Rhizobium tropici*. *Applied Soil Ecology* 40(1): 182–188.

Gao, P., Bai, X., Yang, L., Lv, D, Pan, X., Li, Y. Cai, H., Ji, W., Chen, Q., Zhu, Y. 2011. *OsmiR393*: A salinity-and alkaline stress-related microRNA gene. *Molecular Biology Reports* 38: 237–242.

Gholamhoseini, M., AghaAlikhani, M., Sanavy, S.M., Mirlatifi, S.M. 2013. Interactions of irrigation, weed and nitrogen on corn yield, nitrogen use efficiency and nitrate leaching. *Agricultural Water Management* 126: 9–18.

Ghosh, D., Sen, S., Mohapatra, S. 2017. Modulation of proline metabolic gene expression in Arabidopsis thaliana under water-stressed conditions by a drought-mitigating *Pseudomonas putida* strain. *Annals of Microbiology* 67(10): 655–668.

Gill, S.S., Tuteja, N. 2010. Reactive oxygen species and antioxidant machinery in abiotic stress tolerance in crop plants. *Plant Physiology and Biochemistry* 48(12): 909–930.

Glick, B.R. 2005. Modulation of plant ethylene levels by the bacterial enzyme ACC deaminase. *FEMS Microbiology Letters* 251(1): 1–7.

Glick, B.R. 2012. Plant growth-promoting bacteria: Mechanisms and applications. *Scientifica.* https://doi.org/10.6064/2012/963401

Glick, B.R. 2014. Bacteria with ACC deaminase can promote plant growth and help to feed the world. *Microbiological Research* 169(1): 30–39.

Goswami, D., Thakker, J.N., Dhandhukia, P.C. 2016. Portraying mechanics of plant growth promoting rhizobacteria (PGPR): A review. *Cogent Food & Agriculture* 2(1): 1127500.

Gou, W., Tian, L., Ruan, Z., Zheng, P.E.N.G., Chen, F.U.C.A.I., Zhang, L., Cui, Z., Zheng, P., Li, Z., Gao, M., Shi, W. 2015. Accumulation of choline and glycinebetaine and drought stress tolerance induced in maize (*Zea mays*) by three plant growth promoting rhizobacteria (PGPR) strains. *Pakistan Journal of Botany* 47(2): 581–586.

Gupta, B., Huang, B. 2014. Mechanism of salinity tolerance in plants: Physiological, biochemical, and molecular characterization. *International Journal of Genomics* 701596: 1.

Gupta, S., Schillaci, M., Walker, R., Smith, Penelope, M.C., Watt, M., Roessner, U. 2020. Alleviation of salinity stress in plants by endophytic plant-fungal symbiosis: Current knowledge, perspectives and future directions. *Plant Soil* 461: 219–244.

Hao, Y., Charles, T.C., Glick, B.R. 2011. ACC deaminase activity in a virulent *Agrobacterium tumefaciens* D3. *Canadian Journal of Microbiology* 57(4): 278–286.

Hasegawa, P.M., Bressan, R.A., Zhu, J.K., Bohnert, H.J. 2000. Plant cellular and molecular responses to high salinity. *Annual Review of Plant Biology* 51: 463–499.

Hedden, P., Phillips, A.L. 2000. Manipulation of hormone biosynthetic genes in transgenic plants. *Current Opinion in Biotechnology* 11(2): 130–137.

Holbrook, N.M. and Zwieniecki, M.A. 1999. Embolism repair and xylem tension: Do we need a miracle? *Plant Physiology* 120(1): 7–10.

Hu, Y., Chen, L., Wang, H., Zhang, L., Wang, F., Yu, D. 2013. *Arabidopsis* transcription factor WRKY8 functions antagonistically with its interacting partner VQ9 to modulate salinity stress tolerance. *Plant Journal* 74(5): 730–745.

Huang, Y.M., Srivastava, A.K., Zou, Y.N., Ni, Q.D., Han, Y., Wu, Q.S. 2014. Mycorrhizal-induced calmodulin mediated changes in antioxidant enzymes and growth response of drought-stressed *trifoliate orange*. *Frontiers in Microbiology* 5: 682.

Hussain, M.B., Zahir, Z.A., Asghar, H.N., Asgher, M. 2014. Can catalase and exopolysaccharides producing rhizobia ameliorate drought stress in wheat? *International Journal of Agriculture and Biology* 16(1): 3–13.

Jan, F.G., Hamayun, M., Hussain, A., Jan, G., Iqbal, A., Khan, A. Lee, I.J. 2019. An endophytic isolate of the fungus *Yarrowia lipolytica* produces metabolites that ameliorate the negative impact of salt stress on the physiology of maize. *BMC Microbiology* 19: 1–10.

Jha, C.K., Patel, B., Saraf, M. 2012. Stimulation of the growth of *Jatropha curcas* by the plant growth promoting bacterium *Enterobacter cancerogenus* MSA2. *World Journal of Microbiology and Biotechnology* 28(3): 891–899.

Jha, Y. and Subramanian, R.B. 2014. PGPR regulate caspase-like activity, programmed cell death, and antioxidant enzyme activity in paddy under salinity. *Physiology and Molecular Biology of Plants* 20(2): 201–207.

Jiang, F., Chen, L., Belimov, A.A., Shaposhnikov, A.I., Gong, F., Meng, X., Hartung, W., Jeschke, D.W., Davies, W.J., Dodd, I.C., 2012. Multiple impacts of the plant growth-promoting rhizobacterium *Variovorax paradoxus* 5C-2 on nutrient and ABA relations of *Pisum sativum*. *Journal of Experimental Botany* 63(18): 6421–6430.

Kandasamy, S., Loganathan, K., Muthuraj, R., Duraisamy, S., Seetharaman, S., Thiruvengadam, R., Ponnusamy, B., Ramasamy, S. 2009. Understanding the molecular basis of plant growth promotional effect of *Pseudomonas fluorescens* on rice through protein profiling. *Proteome Science* 7(1): 1–8.

Kang, S.M., Radhakrishnan, R., Khan, A.L., Kim, M.J., Park, J.M., Kim, B.R., Shin, D.H., Lee, I.J. 2014. Gibberellin secreting rhizobacterium, *Pseudomonas putida* H-2-3 modulates the hormonal and stress physiology of soybean to improve the plant growth under saline and drought conditions. *Plant Physiology and Biochemistry* 84: 115–124.

Kasim, W.A., Osman, M.E., Omar, M.N., Abd El-Daim, I.A., Bejai, S., Meijer, J. 2013. Control of drought stress in wheat using plant growth-promoting bacteria. *Journal of Plant Growth Regulation* 32(1): 122–130.

Kaushal, M., Wani, S.P. 2016. Plant growth-promoting rhizobacteria: Drought stress alleviators to ameliorate crop production in drylands. *Annals of Microbiology* 66(1): 35–42.

Keskin, B.C., Sarikaya, A.T., Yuksel, B., Memon, A.R. 2010. Abscisic acid regulated gene expression in bread wheat (*Triticum aestivum* L.). *Australian Journal of Crop Science* 4(8): 617–625.

Khan, A.L., Hamayun, M., Kim, Y.H., Kang, S.M., Lee, J.H., Lee, I.J. 2011. Gibberellins producing endophytic *Aspergillus fumigatus* sp. LH02 influenced endogenous phytohormonal levels, isoflavonoids production and plant growth in salinity stress. *Process Biochemistry* 46: 440–447.

Khan, A.L., Ullah, I., Hussain, J., Kang, S.M., Al-Harrasi, A., Al-Rawahi, A., Lee, I.J. 2016. Regulations of essential amino acids and proteomics of bacterial endophytes S phingomonas sp. L k11 during cadmium uptake. *Environmental Toxicology* 31(7): 887–896.

Khan, M.A., Asaf, S., Khan, A.L., Adhikari, A., Jan, R., Ali, S., Imran, M., Kim, K.M., Lee, I.J. 2019. Halotolerant rhizobacterial strains mitigate the adverse effects of NaCl stress in soybean seedlings. *BioMed Research International*. https://doi.org/10.1155/2019/9530963

Kim, K., Jang, Y., Lee, S.M., Oh, B.T., Chae, J.C., Lee, K.J. 2014. Alleviation of salt stress by *Enterobacter* sp. EJ01 in tomato and arabidopsis is accompanied by up-regulation of conserved salinity responsive factors in plants. *Molecules and Cells* 37(2): 109–117.

Kirkham, M.B. 2005. *Principles of Soil and Plant Water Relations*. Elsevier, Burlington, MA.

Krishna, N., Anjana, J., Roshan Sharma, P., Rupak, T., Shin, P.Y., Eva-Mari, A., Gil, N.H., Lee, C-H. 2013. Towards a critical understanding of the photosystem II repair mechanism and its regulation during stress conditions. *FEBS Letter* 587: 3372–3381.

Kumar, J., Singh, S., Singha, M., Srivastava, P.K., Mishra, R.K., Singh, V.P., Prasad, S.M. 2017. Transcriptional regulation of salinity stress in plants: A short review. *Plant Gene* 11: 160–169.

Kumar, A., Singh, S., Gaurav, A.K., Srivastava, S., Verma, J.P. 2020. Plant growth-promoting bacteria: Biological tools for the mitigation of salinity stress in plants. *Frontiers in Microbiology* 11. doi: 10.3389/fmicb.2021.660075.

Langeroodi, A.R.S., Osipitan, O.A., Radicetti, E., Mancinelli, R., 2020. To what extent arbuscular mycorrhiza can protect chicory (*Cichorium intybus* L.) against drought stress. *Scientia Horticulturae* 263: 109109.

Lei, Z., Zhang, Y-Q. 2015. Effects of phosphate solubilization and phytohormone production of *Trichoderma asperellum* Q1 on promoting cucumber growth under salt stress. *Journal of Integrative Agriculture* 14: 1588–1597.

Li, L., Wang, X., Zhu, P., Wu, H., Qi, S. 2017. Plant growth-promoting endophyte *Piriformospora indica* alleviates salinity stress in *Medicago truncatula*. *Plant Physiology and Biochemistry* 119: 211–223.

Lim, J.H., Kim, S.D. 2013. Induction of drought stress resistance by multi-functional PGPR *Bacillus licheniformis* K11 in pepper. *Plant Pathology Journal* 29(2): 201.

Lindemose, S., O'Shea, C., Jensen, M.K., Skriver, K. 2013. Structure, function and networks of transcription factors involved in abiotic stress responses. *International Journal of Molecular Sciences 14*(3): 5842–5878.

Liu, C.Y., Zhang, F., Zhang, D.J., Srivastava, A.K., Wu, Q.S., Zou, Y.N. 2018. Mycorrhiza stimulates root-hair growth and IAA synthesis and transport in trifoliate orange under drought stress. *Scientific Reports* 8(1): 1–9.

Ma, Y., Rajkumar, M., Zhang, C., Freitas, H. 2016. Inoculation of *Brassica oxyrrhina* with plant growth promoting bacteria for the improvement of heavy metal phytoremediation under drought conditions. *Journal of Hazardous Materials* 320: 36–44.

Maheshwari, D.K., Dheeman, S., Agarwal, M. 2015. Phytohormone-producing PGPR for sustainable agriculture. In *Bacterial Metabolites in Sustainable Agroecosystem*. Springer, Cham, 159–182.

Mahmood-ur-Rahman, Ijaz, M., Qamar, S., Bukhari, S. A., and Malik, K. 2019. Abiotic stress signaling in rice crop. *Advances in Rice Research for Abiotic Stress Tolerance* 551–569. doi: 10.1016/b978-0-12-814332-2.00027-7.

Marulanda, A., Azcón, R., Chaumont, F., Ruiz-Lozano, J.M., Aroca, R. 2010. Regulation of plasma membrane aquaporins by inoculation with a *Bacillus megaterium* strain in maize (*Zea mays* L.) plants under unstressed and salt-stressed conditions. *Planta* 232: 533–543.

Mayak, S., Tirosh, T., Glick, B.R. 2004a. Plant growth-promoting bacteria that confer resistance to water stress in tomatoes and peppers. *Plant Science* 166(2): 525–530.

Mayak, S., Tirosh, T., Glick, B.R. 2004b. Plant growth-promoting bacteria confer resistance in tomato plants to salt stress. *Plant Physiology and Biochemistry* 42(6): 565–572.

Meena, K.K., Sorty, A.M., Bitla, U.M., Choudhary, K. et al. 2017. Abiotic stress responses and microbe-mediated mitigation in plants: The omics strategies. *Frontiers in Plant Science* 8: 172.

Nadeem, S.M., Ahmad, M., Zahir, Z.A., Javaid, A., Ashraf, M. 2014. The role of mycorrhizae and plant growth promoting rhizobacteria (PGPR) in improving crop productivity under stressful environments. *Biotechnology Advances* 32(2): 429–448.

Nathawat, N.S., Kuhad, M.S., Goswami, C.L., Patel, A.L., Kumar, R. 2005. Nitrogen-metabolizing enzymes: Effect of nitrogen sources and saline irrigation. *Journal of Plant Nutrition* 28(6): 1089–1101.

Nautiyal, C.S., Srivastava, S., Chauhan, P.S., Seem, K., Mishra, A., Sopory, S.K. 2013. Plant growth-promoting bacteria *Bacillus amyloliquefaciens* NBRISN13 modulates gene expression profile of leaf and rhizosphere community in rice during salt stress. *Plant Physiology and Biochemistry* 66: 1–9.

Naveed, M., Mitter, B., Reichenauer, T.G., Wieczorek, K., Sessitsch, A. 2014. Increased drought stress resilience of maize through endophytic colonization by *Burkholderia phytofirmans* PsJN and *Enterobacter* sp. FD17. *Environmental and Experimental Botany* 97: 30–39.

Naya, L., Ladrera, R., Ramos, J., González, E.M., Arrese-Igor, C., Minchin, F.R., Becana, M. 2007. The response of carbon metabolism and antioxidant defenses of alfalfa nodules to drought stress and to the subsequent recovery of plants. *Plant Physiology* 144(2): 1104–1114.

Naz, R. and Bano, A. 2015. Molecular and physiological responses of sunflower (*Helianthus annuus* L.) to PGPR and SA under salt stress. *Pakistan Journal of Botany* 47(1): 35–42.

Noreen, S., Ali, B., Hasnain, S. 2012. Growth promotion of *Vigna mungo* (L.) by *Pseudomonas* spp. exhibiting auxin production and ACC-deaminase activity. *Annals of Microbiology* 62(1): 411–417.

Ortiz, N., Armada, E., Duque, E., Roldán, A., Azcon, R. 2015. Contribution of arbuscular mycorrhizal fungi and/or bacteria to enhancing plant drought tolerance under natural soil conditions: Effectiveness of autochthonous or allochthonous strains. *Journal of Plant Physiology* 174: 87–96.

Palaniyandi, S. A., Damodharan, K., Yang, S. H., Suh, J. W. 2014. *Streptomyces* sp. strain PGPA39 alleviates salt stress and promotes growth of 'Micro Tom' tomato plants. *Journal of Applied Microbiology* 117: 766–773.

Pandey, V., Ansari, M.W., Tula, S., Yadav, S., Sahoo, R.K., Shukla, N., Bains, G., Badal, S., Chandra, S., Gaur, A.K., Kumar, A. 2016. Dose-dependent response of *Trichoderma harzianum* in improving drought tolerance in rice genotypes. *Planta 243*(5): 1251–1264.

Pankaj, U., Singh, D.N., Singh, G., Verma, R.K. 2019a. Microbial inoculants assisted growth of *Chrysopogon zizanioides* promotes phytoremediation of salt affected soil. *Indian Journal of Microbiology* 59(2): 137–146.

Pankaj, U., Verma, R.S., Yadav, A., Verma, R.K. 2019b. Effect of arbuscular mycorrhizae species on essential oil yield and chemical composition of commercially grown palmarosa (*Cymbopogon martinii*) varieties in salinity stress soil. *Journal of Essential Oil Research* 31(2): 145–153.

Pankaj, U., Singh, D.N., Mishra, P., Gaur, P., Vivekbabu, C.S., Shanker, K., Verma, R.K. 2020. Autochthonous halotolerant plant growth promoting rhizobacteria promote bacoside A yield of *Bacopa monnieri* (L.) Nash and phytoextraction of salt-affected soil. *Pedosphere* 30(5): 671–683.

Panwar, M., Tewari, R., Nayyar, H. 2016. Native halo-tolerant plant growth promoting rhizobacteria *Enterococcus* and *Pantoea* sp. improve seed yield of Mungbean (*Vigna radiata* L.) under soil salinity by reducing sodium uptake and stress injury. *Physiology and Molecular Biology of Plants* 22(4): 445–459. doi: 10.1007/s12298-016-0376-9

Parida, A.K., Das, A.B., Mittra, B. 2004. Effects of salt on growth, ion accumulation, photosynthesis and leaf anatomy of the mangrove. *Bruguiera Parviflora* 18: 167–174.

Parniske, M. 2008. Arbuscular mycorrhiza: The mother of plant root endosymbioses. *Nature Reviews Microbiology* 6: 763.

Pathak, H., Jain, N., Bhatia, A., Kumar, A., Chatterjee, D. 2016. Improved nitrogen management: A key to climate change adaptation and mitigation. *Indian Journal of Fertilizers* 12(11): 151–162.

Paulucci, N.S., Gallarato, L.A., Reguera, Y.B., Vicario, J.C., Cesari, A.B., García de Lema, M.B., Dardanelli, M.S. 2015. *Arachis hypogaea* PGPR isolated from Argentine soil modifies its lipids components in response to temperature and salinity. *Microbiological Research* 173: 1–9. doi: 10.1016/j.micres.2014.12.01.

Porcel, R., Azcón, R., Ruiz-Lozano, J.M. 2004. Evaluation of the role of genes encoding for Δ1-pyrroline-5-carboxylate synthetase (P5CS) during drought stress in arbuscular mycorrhizal *Glycine max* and *Lactuca sativa* plants. *Physiological and Molecular Plant Pathology* 65(4): 211–221.

Rangarajan, S., Saleena, L.M., Nair, S. 2002. Diversity of *Pseudomonas* spp. isolated from rice rhizosphere populations grown along a salinity gradient. *Microbial ecology* 43(2): 280–289.

Rao, N.K.S., Shivashankara, K.S., Laxman, R.H. 2016. *Abiotic Stress Physiology of Horticultural Crops*. Springer. doi: 10.1007/978-81-322-2725-0.

Rao, N., Dong, Z.Y., Xiao, M., Li, W.-J. 2019. Effect of salt stress on plants and role of microbes in promoting plant growth under salt stress. *Microorganisms in Saline*

Environments: Strategies and Functions, Soil Biology, Vol. 56. Springer, 423–435. doi: 10.1007/978-3-030-18975-4_18.
Rashid, U., Yasmin, H., Hassan, M.N., Naz, R., Nosheen, A., Sajjad, M., Ilyas, N., Keyani, R., Jabeen, Z., Mumtaz, S., Alyemeni, M.N., Ahmad, P. 2021. Drought-tolerant *Bacillus megaterium* isolated from semi-arid conditions induces systemic tolerance of wheat under drought conditions. *Plant Cell Reports* 41(3): 549–569.
Redman, R.S., Kim, Y.O., Woodward, C.J., Greer, C., Espino, L., Doty, S.L., Rodriguez, R.J. 2011. Increased fitness of rice plants to abiotic stress via habitat adapted symbiosis: A strategy for mitigating impacts of climate change. *PLoS ONE* 6(7): e14823.
Robredo, A., Perez-Lopez, U., Miranda-Apodaca, J., Lacuesta, M., Mena-Petite, A., Munoz-Rueda, A. 2011. Elevated CO_2 reduces the drought effect on nitrogen metabolism in barley plants during drought and subsequent recovery. *Environmental and Experimental Botany* 71(3): 399–408.
Rodriguez-Salazar, J., Suárez, R., Caballero-Mellado, J., Iturriaga, G. 2009. Trehalose accumulation in *Azospirillum brasilense* improves drought tolerance and biomass in maize plants. *FEMS Microbiology Letters* 296(1): 52–59.
Rojas-Tapias, D., Moreno-Galván, A., Pardo-Díaz, S., Obando, M., Rivera, D., Bonilla, R. 2012. Effect of inoculation with plant growth-promoting bacteria (PGPB) on amelioration of saline stress in maize (*Zea mays*). *Applied Soil Ecology* 61: 264–272.
Rolli, E., Marasco, R., Vigani, G., Ettoumi, B., Mapelli, F., Deangelis, M.L., Gandolfi, C., Casati, E., Previtali, F., Gerbino, R., Pierotti Cei, F. 2015. Improved plant resistance to drought is promoted by the root-associated microbiome as a water stress-dependent trait. *Environmental Microbiology* 17(2): 316–331.
Sahoo, R.K., Ansari, M.W., Pradhan, M., Dangar, T.K., Mohanty, S.,Tuteja, N. 2014. A novel *Azotobacter vinellandii* (SRI Az 3) functions in salinity stress tolerance in rice. *Plant Signaling & Behavior* 9(7): 511–523.
Saikia, J., Sarma, R.K., Dhandia, R., Yadav, A., Bharali, R., Gupta, V.K., Saikia, R. 2018. Alleviation of drought stress in pulse crops with ACC deaminase producing rhizobacteria isolated from acidic soil of Northeast India. *Scientific Reports* 8(1): 1–16.
Sanaullah, M., Rumpel, C., Charrier, X., Chabbi, A. 2012. How does drought stress influence the decomposition of plant litter with contrasting quality in a grassland ecosystem?. *Plant and Soil* 352(1): 277–288.
Sapre, S., Gontia-Mishra, I., Tiwari, S. 2018. *Klebsiella* sp. confers enhanced tolerance to salinity and plant growth promotion in oat seedlings (*Avena sativa*). *Microbiological Research* 206: 25–32.
Sarkar, A., Pramanik, K., Mitra, S., Soren, T., Maiti, T.K. 2018. Enhancement of growth and salt tolerance of rice seedlings by ACC deaminase-producing *Burkholderia* sp. MTCC 12259. *Journal of Plant Physiology* 231: 434–442.
Sarma, R.K., Saikia, R. 2014. Alleviation of drought stress in mung bean by strain *Pseudomonas aeruginosa* GGRJ21. *Plant and Soil* 377(1): 111–126.
Schellenbaum, L., Müller, J., Boller, T., Wiemken, A., Schüepp, H. 1998. Effects of drought on non-mycorrhizal and mycorrhizal maize: Changes in the pools of non-structural carbohydrates, in the activities of invertase and trehalase, and in the pools of amino acids and imino acids. *New Phytologist* 138(1): 59–66.
Serraj, R., Sinclair, T.R. 2002. Osmolyte accumulation: Can it really help increase crop yield under drought conditions? *Plant, Cell & Environment* 25(2): 333–341.
Shahbaz, M., Ashraf, M. 2013. Improving salinity tolerance in cereals. *Critical Reviews in Plant Sciences* 32: 237–249.
Shahzad, R., Khan, A.L., Bilal, S., Waqas, M., Kang, S.M., Lee, I.J. 2017. Inoculation of abscisic acid-producing endophytic bacteria enhances salinity stress tolerance in *Oryza sativa*. *Environmental and Experimental Botany* 136: 68–77.

Sherameti, I., Tripathi, S., Varma, A., Oelmüller, R. 2008. The root-colonizing endophyte *Pirifomospora indica* confers drought tolerance in arabidopsis by stimulating the expression of drought stress-related genes in leaves. *Molecular Plant-Microbe Interactions* 21(6): 799–807.

Shinozaki, K., Yamaguchi-Shinozaki, K. 2007. Gene networks involved in drought stress response and tolerance. *Journal of Experimental Botany* 58(2): 221–227.

Shukla P. S., Agarwal, P. K., Jha, B. 2012. Improved Salinity Tolerance of *Arachis hypogaea* (L.) by the interaction of halotolerant Plant Growth-Promoting Rhizobacteria. *Journal Plant Growth Regulation* 31: 195–206. doi: 10.1007/s00344-011-9231-y.

Siddiqui, M.H., Khan, M.N., Mohammad, F., Khan, M.M.A. 2008. Role of nitrogen and gibberellin (GA3) in the regulation of enzyme activities and in osmoprotectant accumulation in *Brassica juncea* L. under salt stress. *Journal of Agronomy and Crop Science* 194(3): 214–224.

Singh, R.P., Jha, P.N. 2016. The multifarious PGPR *Serratia marcescens* CDP-13 augments induced systemic resistance and enhanced salinity tolerance of wheat (*Triticum aestivum* L.). *PLoS ONE* 11: e0155026.

Singh, B.K., Trivedi, P., Singh, S., Macdonald, C.A., Verma, J.P. 2018. Emerging microbiome technologies for sustainable increase in farm productivity and environmental security. *Microbiology Australia* 39: 17–23.

Song, S.Y., Chen, Y., Chen, J., Dai, X.-Y., Zhang, W.H. 2011. Physiological mechanisms underlying OsNAC5-dependent tolerance of rice plants to abiotic stress. *Planta* 234: 331–345.

Staudinger, C., Mehmeti-Tershani, V., Gil-Quintana, E., Gonzalez, E.M., Hofhansl, F., Bachmann, G., Wienkoop, S. 2016. Evidence for a rhizobia-induced drought stress response strategy in *Medicago truncatula*. *Journal of Proteomics* 136: 202–213.

Stepien, P., Klobus, G. 2006. Water relations and photosynthesis in *Cucumus sativus* L. leaves under salt stress. *Plant Biology* 50: 610–616.

Sukweenadhi, J., Kim, Y.-J., Choi, E.-S., Koh, S.-C., Lee, S.-W., Kim, Y.-J., Yang, D. C. 2015. *Paenibacillus yonginensis* DCY84T induces changes in *Arabidopsis thaliana* gene expression against aluminum, drought, and salt stress. *Microbiological Research* 172: 7–15. doi: 10.1016/j.micres.2015.01.007

Sun, C., Johnson, J.M., Cai, D., Sherameti, I., Oelmüller, R., Lou, B. 2010. *Piriformospora indica* confers drought tolerance in Chinese cabbage leaves by stimulating antioxidant enzymes, the expression of drought-related genes and the plastid-localized CAS protein. *Journal of Plant Physiology* 167(12): 1009–1017.

Sziderics, A.H., Rasche, F., Trognitz, F., Sessitsch, A., Wilhelm, E. 2007. Bacterial endophytes contribute to abiotic stress adaptation in pepper plants (*Capsicum annuum* L.). *Canadian journal of Microbiology* 53(11): 1195–1202.

Taiz, L., Zeiger, E. 2010. *Photosynthesis: Carbon Reactions. Plant Physiology*. Sinauer Associates, Sunderland, MA, Vol. 5, 782.

Tavakkoli, E., Fatehi, F., Coventry, S., Rengasamy, P., McDonald, G.K. 2011. Additive effects of Na^+ and Cl^- ions on barley growth under salinity stress. *Journal of Experimental Botany* 62: 2189–2203.

Timmusk, S., Abd El-Daim, I.A., Copolovici, L., Tanilas, T., Kännaste, A., Behers, L., Nevo, E., Seisenbaeva, G., Stenström, E., Niinemets, U. 2014. Drought-tolerance of wheat improved by rhizosphere bacteria from harsh environments: Enhanced biomass production and reduced emissions of stress volatiles. *PLoS ONE* 9(5): e96086.

Timmusk, S., Behers, L., Muthoni, J., Aronsson, A.C. 2017. Perspectives and challenges of microbe application for crop improvement. *Frontiers in Plant Science* 8: 49.

Tiwari, S., Lata, C., Chauhan, P.S., Nautiyal, C.S. 2016. *Pseudomonas putida* attunes morphophysiological, biochemical and molecular responses in *Cicer arietinum* L. during drought stress and recovery. *Plant Physiology and Biochemistry* 99: 108–117.

Trindade, I., Capitao, C., Dalmay, T., Fevereiro, M.P., Santos, D.M. 2010. miR398 and miR408 are up-regulated in response to water deficit in *Medicago truncatula. Planta* 231(3): 705–716.

Ullah, A., Nisar, M., Ali, H., Hazrat, A., Hayat, K., Keerio, A.A., Ihsan, M., Laiq, M., Ullah, S., Fahad, S., Khan, A. 2019. Drought tolerance improvement in plants: An endophytic bacterial approach. *Applied Microbiology and Biotechnology* 103(18): 7385–7397.

Umezawa, T., Yoshida, R., Maruyama, K., Yamaguchi-Shinozaki, K., Shinozaki, K. 2004. SRK2C, a SNF1-related protein kinase 2, improves drought tolerance by controlling stress-responsive gene expression in *Arabidopsis thaliana. Proceedings of the National Academy of Sciences* 101(49): 17306–17311.

Upadhyay, S.K., Singh, J.S., Singh, D.P. 2011. Exopolysaccharide-producing plant growth-promoting rhizobacteria under salinity condition. *Pedosphere* 21: 214–222.

Vardharajula, S., Zulfikar Ali, S., Grover, M., Reddy, G., Bandi, V. 2011. Drought-tolerant plant growth promoting *Bacillus* spp.: Effect on growth, osmolytes, and antioxidant status of maize under drought stress. *Journal of Plant Interactions* 6(1): 1–14.

Vargas, L., Santa Brigida, A.B., Mota Filho, J.P. et al. 2014. Drought tolerance conferred to sugarcane by association with *Gluconacetobacter diazotrophicus*: A transcriptomic view of hormone pathways. *PLoS ONE* 9(12): e114744.

Verma, P., Yadav, A.N., Kazy, S.K., Saxena, A.K., Suman, A. 2014. Evaluating the diversity and phylogeny of plant growth promoting bacteria associated with wheat (*Triticum aestivum*) growing in central zone of India. *International Journal of Current Microbiology Applied Sciences* 3(5): 432–447.

Vimal, S.R., Patel, V.K., Singh, J.S. 2018. Plant growth promoting *Curtobacterium albidum* strain SRV4: An agriculturally important microbe to alleviate salinity stress in paddy plants. *Ecological Indicators* 105: 553–562. doi: 10.1016/j.ecolind.2018.05.014

Vurukonda, S.S.K.P., Vardharajula, S., Shrivastava, M., Skz, A. 2016. Enhancement of drought stress tolerance in crops by plant growth promoting rhizobacteria. *Microbiological Research* 184: 13–24.

Wang, Y., Gao, C., Liang, Y., Wang, C., Yang, C., Liu, G. 2010. A novel bZIP gene from *Tamarix hispida* mediates physiological responses to salt stress in tobacco plants. *Journal of Plant Physiology* 167(3): 222–230.

Wang, X., Chen, X., Liu, Y., Gao, H., Wang, Z., Sun, G. 2011. CkDREB gene in *Caragana korshinskii* is involved in the regulation of stress response to multiple abiotic stresses as an AP2/EREBP transcription factor. *Molecular Biology Reports* 38(4): 2801–2811.

Wang, C.J., Yang, W., Wang, C., Gu, C., Niu, D.D., Liu, H.X., Wang, Y.P., Guo, J.H. 2012. Induction of drought tolerance in cucumber plants by a consortium of three plant growth-promoting rhizobacterium strains. *PLoS ONE* 7(12): e52565.

Whipps, J.M. 2001. Microbial interactions and biocontrol in the rhizosphere. *Journal of Experimental Botany* 52: 487–511.

Wu, Q.S., Xia, R.X. 2006. Arbuscular mycorrhizal fungi influence growth, osmotic adjustment and photosynthesis of citrus under well-watered and water stress conditions. *Journal of Plant Physiology* 163(4): 417–425.

Wu, Q.S., Zou, Y.N., He, X.H., Luo, P. 2011a. Arbuscular mycorrhizal fungi can alter some root characters and physiological status in trifoliate orange (*Poncirus trifoliata* L. Raf.) seedlings. *Plant Growth Regulation* 65(2): 273–278.

Wu, Q.S., Zou, Y.N., Wang, G.Y. 2011b. Arbuscular mycorrhizal fungi and acclimatization of micropropagated citrus. *Communications in Soil Science and Plant Analysis* 42(15): 1825–1832.

Yoo, S., Weon, H.Y., Song, J., Sang, M.K. 2019. Induced tolerance to salinity stress by halo-tolerant bacteria *Bacillus aryabhattai* H19-1 and B. mesonae H20-5 in tomato plants. *Journal of Microbiology and Biotechnology* 29(7): 1124–1136.

Yoshiba, Y., Kiyosue, T., Nakashima, K., Yamaguchi-Shinozaki, K., Shinozaki, K. 1997. Regulation of levels of proline as an osmolyte in plants under water stress. *Plant and Cell Physiology* 38(10): 1095–1102.

Zarea, M., Hajinia, S., Karimi, N., Goltapeh, E.M., Rejali, F., Varma, A. 2012. Effect of *Piriformospora indica* and *Azospirillum* strains from saline or non-saline soil on mitigation of the effects of NaCl. *Soil Biology & Biochemistry* 45: 139–146.

Zhang, S., Gan, Y., Xu, B. 2016. Application of plant growth-promoting fungi *Trichoderma longibrachiatum* T6 enhances tolerance of wheat to salt stress through improvement of antioxidative defense system and gene expression. *Frontiers of Plant Science* 7: 1405.

Zhao, B., Ge, L., Liang, R., Li, W., Ruan, K., Lin, H., Jin, Y. 2009. Members of miR-169 family are induced by high salinity and transiently inhibit the NF-YA transcription factor. *BMC Molecular Biology* 10(1): 1–10.

Zhao, X., Wei, P., Liu, Z., Yu, B., Shi, H. 2017. Soybean Na^+/H^+ antiporter GmsSOS1 enhances antioxidant enzyme activity and reduces Na^+ accumulation in *Arabidopsis* and yeast cells under salt stress. *Acta Physiologiae Plantarum* 39: 19.

Zheng, D., Chang, Q., Li, Z., Gao, M., She, Z., Wang, X., Guo, L., Zhao, Y., Jin, C., Gao, F. 2016. Performance and microbial community of a sequencing batch biofilm reactor treating synthetic mariculture wastewater under long-term exposure to norfloxacin. *Bioresource Technology* 222: 139.

3 Role of Arbuscular Mycorrhizal Fungi in Horticultural Crops

Gains and Provocations

Hari Kesh
CCS, Haryana Agricultural University, Hisar, India

Ashutosh Srivastava
Rani Lakshmi Bai Central Agricultural University, Jhansi, India

Prashant Kaushik
Instituto de Conservación y Mejora de la Agrodiversidad, Valencia, Spain
and
Kikugawa Research Station, Shizuoka, Japan

CONTENTS

3.1 INTRODUCTION

Plant important microorganisms are examined based on their features in grow nutrient acquisition, biocontrol growth promotion, and as biocontrol elements for more than three years (Bhardwaj et al. 2014). In India, a selection of microbial formulations having exclusive or perhaps a combination of stresses have been developed and used. The achievements associated with a microbial bio-inoculant within the agriculture field have been impacted by not merely the microorganism's ability to make it. However, the present information on capabilities, ecological adaptations, multitude interactions, and putative beneficial qualities of bio-inoculants in plants entailed are available for few species (Rana et al. 2020). The bulk of the microbiome medical studies are oriented towards characterizing bacterial species earning the eukaryotic bio-inoculants.

DOI: 10.1201/9781003147077-3

Comprehensive metagenomic analyses of the microbiome of harvest vegetation continue to occur in many situations (Suyal et al. 2019). Thus, the normal component of plant microbiome in every crop species is still not exact. It is assumed that some components of plant interacting microorganisms are pathogenic, but a lot of the microorganisms inhabiting plant markets sometimes have neutral or beneficial features. Microorganisms with the fundamental job on the location may not have a simple component or other bio-inoculants. Metagenomic analysis for identification and functional characterization might disclose these fundamental microorganisms' function and their interaction with different microorganisms in the microenvironment (Jovel et al. 2016; Marco and Abram 2019).

Additionally, horizontal transfer of plant crucial genetics of all proteobacteria appears to be discovered (Bulgarelli et al. 2013; Schlaeppi and Bulgarelli, 2015; Singh et al. 2020a, b). Moreover, the microbiome framework of a crop is recognized as being impacted by the genotype, tissue/plant body organ, abiotic environment, and growth stage just where it rises. Like changes/imbalance inside the microbiome framework of the human body triggering disease scenarios, changes in plant parts' microbiome composition could end up in susceptibility to stress/diseases (Gill et al. 2006; Uroz et al. 2016). The compatibility or incompatibility of the released microorganism with the indigenous bio-inoculants is at present generally not familiar. In that case, the issues like marginal and inconsistent plant development promotion, biocontrol of pests, and diseases' additional yield could be solved and enhanced (Chang et al. 2017). This assists in understanding modifications in the microbiome composition because of introduced bio-inoculants.

In the exact same way, changing a procedure for gardening soil microbial tests, furthermore to accessibility of microbial blueprint of every single harvest species, can assist the scientists and region officers to endorse nature favourable and other matters precise (Nair and Raja 2017). Bio-inoculants might remain an effective and sustainable plan to enhance plant growth and productiveness, improving tolerance against abiotic stresses (Khan et al. 2015; Verma et al. 2016; Saini et al. 2019a, b; Pankaj et al. 2020). The growth supportive raw material is a cruicial factor in development of each bio-inoculant due to its changeable complex dynamics and the hetero-geneous combination of elements. The problem is a great deal more complicated by the considerable length of plant life, bio-inoculants, and normally materials bundled into the range of growing bio-inoculants (Rana et al. 2020). For example, two goods obtained by two unique vegetations would belong within an equivalent training. However, the consequences and the method of their activity might be different. Moreover, the full reverse situation might occur; the associated product can create special consequences when utilized in plant life which are different (Rouphael and Colla 2020).

The comfort of any bio-inoculant is not as an outcome related to an individual combination but is the outcome of the synergistic job of huge bioactive molecules (Saini et al. 2021). Since vegetation are sessile organisms that must deal with undesirable outside conditions, most among these techniques are essential for their survival. If they are brought on, in time, these techniques work to establish a defence reaction alongside matter on the eco-friendly adjustments that could impact plant growth irreversibly (Zuccarini and Okurowska 2008). The probable downside between growth and acclimation metabolism results in a type of physical exercise price for plant life, since energy and nutrients are usually destined to grow. Good agronomic management performed to boost the tolerance in crop towards abiotic stresses (created over the centuries due to the technological advancement) which includes the selection of the proper cultivar, the sowing density, and unquestionably the quantity of water as well as fertilizers (Malhi et al. 2021a). Protected raising is truly a cropping method accompanied to shield plant life from unfavourable external conditions. It is primarily ideal for produce and floriculture development in a non-perfect atmosphere through the command of temps, soothing, and possibly atmospheric composition. Another agronomical strategy, mainly used in veggie plant life, is soilless cultivation (Deljou et al. 2014). Thus, bio-inoculants are generally known in these major groups. Bacteria are integrated by this team, yeast, and filamentous fungi, combined with microalgae. They are separated from dirt, plants, water, and composted manure and even some other organic substances. They are set onto earth to enhance harvest results through metabolic pursuits. They assist the uptake of nourishment through the solubilization as well as nitrogen fixation of nourishment; they change a hormonal condition by inducing more stress, e.g., cytokinins. Moreover, they build tolerance to abiotic stresses and produce volatile organic components that can have an immediate effect on vegetation (Malhi et al. 2021b). Positive effects are supplied by microorganisms which produce a protective biofilm on root locations boosting material and fluid uptake. Keeping this overview in mind, this chapter provides suggestions on the use of bio-inoculants, their benefits, and regular strategies of research, together with prospects.

3.2 SOIL METAGENOMICS IN THE CONVERSATION OF MICROBIAL BIODIVERSITY

The soil and RNA analysis has physicochemical characteristics as affected by the growing practice. It's well realized that microorganisms have an excellent amount and variety than alternate organisms on the earth (Delmont et al. 2011; Ranjan et al. 2016; McGee et al. 2019). The microbial variety alongside composition structure is substantially affected by eco-friendly parts. As an outcome, indexing, cataloguing, and evidence of the microorganisms are prerequisites (Fierer and Jackson 2003; Gonzalez et al. 2012). Microbial lots in any habitat are a great deal more linked to the massive measurements of species existing during a particular period. As the planet's microbial society plays an essential role in dirt health management, agro, accessibility of cultivating nourishment, and turnover duties of organic material in soil, they're genuinely influenced by natural and anthropogenic activities (Andersen et al. 2013; Liao et al. 2018). In the most recent past, the utilization of other pesticides, fungicides, herbicides, and artificial fertilizers has led to the degeneration of the grime microflora alongside variety (Keller and Zengler 2004; Saini et al. 2017). Thus, the microorganisms with the changing environment offer a broader picture of how microorganisms are changing the practical choices that come with soils and their flourishing in endangered ecosystems (Prosser et al. 2007). Today, practically all the eco-friendly objectives in farming are on adhering to farming sustainability. Many metagenomic initiatives are finished in agriculture but don't continue some promises to help the marginal farmers. Consequently, practical assessments are needed, which may be worn by the growers' revenue and assistance farming (Souza et al. 2013). Soil metagenomics has pushed the microbial general public's restoration to boost grain yield and dirt health. Metagenomics has the capability to foresee the garden soil microorganisms' construction and undoubtedly the influence on microbial companies of connected niches (Kaushik et al. 2020).

Sustainable farming techniques include a range of microhabitats with superb green variations and genetic biodiversity. Reports from agrarian soils established you're able to have higher microbial stock and insert development promo pursuits (Teuber et al. 2017). Many health-related studies are leading the existing metagenomic developments in agriculture (Karaca and Ince 2019). Soil microorganisms are an important component in triggering the perfect spot development, stress responses, and vegetation defence (De Coninck et al. 2015). For sustainable agricultural production, beneficial microorganisms of agricultural value are able to work as an essential alternative. Metagenomic compensation can deal with fundamental restorative concerns connected with agriculturally main microorganisms (Arora et al. 2018). The rapid development of contemporary molecular biological methods brought a brand-new frontier of science, metagenomics (De Coninck et al. 2015). The information from these solutions will disclose more appropriate biodiversity and more beneficial microorganism capabilities in the earth than those from previously cultured microorganisms. Primarily, soil metagenomic techniques provides us with such helpful information regarding agriculture. Especially, soil microorganisms play important roles in component cycles (carbon, nitrogen, etc.) by decomposing all-natural materials made from plant life, microorganisms, and animals (Bulgarelli et al. 2013).

In farming areas, to reach sustainable production of superior quality plants, vegetables, it's vital to safeguard soil fertility and defeat dirt illness because of repeated cropping. To achieve this goal to assess the physical properties and substance of soils and organic ones and the ability to use their information for gardening soil control, mainly it's essential to elucidate what sort of microorganisms take place in dirt or how soil microorganisms are keen on the phenomena in farming locations, like harvest development, insect infestation as well as suppressiveness (An et al. 2012).

Conversely, metagenome sequencing will provide us with information of biodiversity in the atmosphere, while it takes enormous time and budget (Jovel et al. 2016). Sequencing a metagenome produced from a particular soil sample employs a chance to give us genes which encode novel biocatalysts for biosynthetic or possibly biodegradation procedures like wreckage of deadly elements, synthesis of production, and biofuels of novel drugs. To filter such novel invaluable genes, over 10 metagenomic libraries inside a cloning vector were developed after 2000. Although these stories these days were worthwhile in purchasing several novel genes as well as functions from metagenome, the normal process is generally ineffective and hard. Particularly, knock rates of aim genes are very little in case they occur in non-dominant species. In this context, immediate shotgun sequencing of a metagenome using excessive throughput sequencing methods is starting to be continuously more common (Sboner et al. 2011). A global attempt to mix the worldwide healthcare community's skills to concentrate on sequencing and annotating the grime metagenome was suggested. This worldwide undertaking, the Terra Genome overseas sequencing consortium, has the following main objective: whole sequencing of a 'reference' dirt metagenome.

3.3 ARBUSCULAR MYCORRHIZAL FUNGI AND VEGETABLE PRODUCTION

Vegetables are the herbaceous plant species cultivated for human consumption in which swollen tap roots (carrots), lateral roots (sweet potatos), leaves (lettuce, cabbage), belowground stems (Irish potato), aboveground stems (asparagus), hypocotyls (radish, turnip), flower buds (cauliflower), and petioles (celery) comprise the edible parts (Sharma and Kaushik 2021). Vegetables are valued for their beneficial effects on human health and emphasize the importance of nutraceutical properties (Gruda 2005). Plants are constantly interacting with microorganisms present in their surroundings and are able to create beneficial mutual associations with some that are present in the rhizosphere. Arbuscular mycorrhizal fungi (AMF) positively affect the plants' growth and vigour, fruit yield, nutrient uptake, photosynthetic rates, antioxidant activity, water use efficiency, and tolerance against biotic and abiotic stresses (Table 3.1) (Davies et al. 2005; Yadav et al. 2021a). Application of selected AMF serves as a sustainable alternative for P-fertilizers in tomato and pepper for increasing crop profitability (Conversa et al. 2013; Tanwar et al. 2013a, b). *Glomus intraradices* improves the β-carotene content and accumulation of anthocyanins, carotenoids, phenolics, and mineral nutrients, thus increasing the nutritional profile of lettuce and sweet potato (Baslam et al. 2012). However, the effectiveness depends on the time and type of fungus used to inoculate the vegetables. For example, onions have high P, N, and Zn concentrations and more firmer bulbs when inoculated with *Glomus versiforme* than those inoculated with *G. intraradices* (Charron et al. 2001a, b). Similarly, some vegetables showed increased growth and vigour and are responsive to inoculation whereas some are non-responsive. Growth characteristics of cucumber were significantly improved by *Glomus mosseae*. However, non-significant differences were observed for *G. intraradices* (Chang et al. 2008). *Glomus mosseae* also promote water stress tolerance sustains yield in sunflower plants (Sharma et al. 2021a). AMF inoculation increases the salt tolerance capacity of turmeric by improving nutrient uptake, chlorophyll content, and water use efficiency (Bandi and Kaushik 2022).

Pepper plants inoculated with AMF showed an increased photosynthetic rate and water use efficiency under drought conditions than non-inoculated plants (Jezdinsky et al. 2012; Sharma et al. 2022). Tomato plants colonized with AMF have increased antioxidants like catalase, superoxide dismutase, ascorbate peroxidase in leaves, high accumulation of protein, soluble sugar, and lower Na toxicity under salt stress (He et al. 2010; Latef and ChaoXing 2011). Suppressive effects of AMF on fungal pathogens and disease was seen in many plants (Saini et al. 2020). Inoculation of cucumber

TABLE 3.1
Effect of Arbuscular Mycorrhizal Fungi (AMF) on Growth, Quality, and Stress Tolerance of Vegetable Crops

Crop	Fungus Species	Traits Improved	References
Onion	*Glomus intraradices* and *Glomus versiforme*	High biomass and greater bulb diameter	Charron et al. (2001a)
Asparagus	*Gigaspora margarita*, *Glomus fasciculatum*, and *Glomus* sp. R10	More roots and shoots dry matter, tolerance against *Fusarium oxysporum* f.sp. Asparagi	Matsubara et al. (2001)
Tomato	*Glomus mosseae*	Bioprotector effect against *Phytophthora parasitica*	Pozo et al. (2002)
Lactuca sativa	*Glomus coronatum*, *G. intraradices*, *Glomus claroideum*, *G. mosseae*, *G. constrictum*, and *G. geosporum*	Increased plant water uptake and drought tolerance	Marulanda et al. (2003)
Capsicum annuum	*G. intraradices* and *Glomus* sp.	Enhanced growth and leaf P content	Martin and Stutz (2004)
Potato	*G. intraradices*	High root-to-shoot ratio, low leaf-to-tuber ratio, and high phosphorus use efficiency	Davies et al. (2005)
Tomato	*G. mosseae*	High root and shoot dry matter, more fresh fruit yield, high P, K, Fe, Cu, and Zn content of shoot and tolerance to salt stress	Al-Karaki (2006)
C. annuum	*Glomus deserticola*	Induction of isoforms of acidic chitinases, superoxide dismutase, peroxidase, phenylalanine ammonia-lyase activity, and biocontrol effect against *Verticillium* wilt	Garmendia et al. (2006)
Tomato	*G. mosseae*, *G. geosporum*, *G. claroideum*, *G. fasciculatum*	Increased biomass production and phosphate, potassium, nitrogen uptake	Takacs et al. (2006)
C. annuum	*G. fasciculatum*, *G. constrictum*, *G. geosporum*, *G. tortuosum*, *G. aggregatum*, *G. deserticola*, *G. geosporum*, *G. microaggregatum* and *G. coremioides*	More fruit fresh and dry weight, high chlorophyll, carotenes, and xanthophyll content	Mena-Violante et al. (2006)
Tomato	*G. intraradices*	Increased shoot dry matter, number of flowers and fruits, fruit yield, high ascorbic acid, and total soluble solids, tolerance to drought stress	Subramanian et al. (2006)

(Continued)

TABLE 3.1 *(Continued)*
Effect of Arbuscular Mycorrhizal Fungi (AMF) on Growth, Quality, and Stress Tolerance of Vegetable Crops

Crop	Fungus Species	Traits Improved	References
Lactuca sativa	*G. mosseae*, *G. intraradices*, and *G. coronatum*	Enhanced dry mass production, chlorophyll content and total leaf area, increased K and P uptake, reduced Na and Cl uptake, tolerance to salt stress	Zuccarini (2007)
C. annuum	*G. mosseae*, *G. etuniatum*, *G. fasciculatum*, and *G. margarita*	Increased shoot height, roots and shoots fresh and dry weights and yield, reduced disease severity *Phytophthora capsici*	Ozgonen and Erkilic (2007)
Tomato	*G. intraradices*	Increased leaf area index, marketable yield, improved stem, and fruits P content	Conversa et al. (2007)
L. sativa	*G. intraradices*	Tolerance to water deficit	Aroca et al. (2008)
C. annuum	*G. intraradices* and *G. margarita*	High shoot height, root length, fresh and dry matter of root and shoot, tolerance to salt stress	Turkmen et al. (2008)
Tomato	*Glomus* sp.	High lycopene, β-carotene and phenolic contents, and fresh root weight	Ulrichs et al. (2008)
Cucumber	*G. mosseae*, *G. intraradices*, and *G. versiforme*	N and P content in roots, Mg, Cu and Zn content in shoots, single fruit weight was increased	Wang et al. (2008)
Capsicum annum	*G. clarum*	Enhanced growth and fruit yield by reducing leaf Na^+ concentration and membrane leakage and increasing concentration of N, P, and K	Kaya et al. (2009)
Tomato	AMF	Increased growth, nutrient uptake, and chlorophyll content	Elahi et al. (2010)
Cucumber	*G. rosea*	Increased root fresh weight, shoot dry weight, and photosynthetic efficiency, tolerance to salt stress	Gamalero et al. (2010)
Tomato	*G. mosseae*	Lower Na^+ toxicity in roots and shoots, higher accumulation of protein, proline and soluble sugar, high leaf water potential, tolerance to salt stress	He et al. (2010)
Tomato	*G. mosseae*	Increased activity of SOD and POD in roots and leaves, decreased production of O_2^- and MDA	Zhi et al. (2010)
Tomato	*G. intraradices*	High activity of CAT, APX, POD, SOD, protein and proline, low activity of H_2O_2 and MDA, high K/Na and Ca/Na ratio, improved net assimilation rate and stomatal conductance	Hajiboland et al. (2010)

(Continued)

TABLE 3.1 *(Continued)*
Effect of Arbuscular Mycorrhizal Fungi (AMF) on Growth, Quality, and Stress Tolerance of Vegetable Crops

Crop	Fungus Species	Traits Improved	References
Tomato and carrot	*G. etunicatum*, *G. mosseae*, *A. scrobiculata*, and *K. kentinensis*	Increase in yields and suppression of nematode multiplication *Meloidogyne* sp.	Affokpon et al. (2011)
Carrot and green onion	*G. intraradices* and *G. mosseae*	Increased fresh weight of both root and shoot, enhanced activity of phosphatase, reduction in organophosphorus residues in plant tissues	Wang et al. (2011)
Tomato	*G. fasciculatum*	Improved growth and phosphorus uptake	Tayal et al. (2011)
Cucumber	*G. mosseae*	Enhanced shoot dry weight, protective effect against anthracnose and damping-off pathogen	Saldajeno and Hyakumachi (2011)
Tomato	*G. intraradices*	Reduced damage in roots caused by nematode *Nacobbus aberrans*	Lax et al. (2011)
Tomato	*G. mosseae*	High P and K concentration, enhanced of activity of catalase, peroxidase, superoxide dismutase and ascorbate peroxidase in leaves, lower oxidative damage, tolerance to salt stress	Latef and Chaoxing (2011)
Tomato	*G. mosseae*	Reduced severity of *Botritis cinerea* and lower ABA production	Fiorilli et al. (2011)
Tomato	AMF	Improved fresh weight of stem and root, high chlorophyll a, b, and total chlorophyll content	Chandrasekaran et al. (2021)
Lettuce	AMF	Enhanced concentration of carotenoids, anthocyanins, phenolics, Cu, and Fe	Baslam et al. (2011)
Cucumber	*G. mosseae* and *G. versiforme*	Improved growth, high soluble sugar, proline, N, P, and K content, reduced *Fusarium* wilt severity	Wang et al. (2012)
Pepper	*G. moseea* and *G. etunicatum*	Enhanced P and Zn concentration	Ortas (2012)
Lettuce	AMF	Improved growth and nutritional quality	Baslam et al. (2012)
Lettuce	AMF	Enhanced accumulation of carotenoids and anthocyanins, chlorophylls, and phenolics	Baslam et al. (2012)
C. annuum	*Glomus* sp.	Increased photosynthetic rate and water use efficiency	Jezdinsky et al. (2012)

(Continued)

TABLE 3.1 ***(Continued)***
Effect of Arbuscular Mycorrhizal Fungi (AMF) on Growth, Quality, and Stress Tolerance of Vegetable Crops

Crop	Fungus Species	Traits Improved	References
Tomato	*G. mosseae*	Reduced infection by *Meloidogyne incognita* and *Pratylenchus penetrans*	Vos et al. (2012a, 2012b)
C. annuum	*G. mosseae* and *G. intraradices*	Improved relative water and phosphorus, total chlorophyll, and carotenoid content, low MDA content, tolerance to salt stress	Cekic et al. (2012)
Tomato	*G. intraradices*	High fresh fruit weight, flower number, and marketable fruit number	Conversa et al. (2013)
L. sativa	*G. intraradices*	Improved growth, stomatal conductance and efficiency of photosystem-II, tolerance to salt stress	Aroca et al. (2013)
Lettuce	AMF	Improved level of major chlorophylls, carotenoids, and tocopherols, K, Mg, Cu, Zn, and Fe	Baslam et al. (2013a, 2013b)
Tomato	*G. mosseae*	Lower infection of nematode *M. incognita*	Vos et al. (2013)
C. annuum	*G. intraradices*	High root and shoot biomass and membranes stability	Beltrano et al. (2013)
Cucumber	AMF	Improved plant growth and resistance against *Fusarium oxysporum*	Elwakil et al. (2013)
Sweet potato	*G. intraradices* and *G. mosseae*	Improved β-carotene concentrations	Tong et al. (2013)
Pepper	AMF	Increased chlorophyll index, leaf contents of N, P, Fe, and Zn,	Diaz Franco et al. (2013)
Tomato	*G. mosseae*, *G. clarum*, *G. etunicatum*, *G. intraradices*, and *G. caledonium*	Early flowering, high biomass, enhanced content of phosphorus and zinc	Ortas et al. (2013)
Tomato	AMF	Increased uptake of phosphorus	Watts-Williams et al. (2014)
Onion	AMF	Increased antioxidant activity and nutrient uptake	Mollavali et al. (2015)
Tomato	AMF	Increased root dry matter, uptake rates of N, Mg, P, Fe, Ca, Mn, and tolerance to salt stress	Balliu et al. (2015)
Tomato	AMF	Reduced production of malonaldehyde and hydrogen peroxide, protective effect against Cd stress	Hashem et al. (2016)
Tomato	*Rhizophagus irregularis*	Enhanced shoot FW, leaf area, leaf number, root FW, and levels of growth hormones, tolerance to salt stress	Khalloufi et al. (2017)

(Continued)

TABLE 3.1 ***(Continued)***
Effect of Arbuscular Mycorrhizal Fungi (AMF) on Growth, Quality, and Stress Tolerance of Vegetable Crops

Crop	Fungus Species	Traits Improved	References
Tomato	*G. mosseae*	Increased plant fresh and dry weight, activity of SOD and POD, total chlorophyll, and net photosynthesis rate, reduced incidence of *Cladosporium fulvum*	Wang et al. (2018a)
Cucumber	*G. etunicatum*, *G. intraradices*, and *G. mosseae*	Increased biomass, photosynthetic pigment synthesis, and enhanced antioxidant enzymes like SOD and CAT, tolerance to salinity	Hashem et al. (2018)
Tomato	*Glomus* sp.	Increased growth, dry weight, N, P, K, chlorophyll content and yield, reduced *F. oxysporum* f.sp. *lycopersici*	Kumari and Prabina (2019)
Pea	*Rhizoglomus intraradices*, *Funneliformis mosseae*, *Rhizoglomus fasciculatum*, and *Gigaspora* sp.	High nutrient uptake and chlorophyll synthesis, accumulation of compatible osmolytes, lower cellular electrolyte leakage, enhanced growth, biomass production, and yield	Parihar et al. (2020)
Pea	AMF	Reduced accumulation of arsenic in roots, shoots, and grains	Alam et al. (2020)
Onion	AMF	Increased yield, bulb weight, total sugars, dry matter, phenolics, titratable acidity Ca, Mg, K, and P	Golubkina et al. (2020)

seedlings with *G. mosseae* and *G. versiforme* improve the growth, seedling quality, and dry weight of roots and shoots and reduce the incidence of wilt disease caused by *Fusarium oxysporum* and anthrac-nose disease in cucumber (Saldajeno and Hyakumachi 2011; Chang et al. 2012). Similarly, reduced severity of *Phytophthora* blight caused by *Phytopthora capsici*, and increased capsidiol levels, plant growth, and development, which were seen in pepper plants pre-inoculated with *G. mosseae* (Ozgonen and Erkilic 2007). Infection of the root-knot nematode *Meloidogyne incognita* in mycorrhizal tomato plants was significantly lower as compared to control plants (Vos et al. 2013).

3.4 ARBUSCULAR MYCORRHIZAL FUNGI AND FLOWER PRODUCTION

Flowers are very intimately connected with the many religious and social practices in India. Major flower growers produce high-value cut flowers like roses, gladiolus, orchids, lily, tuberose, gerbera, carnation, and loose flowers like marigolds, chrysanthemums, jasmine, china aster, and gaillardia (Vahoniya et al. 2018). Several previous studies have shown that using biological sources like mycorrhizal application enhances the growth, yield, and tolerance power of flowering plants against various stresses (Table 3.2) (Asrar et al. 2012; Navarro et al. 2012). Many ornamental plants responded positively to AMF inoculation for growth, vigour, and tolerance against biotic and abiotic stresses. Pelargonium (*Pelargonium peltatum*) plants inoculated with three different inocula (TerraVital Hortimix, Endorize-Mix, and AMYkor) enhanced the number of buds, number of flowers, P and K concentration in shoot at low dose of compost, but no changes were seen in shoot biomass and N concentration (Perner et al. 2007). Mixed inoculation of *Gigaspora* and *Scutellospora* sp. in *Callistephus chinesis* and *Impatiens balsamina* improves the number of flowers, shoot height, and dry biomass and saves 30% of the total cost in comparison to chemical fertilizers (Gaur and Adholeya 2000). Chrysanthemums (*Chrysanthemum maximum*) produced a greater number of leaves, root and shoot biomass and accumulated less Cu and Pb in aboveground parts when inoculated with *Funneliformis mosseae* compared to non-inoculated plants (Gonzalez-Chavez and Carillo-Gonzalez 2013). Micropropagated *Gloriosa superba* L. inoculated with *Glomus mosseae*, *Acaulospora laevis*, and a mixed AMF improve leaf number, tuber length, plant height, and survival rate and colchicine content. However, the survival rate was high with *A. laevis* followed by mixed AMF (Yadav et al. 2013). Colonization of *Gerbera jamesonii* with *G. mosseae* significantly enhanced leaf area, number of flowers, high flower longevity, calcium content, and nutrient uptake in soilless culture compared to control plants (Deljou et al. 2014).

Potted snapdragon (*Anthirhinum majus* 'Butterfly') plants colonized with *Glomus deserticola* have enhanced nutrient content such as P, K, N, Mg, and Ca, chlorophyll content, water relations, number, and diameter of flowers, thus alleviating the harmful effect of drought stress (Asrar et al. 2012). The beneficial effect of AMF inoculation under salt stress conditions was also noticed in several ornamental plant species. For example, effectiveness of *Rhizoglomus intraradices* and *G. iranicum*

TABLE 3.2
Effect of Arbuscular Mycorrhizal Fungi (AMF) on Growth, Quality, and Stress Tolerance of Ornamental Crops

Crop	AM Fungus Species	Traits Improved	References
Lilium sp.	*Glomus*, *Gigaspora*, *Scutellospra* sp. and *G. intraradices*	Increased shoot length, bulblet size, and weight and P content, early flowering	Varshney et al. (2002)
Chrysanthemum morifolium	AMF	Increased rooting rate, plant height, leaf area, fresh, and dry weight of shoots, early flowering	Sohn et al. (2003)
Rosa multiflora	*Glomus albidum*, *G. claroideum*, and *G. diaphanum*	Enhanced leaf and stem dry weight, leaf area, N, P, K, Ca, Fe, Zn, Al, and B concentration, chlorophyll content, tolerance to bicarbonate	Cartmill et al. (2007)
Pelargonium peltatum	*Glomus mosseae*, *G. intraradices*, *G. sclaroideum*, *G. microaggregatum*, *G. etunicatum*, and *Glomus* sp.	Increased number of buds, flowers, P and K concentration	Perner et al. (2007)
Catharanthus roseus	*G. albidum*, *G. claroideum*, and *G. diaphanum*	Enhanced growth and content of P, Zn, Cu, Mn, B, and Mo and increased antioxidant activity	Cartmill et al. (2008)
Eustoma grandiflorum	AMF	Increased yield, stem length, and number of flowering stems	Meir et al. (2010)
Tagetes erecta	*G. constrictum*	Increased flowers' carotene, leaves chlorophylls a and b, phosphorous content, and tolerance to water stress	Asrar and Elhindi (2011)
Antirhinum majus	*G. deserticola*	High root and shoot dry mass, water use efficiency, flower yield, P, N, K, Mg, Ca, and chlorophyll content and tolerance to water stress	Asrar et al. (2012)
Rauwolfia serpentina	*G. mosseae* and *Acaulospora laevis*	Improved height, root and shoot P content, chlorophyll content, and stomatal conductivity	Kaushish et al. (2012)
Dianthus caryophyllus	*G. mosseae*, *A. laevis*, and *Gigaspora* sp.	Improved fresh and dry root and shoot weight, chlorophyll a and b content, shoot and root P content	Bhatti et al. (2013)
Hibiscus sabdariffa	*Glomus* sp., *Gigaspora* sp., and *Scutellospora* sp.	Improved vegetative growth and anthocyanin concentration	Sembok et al. (2015)

(Continued)

TABLE 3.2 ***(Continued)***
Effect of Arbuscular Mycorrhizal Fungi (AMF) on Growth, Quality, and Stress Tolerance of Ornamental Crops

Crop	AM Fungus Species	Traits Improved	References
Calendula officinalis, *Origanum majorana*, and *Melissa officinalis*	AMF	Increased biomass, number of flowers, yield of rosmarinic acid, and lithospermic acid	Engel et al. (2016)
H. sabdariffa	*Glomus versiforme* and *Rhizophagus irregularis*	Increased number of fruits, plant dry weight, economical yield, biological yield	Fallahi et al. (2016)
Hibicus rosa sinensis	*G. mosseae*	Increased shoot length, number of leaves and N, P, K, and Zn concentration	Kasliwal and Srinivasamurthy (2016)
Osmium basilicum	*G. deserticola*	Improved growth, water use and photosynthetic efficiency, K/Na and Ca/Na ratio, and tolerance to salt stress	Elhindi et al. (2017)
Rosa damascena	AMF	Improved flower yield, macronutrient content, chlorophyll and carotenoid content, tolerance to drought stress	Abdel-Salam et al. (2018)
C. morifolium	*Funneliformis mosseae* and *Diversispora versiformis*	High root length, root and shoot dry weight, and root N concentration	Wang et al. (2018b)
Zelkova serrata	*F. mosseae*	Increased photosynthetic and biomass, increased P, K^+, and Mg^{2+} content, and stomatal conductance of leaves	Wang et al. (2019)
Hibiscus sabdariffa	*G. mosseae*	Increased chlorophyll a and b content, calyx parameters, i.e., total proteins, lipids, and crude fibre, phenolics, flavanoids, and calyx mineral content	El-Kinany et al. (2020)

improved the growth and quality of carnations (*Dianthus caryophyllus*) and euonymus (*Eriobotrya japonica*) under salt stress, enhancing the uptake of P, K, Mg, and Ca and by reducing the uptake of Na^+ and Cl^- to the shoot. This indicates that Na^+ and Cl^- might be retained in intra-radical hyphae of AMF or compartmentalized in the vacuoles of root cells without moving into the cytoplasm (Navarro et al. 2012; Gomez-Bellot et al. 2015). Colonization of *Tagetes erecta* with AMF *R. mosseae, R. intraradices*, and *G. constrictum* enhance the root and shoot biomass, the activity of antioxidants such as catalase, superoxide dismutase and peroxidase, reduction in the production of reactive oxygen species (ROS), and tolerance to cadmium metal stress compared to non-mycorrhizal plants (Liu et al. 2011). Mixed inoculation of *Glomus albidum*, *Claroideoglomus claroideum*, and *Glomus diaphanum* enhance the tolerance of *Rosa multiflora* and *Catharantus roseus* to high alkalinity by improving the growth, leaf concentration of N, P, K, Ca, Zn, Fe, B, Mn, and Mo, chlorophyll synthesis and nutrient uptake and translocation, as well as low iron reductase and soluble alkaline and phosphate activities and tolerance to alkalinity (Cartmill et al. 2007, 2008).

3.5 ARBUSCULAR MYCORRHIZAL FUNGI AND FRUIT TREES PRODUCTION

Fruit trees are historically and intensively used for fruit, fodder, seeds, and medicine by human beings (Ambe 2001). They contribute to food security, solve nutritional issues, and are an essential source of revenue for fruit growers. Due to their nutritive and economic value, fruit trees serve as an alternative to agroforestry and field crops (Leakey et al. 2005). Mycorrhizal fungal root colonization affects the growth, vigour, and health of host plants with better nutrient uptake, tolerance to drought, salt and heavy metals, and more excellent resistance against pathogens (Sharma et al. 2021b; Yadav et al. 2021b). AMF are known to help in the absorption and assimilation of macro- and microelements such as P, N, Ca, and Zn (Wu and Zou 2009). Single and mixed inoculation effects of two AMF, namely, *G. mosseae* and *Glomus intraradices*, were evaluated in *Morus alba* L. seedlings under a greenhouse (Table 3.3). The inoculated plants exhibited a greater number of leaves, plant height, longer roots, high chlorophyll content, and increased nitrogen and phosphorus content (Lu et al. 2015).

Similarly, single or mixed inoculation of strawberries (*Fragaria* × *ananassa*) with *F. mosseae* and *F. geosporus* increased growth, yield, and water use efficiency under water-deficient conditions (Boyer et al. 2015). Enhanced watermelon efficiency has also been observed when inoculated with AMF suggesting the role of fungus in enhancing water uptake and its efficient utilization (Omirou et al. 2013). Growth and yield characteristics of strawberries such as size and number of fruits were significantly improved by *R. intraradices*, *G. ageratum*, *G. viscosum*, *C. etunicatum*, and *C. claroideum* with 70% of the conventional fertilization compared to non-mycorhizal plants with conventional fertilization (Bona et al. 2015). *G. versiforme* improves the mineral composition under drought stress, including P, K, N, Mn, Fe, and Zn of trifoliate orange (*Poncirus trifoliata* L.) in comparison to

TABLE 3.3
Effect of Arbuscular Mycorrhizal Fungi (AMF) on Growth, Quality, and Stress Tolerance of Fruit Crops

Crop	Fungus Species	Traits Improved	References
Ziziphus mauritiana	*Gigaspora margarita*, *Glomus constrictum*, *Glomus fasciculatum*, *Glomus mosseae*, *Sclerocystis rubiformis*, and *Scutellospora calospora*	Increased nutrient uptake of N, P, K, Ca, and Mg, increased accumulation of free proline, amino acids, soluble protein, total chlorophyll, and reducing sugars	Mathur and Vyas (2000)
Psidium guajava	*Glomus diaphanum*, *Glomus albidum*, and *Glomus claroides*	Improved shoot growth and leaf production, increased content of P, Mg, Cu, and Mo in leaves	Estrada-Luna et al. (2000)
Persea sp.	*Glomus clarum*, *Scutellospora heterogama*, *Glomus etunicatum*, *Acaulospora scrobiculata*, and *Glomus manihotis*	Better vegetative growth, improved carbohydrate content, and good nutrition	Silveira et al. (2002)
Banana	VAM	Improved root length, shoot length, root weight, shoot weight, and chlorophyll content	Thaker and Jasrai (2002)
Banana	*Glomus* sp., *G. proliferum*, *G. intraradices*, and *G. versiformes*	Improvement in growth, shoot P content, and lower disease severity caused by *Cylindrocladium spathiphylli*	Declerck et al. (2002)
Strawberry	AMF	High biomass production, leaf area, shoot-to-root ratio, good efficiency of photosystems, tolerance to water stress	Borkowska (2002)
Plantago lanceolata	AMF	High P content and plant dry mass	Paradi et al. (2003)
Grapevine	*G. etunicatum*, *Glomus caledonium*, *G. clarum*, and *Glomus mosseae*	Increased leaf area, P content, and total sucrose concentration	Caglar and Bayram (2006)
Grapevine	*Glomus* sp.	Increased uptake of N and K, improved vegetative growth	Ocete et al. (2015)
Peach	AMF	Increased N, Ca, and Mg content of leaves	Borkowska et al. (2008)
Peach	*Acaulospora* sp., *G. clarum*, and *G. etunicatum*	Increased absorption of N, P, and K, stem height, diameter, foliage area, stem fresh, and dry weight	Nunes et al. (2008)
Peach	*Acaulospora* sp., *G. clarum*, *G. etunicatum*, and *S. heterogama*	Increased N, P, and K tissue content and good vegetative growth	Nunes et al. (2009)

(Continued)

TABLE 3.3 *(Continued)*

Effect of Arbuscular Mycorrhizal Fungi (AMF) on Growth, Quality, and Stress Tolerance of Fruit Crops

Crop	Fungus Species	Traits Improved	References
Strawberry	VAM	Increased antioxidant activity of CAT and SOD, free proline, soluble protein, H^+-ATPase activity, decreased MDA content, and membrane leakage	Yin et al. (2010)
Cherry	*G. clarum*, *G. caledonium*, *G. etunicatum*, *Glomus intraradices*, and *G. mosseae*	Healthy growth and high P and Zn content	Aka-Kacar et al. (2010)
Plum and cherry	AMF	High fruit yield	Świerczyński and Stachowiak (2010)
Morus alba	*G. mosseae* and *G. intraradices*	Longer roots, more leaves and biomass production, high chlorophyll content, high seedlings N and P content	Lu et al. (2015)
Parkia biglobosa, *Tamarindus indica*, and *Z. mauritiana*	*Glomus aggregatum*	Increased N, P, and K content, shoot height and total dry weight	Sidibé et al. (2012)
Carica papaya	*Glomus* sp. and *Acaulospora* sp.	Improved root length and shoot height	Sankaralingam et al. (2016)

non-mycorhizal plants (Wu and Zou 2009). Plants of pistachio cultivars (Qazvini and Badami-Riz-Zarand) had increased P and Zn uptake and favourable leaf water status when inoculated with *F. mosseae* and *R. intraradices* under water-deficit conditions (Bagheri et al. 2012).

Similarly, in greenhouse-grown melons (*Cucumis melo*) when inoculated with *G. versiforme*, *R. intraradices*, and *F. mosseae* showed improved root lengths, plant heights, net photosynthetic rates, biomass production, and tolerance to drought stress compared to non-inoculated plants (Huang et al. 2011). Citrus seedlings and grapevine rootstocks (*Vitis vinifera* L.), namely, Dogridge, 1103, Paulsen and Harmony when inoculated with *F. mosseae*, *Paraglomus occultum*, and *R. intraradices*, respectively, had improved stem diameter, plant height, root and shoot biomass, low concentration of Na and Cl, high K, and Mg concentration of leaves and high K/Na ratio compared to the non-inoculated plants (Wu et al. 2010; Khalil 2013). Colonization of olive (*Olea europea* L.) seedlings with *R. intraradices*, *F. mosseae*, and *Claroideoglomus claroideum* increases the root and shoot biomass, nutrient uptake, and tolerance against salt stress, with *F. mosseae* being the most efficient (Porras-Soriano et al. 2009). AMF *R. irregularis* alleviate the salt stress to a high degree compared to *F. caledonius* and *F. mosseae*, indicating the selection of AMF should be based on a genotype and condition when inoculating the three cultivars of strawberry, viz., Albion, Charlotte, and Seascape under different salt concentrations (Sinclair et al. 2014). Forty days after inoculation of potted banana (*Musa acuminata*) with AMF, *R. intraradices* showed high shoot biomass and reduced Al concentration in both roots and shoots compared to the non-inoculated plants (Rufyikiri et al. 2000). Magnesium concentration and CO_2 assimilation rates of two citrus cultivars, namely, Newhall and Ponkan, were enhanced under Mg-stressed conditions when inoculated with mycorrhiza *G. versiforme* (Xiao et al. 2014).

3.6 CONCLUSIONS AND FUTURE PROSPECTS

AMF may successfully fight a variety of environmental signals, including salinity, drought, nutrient stress, alkali stress, cold stress, and high temperatures, and so assist in enhancing the per hectare output of a wide range of horticultural crops. Mineral elements, food supplements, amino acids, along with poly oligosaccharides, and the trace of organic plant hormones are likely the most recognized parts. Nevertheless, it is essential to underline that the bio-inoculants exercise must not rely on the product's nourishment or perhaps typically established plant hormone content. The systems caused by bio-inoculants are tough to find and remain under investigation. High-throughput and omics technology phenotyping seem to be useful methods to realize bio-inoculants' activity and hypothesize a training method. They might act on plant physiology alongside metabolism by boosting dirt scenarios. They are inside a location to change some molecular activities that enable water to be enhanced by you and nutrient use effectiveness of vegetation, promote plant development, and balance abiotic stresses by increasing primary and secondary metabolic rate. Of all ideas on the dialogue is about the usage of these things in the role and traumatic factors of theirs as nourishment, not producing a curative feature.

REFERENCES

Abdel-Salam, E., Alatar, A., El-Sheikh, M.A. 2018. Inoculation with arbuscular mycorrhizal fungi alleviates harmful effects of drought stress on Damask rose. *Saudi Journal of Biological Sciences* 25:1772–1780.

Affokpon, A., Coyne, D.L., Lawouin, L., Tossou, C., Agbede, R.D., Coosemans, J. 2011. Effectiveness of native West African arbuscular mycorrhizal fungi in protecting vegetable crops against root-knot nematodes. *Biology and Fertility of Soils* 47:207–217.

Aka-Kacar, Y., Akpinar, C., Agar, A., Yalcin-Mendi, Y., Serce, S., Ortas, I. 2010. The effect of mycorrhiza in nutrient uptake and biomass of cherry rootstocks during acclimatization. *Romanian Biotechnological Letters* 15:5246–5252.

Al-Karaki, G.N. 2006. Nursery inoculation of tomato with arbuscular mycorrhizal fungi and subsequent performance under irrigation with saline water. *Scientia Horticulturae* 109:1–7.

Alam, M.Z., Hoque, M.A., Ahammed, G.J., Carpenter-Boggs, L. 2020. Effects of arbuscular mycorrhizal fungi, biochar, selenium, silica gel, and sulfur on arsenic uptake and biomass growth in *Pisum sativum* L. *Emerging Contaminants* 6:312–322.

Ambe, A.G. 2001. Les fruitiers sauvages comestibles des savanes guineennes de Cote d'Ivoire: Etat des connaissances par une population locale, les Malinke. *Biotechnology, Agronomy Society and Environment* 5:43–58.

An, S.S., Cheng, Y., Huang, Y.M., Liu, D. 2012. Effects of revegetation on soil microbial biomass, enzyme activities, and nutrient cycling on the Loess Plateau in China. *Restoration Ecology* 21:600–607.

Andersen, R., Chapman, S.J., Artz, R.R.E. 2013. Microbial communities in natural and disturbed peatlands: A review. *Soil Biology and Biochemistry* 57:979–994.

Aroca, R., Verniery, P., Ruiz-Lozano, J.M. 2008. Mycorrhizal and non-mycorrhizal *Lactuca sativa* plants exhibit contrasting responses to exogenous ABA during drought stress and recovery. *Journal of Experimental Botany* 59:2029–2041.

Aroca, R., Ruiz-Lozano, J.M., Zamarreño, Á.M., Paz, J.A., García-Mina, J.M., Pozo, M.J., López-Ráez, J.A. 2013. Arbuscular mycorrhizal symbiosis influences strigolactone production under salinity and alleviates salt stress in lettuce plants. *Journal of plant physiology* 170:47–55.

Arora, N.K., Fatima, T., Mishra, I., Verma, M., Mishra, J., Mishra, V. 2018. Environmental sustainability: Challenges and viable solutions. *Environmental Sustainability* 1:309–340.

Asrar, A.-W.A., Elhindi, K.M. 2011. Alleviation of drought stress of marigold (*Tagetes erecta*) plants by using arbuscular mycorrhizal fungi. *Saudi journal of biological sciences* 18:93–98.

Asrar, A.A., Abdel-Fattah, G.M., Elhindi, K.M. 2012. Improving growth, flower yield, and water stress of snapdragon (*Antirhinum majus* L.) plants grown under well-watered and water stress conditions using arbuscular mycorrhizal fungi. *Photosynthetica* 50:305–316.

Bagheri, V., Shamshiri, M.H., Shirani, H., Roosta, H. 2012. Nutrient uptake and distribution in mycorrhizal pistachio seedlings under drought stress. *Journal of Agriculture, Science and Technology* 14:1591–1604.

Balliu, A., Sallaku, G., Rewald, B. 2015. AMF inoculation enhances growth and improves the nutrient uptake rates of transplanted, salt-stressed tomato seedlings. *Sustainability* 7:15967–15981.

Bandi, A., Kaushik, P. 2022. Mechanisms and approaches for salt tolerance in turmeric: A breeding perspective. *OBM Genetics*, 6 (2):1.

Baslam, M., Garmendia, I., Goicoechea, N. 2011. Arbuscular mycorrhizal fungi (AMF) improved growth and nutritional quality of greenhouse-grown lettuce. *Journal of Agricultural and Food Chemistry* 59:5504–5515.

Baslam, M., Garmendia, I., Goicoechea, N. 2012. Elevated CO_2 may impair the beneficial effect of arbuscular mycorrhizal fungi on the mineral and phytochemical quality of lettuce. *Annals of Applied Biology* 161:180–191.

Baslam, M., Esteban, R., Garcia-Plazaola, J.I., Goicoechea, N. 2013a. Effectiveness of arbuscular mycorrhizal fungi (AMF) for inducing the accumulation of major carotenoids, chlorophylls and tocopherol in green and red leaf lettuces. *Applied Microbiology and Biotechnology* 97:3119–3128.

Baslam, M., Garmendia, I., Goicoechea, N. 2013b. The arbuscular mycorrhizal symbiosis can overcome reductions in yield and nutritional quality in green house-lettuces cultivated at inappropriate growing seasons. *Scientia Horticulturae* 164:145–154.

Beltrano, J., Ruscitti, M., Arango, M.C., Ronco, M. 2013. Effects of arbuscular mycorrhiza inoculation on plant growth, biological and physiological parameters and mineral nutrition in pepper grown under different salinity and P levels. *Journal of Plant Nutrition and Soil Science* 13:123–141.

Bhardwaj, D., Ansari, M.W., Sahoo, R.K., Tuteja, N. 2014. Biofertilizers function as key player in sustainable agriculture by improving soil fertility, plant tolerance and crop productivity. *Microbial Cell Factories* 13:66.

Bhatti, S.K., Kumar, A., Rana, T., Kaur, N. 2013. Influence of AM fungi (*Glomus mosseae, Acaulospora laevis* and *Gigaspora* sp.) alone and in combination with *Trichoderma viride* on growth responses and physiological parameters of *Dianthus caryophyllus* Linn. *Advances in Bioresearch* 4.

Bona, E., Lingua, G., Manassero, P. 2015. AM fungi and PGP pseudomonads increase flowering, fruit production, and vitamin content in strawberry grown at low nitrose and phosphorus levels. *Mycorrhiza* 25:181–193.

Borkowska, B. 2002. Growth and photosynthetic activity of micropropagated strawberry plants inoculated with endomycorrhizal fungi (AMF) and growing under drought stress. *Acta physiologiae plantarum* 24:365–370.

Borkowska, B., Balla, I., Szucs, E., Michaczuk, B. 2008. Evaluation of the response of micropropagated peach rootstock 'Cadaman' and cv. 'Cresthaven' to mycorrhization using chlorophyll a fluorescence method. *Journal of Fruit and Ornamental Plant Research* 16:243–260.

Boyer, L.R., Brain, P., Xu, X.M., Jeffries, P. 2015. Inoculation of drought-stressed strawberry with a mixed inoculum of two arbuscular mycorrhizal fungi: Effects on population dynamics of fungal species in roots and consequential plant tolerance to water deficiency. *Mycorrhiza* 25:215–227.

Bulgarelli, D., Schlaeppi, K., Spaepen, S., Themaat, E.V.L., Schulze-Lefert, P. 2013. Structure and functions of the bacterial microbiota of plants. *Annual Review of Plant Biology* 64:807–838.

Caglar, S., Bayram, A. 2006. Effects of vesicular-arbuscular mycorrhizal (VAM) fungi on the leaf nutritional status of four grapevine rootstocks. *European Journal of Horticultural Science* 71:109.

Cartmill, A.D., Alarcon, A., Valdez-Aguilar, L.A. 2007. Arbuscular mycorrhizal fungi enhance tolerance of *Rosa multiflora* cv. Burr to bicarbonate in irrigation water. *Journal of Plant Nutrition* 30:1517–1540.

Cartmill, A.D., Valdez-Aguilar, L.A., Bryan, D.L., Alarcon, A. 2008. Arbuscular mycorrhizal fungi enhance tolerance of *Vinca* to high alkalinity in irrigation water. *Scientia Horticulturae* 115: 275–284.

Cekic, F.O., Unyayar, S., Ortas, I. 2012. Effects of arbuscular mycorrhizal inoculation on biochemical parameters in *Capsicum annuum* grown under long term salt stress. *Turkish Journal of Botany* 36:63–72.

Chandrasekaran, M., Boopathi, T., Manivannan, P. 2021. Comprehensive assessment of ameliorative effects of AMF in alleviating abiotic stress in tomato plants. *Journal of Fungi* 7:303.

Chang, X.W., Xiao Lin, L., Jian Chao, Z., Gui Qiang, W., Yong Yi, D. 2008. Effects of arbuscular mycorrhizal fungi on growth and yield of cucumber plants. *Communications in Soil Science and Plant Analysis* 39:499–509.

Chang, X.W., Xiao Lin, L., Fu Qiang, S. 2012. Protecting cucumber from *Fusarium* wilt with arbuscular mycorrhizal fungi. *Communications in Soil Science and Plant Analysis* 43:2851–2864.

Chang, H-X., Haudenshield, J.S., Bowen, C.R., Hartman, G.L. 2017. Metagenome wide association study and machine learning prediction of bulk soil microbiome and crop productivity. *Frontiers in Microbiology* 8:519.

Charron, G., Furlan, V., Bernier-Cardou, M., Doyon, G. 2001a. Response of onion plants to arbuscular mycorrhizae. 2. Effects of nitrogen fertilization on biomass and bulb firmness. *Mycorrhiza* 11:145–150.

Charron, G., Furlan, V., Bernier-Cardou, M., Doyon, G. 2001b. Response of onion plants to arbuscular mycorrhizae. 1. Effects of inoculation method and phosphorus fertilization on biomass and bulb firmness. *Mycorrhiza* 11:187–197.

Conversa, G., Elia, A., Rotonda, P. 2007. Mycorrhizal inoculation and phosphorus fertilization effect on growth and yield of processing tomato. *Acta Horticulturae* 75:333–338.

Conversa, G., Lazzizera, C., Bonasia, A., Elia, A. 2013. Yield and phosphorus uptake of a processing tomato crop grown at different phosphorus levels in a calcareous soil as affected by mycorrhizal inoculation under field conditions. *Biology and Fertility of Soils* 49:691–703.

Davies, F.T., Calderón, C.M., Huaman, Z. 2005. Influence of arbuscular mycorrhizae indigenous to Peru and a flavonoid on growth, yield, and leaf elemental concentration of 'Yungay' potatoes. *Horticultural Science* 40:381–385.

De Coninck, B., Timmermans, P., Vos, C., Cammue, B.P.A., Kazan, K. 2015. What lies beneath: Belowground defense strategies in plants. *Trends in Plant Science* 20:91–101.

Declerck, S., Risède, J.M., Rufyikiri, G., Delvaux, B. 2002. Effects of arbuscular mycorrhizal fungi on severity of root rot of bananas caused by *Cylindrocladium spathiphylli*. *Plant Pathology* 51:109–115.

Deljou, M.J.N., Marouf, A., Hamedan, H.J. 2014. Effect of inoculation with arbuscular mycorrhizal fungi (AMF) on Gerbera cut flower (*Gerbera jamesonii*) production in soilless cultivation. *Acta Horticulturae* 1034:417–422.

Delmont, T.O., Robe, P., Clark, I., Simonet, P., Vogel, T.M. 2011. Metagenomic comparison of direct and indirect soil DNA extraction approaches. *Journal of Microbiological Methods* 86:397–400.

Diaz Franco, A., Alvarado Carrillo, M., Ortiz Chairez, F., Grageda Cabrera, O. 2013. Plant nutrition and fruit quality of pepper associated with arbuscular mycorrhizal in greenhouse. *La Revista Mexicana de Ciencias Agrícolas* 4:315–321.

Elhindi, K.M., El-Din, A.S., Elgorban, A.M. 2017. The impact of arbuscular mycorrhizal fungi in mitigating salt-induced adverse effects in sweet basil (*Ocimum basilicum* L.). *Saudi Journal of Biological Sciences* 24:170–179.

El-Kinany, R.G., Salama, Y.E., Rozan, M.A., Bayom, H.M., Nassar, A.M. 2020. Impacts of humic acid, indole butyric acid (IBA) and arbuscular mycorrhizal fungi (*Glomus mosseae*) as growth promoters on yield and phytochemical characteristics of *Hibiscus sabdariffa* (*roselle*). *Alexandria Science Exchange Journal* 41:29–41.

Elahi, F.E., Aminuzzaman, F.M., Mridha, M.A., Begum, B., Harun, A.K. 2010. AMF inoculation reduced arsenic toxicity and increased growth, nutrient uptake and chlorophyll content of tomato grown in arsenic amended soil. *Advances in Environmental Biology* 4:194–200.

Elwakil, M.A., Baka, Z.A., Soliman, H.M., Sadek, M.S. 2013. A modern tactic for reducing the biotic stress on cucumber plants caused by *Fusarium oxysporum*. *Plant Pathology Journal* 12:6–31.

Engel, R., Szabo, K., Abranko, L., Rendes, K., Füzy, A., Takács, T. 2016. Effect of arbuscular mycorrhizal fungi on the growth and polyphenol profile of marjoram, lemon balm, and marigold. *Journal of agricultural and food chemistry* 64:3733–3742.

Estrada-Luna, A.A., Davies Jr, F.T., Egilla, J.N. 2000. Mycorrhizal fungi enhancement of growth and gas exchange of micropropagated guava plantlets (*Psidium guajava L.*) during ex vitro acclimatization and plant establishment. *Mycorrhiza* 10:1–8.

Fallahi, H.-R., Ghorbany, M., Samadzadeh, A., Aghhavani-Shajari, M., Asadian, A.-H. 2016. Influence of arbuscular mycorrhizal inoculation and humic acid application on growth and yield of Roselle (*Hibiscus sabdariffa* L.) and its mycorrhizal colonization index under deficit irrigation. *International Journal of Horticultural Science and Technology* 3:113–128.

Fierer, N., Jackson, R.B. 2003. The diversity and biogeography of soil bacterial communities. *Proceedings of the National Academy of Sciences of the United States of America* 103:626–631.

Fiorilli, V., Catoni, M., Francia, D., Cardinale, F., Lanfranco, L. 2011. The arbuscular mycorrhizal symbiosis reduces disease severity in tomato plants infected by *Botrytis cinerea. Journal of Plant Pathology* 93:237–242.

Gamalero, E., Berta, G., Massa, N., Glick, B.R., Lingua, G. 2010. Interactions between *Pseudomonas putida* UW4 and *Gigaspora rosea* BEG9 and their consequences for the growth of cucumber under salt-stress conditions. *Journal of Applied Microbiology* 108:236–245.

Garmendia, I., Aguirreolea, J., Goicoechea, N. 2006. Defence related enzymes in pepper roots during interactions with arbuscular mycorrhizal fungi and/or *Verticillium dahliae. Biocontrol* 51:293–310.

Gaur, A., Adholeya, A. 2000. Growth and flowering in Petunia hybrid, *Callistephus chinensis* and *Impatiens balsamina* inoculated with mixed AM inocula or chemical fertilizers in a soil of low P fertility. *Scientia Horticulturae* 84:151–162.

Gill, S.R., Pop, M., DeBoy, R.T., Eckburg, P.B., Eckburg, P.B., Turnbaugh, P.J., Samuel, B.S., Gordon, J.I., Relman, D.A., Fraser-Liggett, C.M., Nelson, K.E. 2006. Metagenomic analysis of the human distal gut microbiome. *Science* 312:1355–1359.

Golubkina, N., Amagova, Z., Matsadze, V., Zamana, S., Tallarita, A., Caruso, G. 2020. Effects of arbuscular mycorrhizal fungi on yield, biochemical characteristics, and elemental composition of garlic and onion under selenium supply. *Plants* 9:84.

Gomez-Bellot, M.J., Ortuno, M.F., Nortes, P.A., Vicente-Sánchez, J., Banon, S., Sanchez-Blanco, M.J. 2015. Mycorrhizal euonymus plants and reclaimed water biomass, water status and nutritional responses. *Scientia Horticulturae* 186:61–69.

Gonzalez, A., King, A., Robeson, M.S., Song, S., Shade, A., Metcalf, J.L., Knight, R. 2012. Characterizing microbial communities through space and time. *Current Opinion in Biotechnology* 23:431–436.

Gonzalez-Chavez, M.D.C.A., Carillo-Gonzalez, R. 2013. Tolerance of *Chrysantemum maximum* to heavy metals: The potential for its use in the re-vegetation of tailing heaps. *Journal of Environmental Sciences* 25:367–375.

Gruda, N. 2005. Impact of environmental factors on product quality of greenhouse vegetables for fresh consumption. *Critical Reviews in Plant Sciences* 24:227–247.

Hajiboland, R., Aliasgharzadeh, N., Laiegh, S.F., Poschenrieder, C. 2010. Colonization with arbuscular mycorrhizal fungi improves salinity tolerance of tomato (*Solanum lycopersicum* L.) plants. *Plant Soil* 331:313–327.

Hashem, A., Abd_Allah, E.F., Alqarawi, A.A., Al Huqail, A.A., Egamberdieva, D., Wirth, S. 2016. Alleviation of cadmium stress in *Solanum lycopersicum* L. by arbuscular mycorrhizal fungi via induction of acquired systemic tolerance. *Saudi Journal of Biological Sciences* 23:272–281.

Hashem, A., Alqarawi, A.A., Radhakrishnan, R., Al-Arjani, A.-B.F., Aldehaish, H.A., Egamberdieva, D., Abd_Allah, E.F. 2018. Arbuscular mycorrhizal fungi regulate the oxidative system, hormones and ionic equilibrium to trigger salt stress tolerance in *Cucumis sativus* L. *Saudi Journal of Biological Sciences* 25:1102–1114.

He, Z.Q., Tang, H.R., Li, H.X., He, C.X., Zhang, Z.B., Wang, H.S. 2010. Arbuscular mycorrhizal alleviated ion toxicity, oxidative damage and enhanced osmotic adjustment in tomato subjected to NaCl stress. *American-Eurasian Journal of Agricultural & Environmental Sciences* 7:676–683.

Huang, Z., Zou, Z.R., He, C.X., He, Z.Q., Zhang, Z.B., Li, J.M. 2011. Physiological and photosynthetic responses of melon (*Cucumis melo* L.) seedlings to three *Glomus* species under water deficit. *Plant Soil* 339:391–399.

Jezdinsky, A., Vojtiskova, J., Slezak, K., Petrikova, K., Pokluda, R. 2012. Effect of drought stress and *Glomus* inoculation on selected physiological processes of sweet pepper (*Capsicum annuum* L. cv. 'Slavy'). *Acta Universitatis Agriculturae et Silviculturae Mendelianae Brunensis* 60:69–76.

Jovel, J., Patterson, J., Wang, W., Hotte, N., O'Keefe, S., Mitchel, T., Perry, T., Kao, D., Mason, A.L., Madsen, K.L., Wong, G.K.S. 2016. Characterization of the gut microbiome using 16S or shotgun metagenomics. *Frontiers in Microbiology* 7:459.

Karaca, M., Ince, A.G. 2019. Conservation of biodiversity and genetic resources for sustainable agriculture. In *Innovations in Sustainable Agriculture*, eds. M. Farooq and M. Pisante, 363:410. Cham: Springer.

Kasliwal, S., Srinivasamurthy, K.M. 2016. Influence of Arbuscular mycorrhizae inoculation on growth and development of *Hibiscus rosa sinensis*. *International Journal of Current Microbiology and Applied Sciences* 5:659–666.

Kaushik, P., Sandhu, O.S., Brar, N.S., Kumar, V., Malhi, G.S., Kesh, H., Saini, I. 2020. *Soil Metagenomics: Prospects and Challenges*. IntechOpen. doi: 10.5772/intechopen.93306. https://www.intechopen.com/online-first/soil-metagenomics-prospects-and-challenges

Kaushish, S., Kumar, A., Aggarwal, A., Parkash, V. 2012. Influence of inoculation with the endomycorrhizal fungi and *Trichoderma viride* on morphological and physiological growth parameters of *Rauwolfia serpentina* Benth. Ex. Kurtz. *Indian Journal of Microbiology* 52:295–299.

Kaya, C., Ashraf, M., Sonmez, O., Aydemir, S., Tuna, A.L., Cullu, M.A. 2009. The influence of arbuscular mycorrhizal colonization on key growth parameters and fruit yield of pepper plants grown at high salinity. *Scientia Horticulturae* 121:1–6.

Keller, M., Zengler, K. 2004. Tapping into microbial diversity. *Nature Reviews Microbiology* 2:141–150.

Khalil, H.A. 2013. Influence of vesicular-arbuscula mycorrhizal fungi (*Glomus* spp.) on the response of grapevines rootstocks to salt stress. *Asian Journal of Crop Science* 5:393–404.

Khalloufi, M., Martínez-Andújar, C., Lachaâl, M., Karray-Bouraoui, N., Pérez-Alfocea, F., Albacete, A. 2017. The interaction between foliar GA3 application and arbuscular mycorrhizal fungi inoculation improves growth in salinized tomato (*Solanum lycopersicum* L.) plants by modifying the hormonal balance. *Journal of Plant Physiology* 214:134–144.

Khan, K., Pankaj, U., Verma, S.K., Gupta, A.K., Singh, R.P., Verma, R.K. 2015. Bio-inoculants and vermicompost influence on yield, quality of *Andrographis paniculata*, and soil properties. *Industrial Crops and Products* 70:404–409.

Kumari, S.M.P., Prabina, B.J. 2019. Protection of tomato, *Lycopersicon esculentum* from wilt pathogen, *Fusarium oxysporum f. sp. lycopersici* by arbuscular mycorrhizal fungi, *Glomus sp. International Journal of Current Microbiology and Applied Sciences* 8:1368–1378.

Latef, A.A.H.A, Chaoxing, H. 2011. Effect of arbuscular mycorrhizal fungi ongrowth, mineral nutrition, antioxidiant enzymes activity and fruit yield of tomato grown under salinity stress. *Scientia Horticulturae* 127:228–233.

Lax, P., Becerra, A.G., Soteras, F., Cabello, M., Doucet, M.E. 2011. Effect of the arbuscular mycorrhizal fungus *Glomus intraradices* on the false root-knot nematode *Nacobbus aberrans* in tomato plants. *Biology and Fertility of Soils* 47:591–597.

Leakey, R.R.B., Tchoundjeu, Z., Schreckenberg, K., Shackleton, S.E., Shackleton, C.M. 2005. Agroforestry tree products (AFTPs): Targeting poverty reduction and enhanced livelihoods. *International Journal of Agricultural Sustainability* 3:1–23.

Liao, K., Bai, Y., Huo, Y., Jian, Z., Hu, W., Zhao, C., Jiuhui, Q. 2018. Integrating microbial biomass, composition and function to discern the level of anthropogenic activity in a river ecosystem. *Environment International* 116:147–155.

Liu, L.Z., Gong, Z.Q., Zhang, Y.L., Li, P.J. 2011. Growth, cadmium accumulation and physiology of marigold (*Tagetes erecta* L.) as affected by arbuscular mycorrhizal fungi. *Pedosphere* 21:319–327.

Lu, N., Zhou, X., Cui, M., Yu, M., Zhou, J., Qin, Y., Li, Y. 2015. Colonization with arbuscular mycorrhizal fungi promotes the growth of *Morus alba* L. seedlings under greenhouse conditions. *Forests* 6:734–747.

Malhi, G.S., Kaur, M., Kaushik, P. 2021a. Impact of climate change on agriculture and its mitigation strategies: A review. *Sustainability* 13:1318.

Malhi, G.S., Kaur, M., Kaushik, P., Alyemeni, M.N., Alsahli, A.A., Ahmad, P. 2021b. Arbuscular mycorrhiza in combating abiotic stresses in vegetables: An eco-friendly approach. *Saudi Journal of Biological Sciences* 28:1465.

Marco, D.E., Abram, F. 2019. Editorial: Using genomics, metagenomics and other 'omics' to assess valuable microbial ecosystem services and novel biotechnological applications. *Frontiers in Microbiology* 10:151.

Martin, C.A., Stutz, J.C. 2004. Interactive effects of temperature and arbuscular mycorrhizal fungi on growth, P uptake and root respiration of *Capsicum annuum* L. *Mycorrhiza* 14:241–244.

Marulanda, A., Azcón, R., Ruiz-Lozano, J.M. 2003. Contribution of six arbuscular mycorrhizal fungal isolates to water uptake by *Lactuca sativa* L. plants under drought stress. *Physiol. Plant* 119:526–533.

Mathur, N., Vyas, A. 2000. Influence of arbuscular mycorrhizae on biomass production, nutrient uptake and physiological changes in *Ziziphus mauritiana Lam.* under water stress. *Journal of Arid Environments* 45:191–195.

Matsubara, Y., Ohba, N., Fukui, H. 2001. Effect of arbuscular mycorrhizal fungus infection on the incidence of *Fusarium* root rot in asparagus seedlings. *Japanese Society for Horticultural Science* 70:202–206.

McGee, K.M., Robinson, C.V., Hajibabaei, M. 2019. Gaps in DNA-based biomonitoring across the globe. *Frontiers in Ecology and Evolution* 7:1–7.

Meir, D., Pivonia, S., Levita, R., Dori, I., Ganot, L. 2010. Application of mycorrhizae to ornamental horticultural crops: *Lisianthus* (*Eustoma gradiflorum*) as a test case. *Spanish Journal of Agricultural Research* 5–10.

Mena-Violante, H.G., Ocampo-Jiménez, O., Dendooven, L., Martínez-Soto, G., González-Casta˜neda, J., Davies, F.T., Olalde-Portugal, V. 2006. Arbuscular mycorrhizal fungi enhance fruit growth and quality of chile ancho (*Capsicum annuum* L. cv San Luis) plants exposed to drought. *Mycorrhiza* 16:261–267.

Mollavali, M., Bolandnazar, S., Nazemieh, H., Zare, F., Aliasgharzad, N. 2015. The effect of mycorrhizal fungi on antioxidant activity of various cultivars of onion (*Allium cepa* L). *International Journal of Biosciences (IJB)* 6:66–79.

Nair, G.R., Raja, S.S.S. 2017. Decoding complex soil microbial communities through new age "omics". *Journal of Microbial & Biochemical Technology* 9:301–309.

Navarro, A., Elia, A., Conversa, G., Campi, P., Mastrorilli, M. 2012. Potted mycorrhizal carnation plants and saline stress: Growth, quality and nutritional plant responses. *Scientia Horticulturae* 140:131–139.

Nunes, J.L. da S., Souza, P.V.D. de, Marodin, G.A.B., Fachinello, J.C. 2008. Inoculation of arbuscular mycorrhizal fungi in peach rootstock cv Okinawa. *Revista Brasileira de Fruticultura* 30:1100–1106.

Nunes, J.L. da S., Souza, P.V.D. de, Marodin, G.A.B., Fachinello, J.C. 2009. Efficiency of arbuscular mycorrhizal fungi on growth of 'aldrighi' peach tree rootstock. *Bragantia* 68:931–940.

Ocete, R., Armendáriz, I., Cantos, M., Álvarez, D., Azcón, R. 2015. Ecological characterization of wild grapevine habitats focused on arbuscular mycorrhizal symbiosis. *VITIS-Journal of Grapevine Research* 54:207–211.

Omirou, M., Ioannides, I.M., Ehaliotis, C. 2013. Mycorrhizal inoculation affects arbuscular mycorrhizal diversity in watermelon roots, but leads to improved colonization and plant response under water stress only. *Applied Soil Ecology* 63:112–119.

Ortas, I. 2012. Do maize and pepper plants depend on mycorrhizae in terms of phosphorus and zinc uptake? *Journal of Plant Nutrition* 35:1639–1656.

Ortas, I., Sari, N., Akpinar, C., Yetisir, H. 2013. Selection of arbuscular mycorrhizal fungi species for tomato seedling growth, mycorrhizal dependency and nutrient uptake. *European Journal of Horticultural Science* 78:209–218.

Ozgonen, H., Erkilic, A. 2007. Growth enhancement and *Phytophthora* blight (*Phytophthora capsici* Leonian) control by arbuscular mycorrhizal fungal inoculation in pepper. *Crop Protection* 26:1682–1688.

Pankaj, U., Singh, D.N., Mishra, P., Gaur, P., Vivekbabu, C.S., Shanker, K., Verma, R.K. 2020. Autochthonous halotolerant plant growth promoting rhizobacteria promote bacoside A yield of *Bacopa monnieri* (L) Nash and phytoextraction of salt-affected soil. *Pedosphere* 30(5):671–683.

Paradi, I., Bratek, Z., Lang, F. 2003. Influence of arbuscular mycorrhiza and phosphorus supply on polyamine content, growth and photosynthesis of *Plantago lanceolata*. *Biologia Plantarum* 46:563–569.

Parihar, M., Rakshit, A., Rana, K., Tiwari, G., Jatav, S.S. 2020. The effect of arbuscular mycorrhizal fungi inoculation in mitigating salt stress of pea (*Pisum Sativum* L.). *Communications in Soil Science and Plant Analysis* 51:1545–1559.

Perner, H., Schwarz, D., Bruns, C., Mäder, P., George, E. 2007. Effect of arbuscular mycorrhizal colonization and two levels of compost supply on nutrient uptake and flowering of pelargonium plants. *Mycorrhiza* 17:469–474.

Porras-Soriano, A., Sorano-Marintin, M.L., Porras-Piedra, A., Azcon, P. 2009. Arbuscular mycorrhizal fungi increased growth, nutrient uptake and tolerance to salinity in olive trees under nursery conditions. *Journal of Plant Physiology* 166:1350–1359.

Pozo, M.J., Cordier, C., Dumas-Gaudot, E., Gianinazzi, S., Barea, J.M., Azcón-Aguilar, C. 2002. Localized vs. systemic effect of arbuscular mycorrhizal fungi on defence responses to *Phytophthora* infection in tomato plants. *Journal of Experimental Botany* 53:525–534.

Prosser, J.I., Bohannan, B.J., Curtis, T.P. et al. 2007. The role of ecological theory in microbial ecology. *Nature Reviews Microbiology* 5:384–392.

Rana, K.L., Kour, D., Kaur, T., Devi, R., Yadav, A.N., Yadav, N., Dhaliwal, H.S., Saxena, A.K. 2020. Endophytic microbes: Biodiversity, plant growth-promoting mechanisms and potential applications for agricultural sustainability. *Antonie van Leeuwenhoek* 113:1075–1107.

Ranjan, R., Rani, A., Metwally, A., McGee, H.S., Perkins, D.L. 2016. Analysis of the microbiome: Advantages of whole genome shotgun versus 16S amplicon sequencing. *Biochemical and Biophysical Research Communications* 469:967–977.

Rouphael, Y., Colla, G. 2020. Toward a sustainable agriculture through plant biostimulants: From experimental data to practical applications. *Agronomy* 10:1461. doi: 10.3390/agronomy10101461.

Rufyikiri, G., Declerck, S., Dufey, J.E., Delvaux, B. 2000. Arbuscular mycorrhizal fungi might alleviate aluminium toxicity in banana plants. *New Phytologist* 148:343–352.

Saini, I., Yadav, K., Esha, E., Aggarwal, A. 2017. Effect of bio-inoculants on morphological and biochemical parameters of *Zinnia elegans* Jacq. *Journal of Applied Horticulture* 19:167–172.

Saini, I., Aggarwal, A., Kaushik, P. 2019a. Inoculation with mycorrhizal fungi and other microbes to improve the morpho-physiological and floral traits of *Gazania rigens* (L.) Gaertn. *Agriculture* 9:51.

Saini, I., Aggarwal, A., Kaushik, P. 2019b. Influence of biostimulants on important traits of *Zinnia elegans* Jacq. under open field conditions. *International Journal of Agronomy* 3082967.

Saini, I., Rani, K., Gill, N., Sandhu, K., Bisht, N., Kumar, T., Kaushik, P. 2020. Significance of arbuscular mycorrhizal fungi for *Acacia*: A review. *Pakistan Journal of Biological Sciences* 23:1231–1236.

Saini, I., Kaushik, P., Al-Huqail, A.A., Khan, F., Siddiqui, M.H. 2021. Effect of the diverse combinations of useful microbes and chemical fertilizers on important traits of potato. *Saudi Journal of Biological Sciences* 28:2641–2648.

Saldajeno, M.G.B., Hyakumachi, M. 2011. The plant growth-promoting fungus *Fusarium equiseti* and the arbuscular mycorrhizal fungus *Glomus mosseae* stimulate plant growth and reduce severity of anthracnose and damping-off diseases in cucumber (*Cucumis sativus*) seedlings. *Annals of Applied Biology* 159:28–40.

Sankaralingam, S., Kathiresan, D., Harinathan, B., Palpperumal, S., Shankar, T., Ramya, B., Prabhu, D. 2016. Effect of arbuscular mycorrhizae fungi on growth and development of Carica papaya. *Scientific Journal of Seoul Sciences* 4: 1–7.

Sboner, A., Mu, X.J., Greenbaum, D., Auerbach, R.K., Gerstein, M.B. 2011. The real cost of sequencing: Higher than you think! *Genome Biology* 12:125.

Schlaeppi, K., Bulgarelli, D. 2015. The plant microbiome at work. *Molecular Plant Microbe Interaction* 28:212–217.

Sharma, M., Kaushik, P. 2021. Vegetable phytochemicals: An update on extraction and analysis techniques. *Biocatalysis and Agricultural Biotechnology* 36:102149.

Sharma, M., Delta, A.K., Kaushik, P. 2021a. *Glomus mosseae* and *Pseudomonas fluorescens* application sustains yield and promote tolerance to water stress in *Helianthus annuus* L. *Stresses* 1:305–316.

Sharma, M., Saini, I., Kaushik, P., Al Dawsari, M.M., Al Balawi, T., Alam, P. 2021b. Mycorrhizal fungi and *Pseudomonas fluorescens* application reduces root-knot nematode (*Meloidogyne javanica*) infestation in eggplant. *Saudi Journal of Biological Sciences* 2814(9):3685–3691.

Sharma, M., Sharma, V., Delta, A.K., and Kaushik, P. 2022. *Rhizophagus irregularis* and nitrogen fixing azotobacter with a reduced rate of chemical fertilizer application enhances pepper growth along with fruits biochemical and mineral composition. *Sustainability*, 14(9):5653.

Sidibé, D., Sanou, H., Teklehaimanot, Z., Mahamadi, D., Koné, S. 2012. The use of mycorrhizal inoculation in the domestication of *Ziziphus mauritiana* and *Tamarindus indica* in Mali (West Africa). *Agroforestry Systems* 85:519–528.

Silveira, S.V. da, Souza, P.V.D. de, Koller, O.C. 2002. Influence of arbuscular mycorrhizal fungi on vegetative growth of avocado rootstocks. *Pesquisa Agropecuária Brasileira* 37:303–309.

Sinclair, G., Charest, C., Dalpé, Y., Khanizadeh, S. 2014. Influence of colonization by arbuscular mycorrhizal fungi on three strawberry cultivars under salty conditions. *Agricultural and Food Science* 23:146–158.

Singh, A., Kumar, M., Verma, S., Choudhary, P., Chakdar, H. 2020a. Plant microbiome: Trends and prospects for sustainable agriculture. In *Plant Microbe Symbiosis*, eds. A. Varma, S. Tripathi and R. Prasad, 129–151. New York City: Springer.

Singh, H., Sethi, S., Kaushik, P., Fulford, A. 2020b Grafting vegetables for mitigating environmental stresses under climate change: A review. *Journal of Water and Climate Change* 11:1784–1797.

Sohn, B.K., Kim, K.Y., Chung, S.J., Kim, W.S., Park, S.M., Kang, J.G., Rim, Y.S., Cho, J.S., Kim, T.H., Lee, J.H. 2003. Effect of the different timing of AMF inoculation on plant growth and flower quality of *chrysanthemum. Scientia Horticulturae* 98:173–183.

Sembok, W.W., Abu-Kassim, N., Hamzah, Y., Rahman, Z.A. 2015. Effect of mycorrhizal inoculation on growth and quality of roselle (*Hibiscus sabdariffa* L.) grown in soiless culture system. *Malaysian Applied Biology* 44:57–62.

Souza, R.C., Cantao, M.E., Vasconcelos, A.T.R., Nogueira, M.A., Hungria, M. 2013. Soil metagenomics reveals differences under conventional and no-tillage with crop rotation or succession. *Applied Soil Ecology* 72:49–61.

Subramanian, K.S., Santhanakrishnan, P., Balasubramanian, P. 2006. Responses of field grown tomato plants to arbuscular mycorrhizal fungal colonization under varying intensities of drought stress. *Scientia Horticulturae* 107:245–253.

Suyal, D.C., Joshi, D., Debbarma, P., Soni, R., Das, B., Goel, R. 2019. Soil metagenomics: Unculturable microbial diversity and its function. In *Mycorrhizosphere and Pedogenesis*, eds. A. Varma and D. Choudhary, 355–362. Singapore: Springer.

Świerczyński, S., Stachowiak, A. 2010. The influence of mycorrhizal fungi on the growth and yielding of plum and sour cherry trees. *Journal of Fruit and Ornamental Plant Research* 18:71–77.

Takacs, T., Biro, I., Anton, A., Xing, H.C. 2006. Inter- and intraspecific variability in infectivity and effectiveness of five *Glomus* sp. strains and growth response of tomato host. *Agrokemia es Talajtan* 55:251–260.

Tanwar, A., Aggarwal, A., Kadian, N., Gupta, A. 2013a. Arbuscular mycorrhizal inoculation and super phosphate application influence plant growth and yield of *Capsicum annuum. Journal of Soil Science and Plant Nutrition* 13:55–66.

Tanwar, A., Aggarwal, A., Kaushish, S., Chauhan, S. 2013b. Interactive effects of AM fungi with *Trichoderma viride* and *Pseudomonas fluorescens* on growth and yield of broccoli. *Plant Protection Science* 49:137–145.

Tayal, P., Kapoor, R., Bhatnagar, A.K. 2011. Functional synergism among *Glomus fasciculatum*, *Trichoderma viride* and *Pseudomonas fluorescens* on *Fusarium* wilt in tomato. *Journal of Plant Pathology* 93:745–750.

Teuber, S., Ahlrichs, J., Henkner, J., Knopf, T., Kühn, P., Scholten, T. 2017. Soil cultures – The adaptive cycle of agrarian soil use in Central Europe: An interdisciplinary study using soil scientific and archaeological research. *Ecology and Society* 22:13.

Thaker, M.N., Jasrai, Y.T. 2002. Increased growth of micropropagated banana (*Musa paradisiaca*) with VAM symbiont. *Plant Tissue Cult* 12:147–154.

Turkmen, O., Sensoy, S., Demir, S., Erdinc, C. 2008. Effects of two different AMF species on growth and nutrient content of pepper seedlings grown under moderate salt stress. *African Journal of Biotechnology* 7:392–396.

Ulrichs, C., Fischer, G., Büttner, C., Mewis, I. 2008. Comparison of lycopene, β-carotene and phenolic contents of tomato using conventional and ecological horticultural practices, and arbuscular mycorrhizal fungi (AMF). *Agronomia Colombiana* 26:40–46.

Uroz, S., Buée, M., Deveau, A., Mieszkin, S., Martin, F. 2016. Ecology of the forest microbiome: Highlights of temperate and boreal ecosystems. *Soil Biology and Biochemistry* 103:471–488.

Vahoniya, D., Panigrahy, S.R., Patel, D., Patel, J. 2018. Status of floriculture in India: With special focus to marketing. *Indian Journal of Pure & Applied Biosciences* 6:1434–1438.

Varshney, A., Sharma, M.P., Adholeya, A., Dhawan, V., Srivastava, P.S. 2002. Enhanced growth of micropropagated bulblets of *Lilium* sp. inoculated with arbuscular mycorrhizal fungi at different P fertility levels in an alfisol. *Journal of Horticultural Science and Biotechnology* 77:258–263.

Verma, S.K., Pankaj, U., Khan, K., Singh, R., Verma, R.K. 2016. Bio-inoculants and vermicompost improve *Ocimum basilicum* yield and soil health in a sustainable production system. *Clean – Soil, Air, Water* 44 (9999):1–8.

Vos, C., Claerhout, S., Mkandawire, R., Panis, B., Waele, D.D., Elsen, A. 2012a. Arbuscular mycorrhizal fungi reduce root-knot nematode penetration through altered root exudation of their host. *Plant Soil* 354:335–345.

Vos, C.M., Tesfahun, A.N., Panis, B., de Waele, D., Elsen, A. 2012b. Arbuscular mycorrhizal fungi induce systemic resistance in tomato against the sedentary nematode *Meloidogyne incognita* and the migratory nematode *Pratylenchus penetrans. Applied Soil Ecology* 61:1–6.

Vos, C., Schouteden, N., Tuinen, D., van Chatagnier, O., Elsen, A., Waele, D., dePanis, B., Gianinazzi-Pearson, V. 2013. Mycorrhiza-induced resistance against the root-knot nematode Meloidogyne incognita involves priming of defense gene responses in tomato. *Soil Biology and Biochemistry* 60:45–54.

Wang, C., Li, X., Zhou, J., Wang, G., Dong, Y. 2008. Effects of arbuscular mycorrhizal fungi on growth and yield of cucumber plants. *Communications in Soil Science and Plant Analysis* 39:499–509.

Wang, F.Y., Tong, R.J., Shi, Z.Y., Xu, X.F., He, X.H. 2011. Inoculation with arbuscular mycorrhizal fungi increase vegetable yields and decrease phoxim concentrations in carrot and green onion and their soil. *PLoS One* 6:e16949.

Wang, C., Li, X., Song, F. 2012. Protecting cucumber from *Fusarium wilt* with arbuscular mycorrhizal fungi. *Communications in soil science and plant analysis* 43:2851–2864.

Wang, Y.-Y., Yin, Q.-S., Qu, Y., Li, G.-Z., Hao, L. 2018a. Arbuscular mycorrhiza-mediated resistance in tomato against *Cladosporium fulvum*-induced mould disease. *Journal of Phytopathology* 166:67–74.

Wang, Y., Wang, M., Li, Y., Wu, A., Huang, J. 2018b. Effects of arbuscular mycorrhizal fungi on growth and nitrogen uptake of *Chrysanthemum morifolium* under salt stress. *PLoS One* 13: e0196408.

Wang, J., Fu, Z., Ren, Q., Zhu, L., Lin, J., Zhang, J., Cheng, X., Ma, J., Yue, J. 2019. Effects of arbuscular mycorrhizal fungi on growth, photosynthesis, and nutrient uptake of *Zelkova serrata* (Thunb.) Makino seedlings under salt stress. *Forests* 10:186.

Watts-Williams, S.J., Turney, T.W., Patti, A.F., Cavagnaro, T.R. 2014. Uptake of zinc and phosphorus by plants is affected by zinc fertilizer material and arbuscular mycorrhizas. *Plant Soil* 376:165–175.

Wu, Q.S., Zou, Y.N. 2009. Mycorrhizal influence on nutrient uptake of citrus exposed to drought stress. *Philippine Agricultural Scientist* 92:33–38.

Wu, Q.S., Zou, Y.N., He, X.H. 2010. Contricutions of arbuscular mycorrhizal fungi to growth, photosynthesis, root morphology and ionic balance of citrus seedlings under salt stress. *Acta Physiologiae Plantarum* 32:297–304.

Xiao, J.X., Hu, C.Y., Chen, Y.Y., Yang, B., Hua, J. 2014. Effects of low magnesium and arbuscular mycorrhizal fungus on the growth, magnesium distribution and photosynthesis of two citrus cultivars. *Scientia Horticulturae* 177:14–20.

Yadav, K., Aggarwala, A., Singh, N. 2013. Arbuscular mycorrhizal fungi (AMF) induced acclimatization, growth enhancement and colchicine content of micropropagated *Gloriosa superba* L. plantlets. *Industrial Crops Production* 45:88–93.

Yadav, A., Saini, I., Kaushik, P., Ansari, M.A., Khan, M.R., Haq, N. 2021a. Effects of arbuscular mycorrhizal fungi and P-solubilizing *Pseudomonas fluorescens* (ATCC-17400) on morphological traits and mineral content of sesame. *Saudi Journal of Biological Sciences* 28:2649–2654.

Yadav, V.K., Jha, R.K., Kaushik, P., Altalayan, F.H., Al Balawi, T., Alam, P. 2021b. Traversing arbuscular mycorrhizal fungi and *Pseudomonas fluorescens* for carrot production under salinity. *Saudi Journal of Biological Sciences* 28:4217–4223.

Yin, B., Wang, Y., Liu, P., Hu, J., Zhen, W. 2010. Effects of vesicular-arbuscular mycorrhiza on the protective system in strawberry leaves under drought stress. *Frontiers of Agriculture in China* 4:165–169.

Zhi, H., ChaoXing, H., Zhong Qun, H., Zhi Rong, Z., Zhi Bin, Z. 2010. The effects of arbuscular mycorrhizal fungi on reactive oxy radical scavenging system of tomato under salt tolerance. *Agricultural Sciences in China* 9:1150–1159.

Zuccarini, P. 2007. Mycorrhizal infection ameliorates chlorophyll content and nutrient uptake of lettuce exposed to saline irrigation. *Plant Soil and Environment* 53:283–289.

Zuccarini, P., Okurowska, P. 2008. Effects of mycorrhizal colonization and fertilization on growth and photosynthesis of sweet basil under salt stress. *Journal of Plant Nutrition* 31:497–513.

4 Endophytes and Their Role in Managing Degraded Land and Enhancing Food, Fodder, Fuel and Fibre Production

Deepamala Maji
CSIR, Lucknow, India

Suman Singh
University of Lucknow, Lucknow, India

Ashutosh Awasthi
Teerthanker Mahaveer University, Moradabad, India

CONTENTS

DOI: 10.1201/9781003147077-4

4.1 INTRODUCTION

Demand for more food production due to the increasing population across the globe has led to the indiscriminate use of chemicals in agricultural practices. Chemicals in the form of fertilizers, insecticides, fungicides, pesticides, etc., have led to degradation of land in terms of vitality and productivity (Stocking 2001). Greenhouse gases released as a result of these chemical applications have resulted in change in climatic conditions leading to floods or droughts, rise in salinity and temperature fluctuations. This shift from the normal natural climatic routine had led to a drastic deceleration of agricultural outputs. Moreover, problems like water unavailability, fragmentation of land and other anthropogenic activities like construction of buildings, industries and roadways, and use of machines that emit greenhouse gases enhance the above problems (Stocking 2001). Land fertility is again jeopardised and agricultural land is now known to be degraded (O'Riordan 2000).

Shortage of fertile land for agriculture creates constrain on the production of important elements of human survival, namely, food, fodder, fuel and fibre. All the 4Fs being plant products directly depend on land available for cultivation. Reduced available land for food (rice, wheat, pulses besides, vegetables and oilseeds), pastures for cattle feed, fuel crop (*Jatropha*, *Castor*, etc.) for various energy needs and fibre crops (jute, hemp, coconut, etc.) makes it a matter of immediate concern to expand land or think of ways to improve productivity of the existing land or even explore other sources of the 4Fs, if possible.

Amongst the numerous microbial alternatives being exploited today to face the discussed challenges, endophytes have emerged as an extremely important and suitable candidate. Endophytes have been bestowed with properties such as fixation of atmospheric nitrogen, phosphate solubilization, IAA and other phytohormone production and siderophore production to combat environmental and nutrient stress (Card et al. 2016), and induced systemic resistance (ISR) mode of defence against biotic attack that helps them to promote plant growth and health (Ryan et al. 2008). Endophytes by means of their nutrient recycling properties improve the fertility of soil, thereby reclaiming degraded land (Ajilogba and Babalola 2019).

The initial part of this chapter deals with the definition, causes and problems associated with land degradation. Further, endophytes (both bacterial and fungal) and the different strategies they use to promote plant growth are discussed. Finally, the chapter discusses the endophyte-mediated enhancement in the production of the 4Fs with examples. The role of endophytes in increasing food and fibre production in the available cultivable is discussed. Microbial fuel production, and not plant-based biofuel, is the call of the hour as it may save more land for food production and constitutes another part of the chapter. Endophytes present in pastures have an inherent quality of damaging pests and insects and accidently harm cattle. Strains with the ability to destroy pests but not cattle that have been developed are also discussed. Thus, endophytes and their role in enhancing production of food, fodder, fuel and fibre under degraded land conditions are the topics discussed here (Figure 4.1).

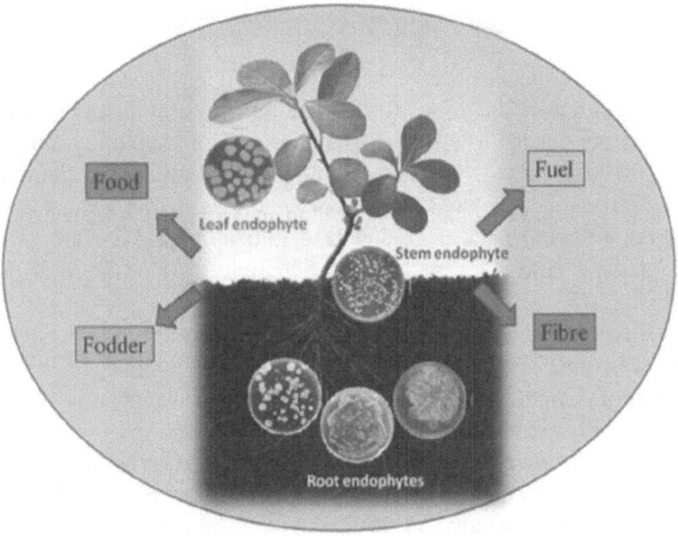

FIGURE 4.1 Endophytes in enhancing food, fodder, fuel and fibre production.

4.2 DEGRADED LAND

Today one of the most pressing problems faced by our planet is land degradation. As addressed by O'Riordan (2000), "land degradation is An Old problem with a New Urgency." According to the United Nations Convention to Combat Desertification (UNCCD), land degradation can be defined as the "reduction or loss of biological or economic productivity resulting from land uses or from a process or combination of processes, including human activities" (Stocking 2001). Both climate changes as well as anthropogenic activities serve as drivers for land degradation. The nomadic form of human civilization where animal and crop husbandry led to destruction of natural vegetation gave way to present-day agricultural lands that initiated the process of land degradation. Thus, land degradation is as old as agriculture itself and can be seen as a side effect of agricultural practices. Anthropogenic degradation of agricultural lands occurs to this date where land is getting converted to factories, industries, cities and roadways. Intense desertification is caused by land degradation in dryland areas. Owing to unsustainable agriculture practices, about one-fourth of the total land area across the globe has been degraded and about 24 billion tons of fertile land is being lost annually (UN News 2019). According to a report of the Intergovernmental Science-Policy Platform on Biodiversity and Ecosystem Services – 2018, this continuous trend may convert about 95% of the planet's land areas into degraded lands by 2050. This phenomenon is largely hampering human life and it is estimated that land degradation affects about 3.2 billion people globally, especially smallholder farmers, rural communities and the very poor. The rising population of the world is believed to increase by about 35% and amount to 9.7 billion in 2050, thus causing higher demands for agricultural products including food, feed, fibre and fuel (UN Reports 2017).

Such land destruction was within the limits until a few decades back when agriculture was sustainable. But presently, the excessive use of chemicals to improve agricultural output has led to diminished soil fertility and viability, thus christening the lands to be degraded. The new challenge is to convert these degraded lands into useful land to serve the dual purpose of land reclamation, i.e., using management practices to bring back degraded land into its previous state and thereby increasing land for cultivation and food productivity to feed the ever-increasing population globally.

Today, we face constraints that very adversely affect agricultural productivity which once was done in a sustainable and eco-friendly manner. Such constraints include meage/unreliable rainfall, improper supply of water and water use efficiency and over-utilization of land agricultural purposes, urbanization and land degradation. The overuse of synthetic chemicals in agro-practices, the attack of plant pathogens and pests, and different abiotic stress have also resulted in decline in soil nutrients and ultimately soil fertility. The current chemical-based agricultural practices, differences in the abiotic components such as solar radiation, rainfall, moisture content and temperature as well as various anthropogenic activities have further led to detrimental environmental health and climatic change. The intense chemical-based agriculture has negatively affected the ecological balance, reducing the fertility of the soil, causing groundwater pollution and contaminating the food chain which has ultimately resulted in reduced diversity of the microbial community, pH of soil and improved resistance of microorganisms towards chemical inputs. Throughout the globe, agriculturists and scientists are engaged in establishing new and innovative alternatives for improving agriculture productivity and maintaining ecological health and balance. The expected result includes higher crop yield due to plant resilience and adaptability to altering climatic conditions (abiotic) and biotic constraints. Nowadays, microbial inoculants are being used in the form of biopesticides, bioflocculants, biofertilizers, bioremediation agents and biostimulants in contemporary agriculture to improve soil fertility and boost nutrient cycling and to improve vigour, yield and productivity of crops (Enebe and Babalola 2018; Igiehon and Babalola 2018; Ajilogba and Babalola 2019; Pankaj 2020).

The beneficial roles of microorganisms especially from the rhizosphere, rhizoplane and phylloplane, such as plant growth-promoting rhizobacteria/microorganisms (PGPR/PGPM), have been well documented (Aremu et al. 2017; Igiehon et al. 2019; Uzoh et al. 2019). Moreover, the application of such organisms in consortia has been proved to be of more importance than single strain application (Awasthi et al. 2014; Verma et al. 2016; Pankaj et al. 2019). Furthermore, imperative invasion of plants by pathogenic agents and plants in the susceptible environment may be a result of a reduction in the diversity of the beneficial microorganisms and unevenness in the microbial community of the plant rhizosphere and soil. Plant pathogens are detrimentally affected by the enzymes and/or metabolites synthesized by the plant microbiome; thus contributing to plant and soil health (Le Cocq et al. 2017). Recently, focus has been shifted to not only rhizospheric or exophytes but on endophytes; some important candidates of the plant microbial community that take possession of the internal host plant tissues with no apparent disease symptoms.

4.3 ENDOPHYTES

Endophyte meaning (Greek-endon = within, phyton = plant) was coined by De Bary in 1866 as "Any organism occurring within plant tissues". Endophytes include a plethora of microorganisms both bacteria and fungi (Raghukumar 2008). Plants growing in diverse environmental conditions such as Arctic tundra, temperate, hot deserts, forests and grasslands as well as croplands harbour a wide range of endophytes making them omnipresent (Arnold 2007; Arnold and Lutzoni 2007; Jain et al. 2021). Besides, endophytes also thrive in gymnosperms and angiosperms, vascular and non-vascular plants, like ferns and mosses (Arnold 2007), thus contributing to the enormous fungal diversity on earth (Jain et al. 2021).

Endophytes are associated with intercellular spaces of the specific host with no antagonistic effects exhibiting a mutualistic, symbiotic, communalistic and trophobiotic relationships with the host tissue (Swarnalatha et al. 2015) and are also responsible for the production of a wide range of secondary metabolites. These molecules with bioactive capacities include growth regulators, antiviral, antibacterial, antifungal and also insecticidal compounds that cause growth improvement and confer resistance to the host against a number of diseases and environmental stresses (Shentu et al. 2014; Hardoim et al. 2015; Card et al. 2016). Such bioactive compounds may serve as reliable agriculturally and pharmaceutically potent molecules (Nisa et al. 2015). Dead materials of both animal and plants' origin are also acted upon by endophytes that act as initial degraders, thereby helping nutrient cycling in the biosphere (Hungria et al. 2010). Internal tissue system of healthy plants serves as a place, where endophytes establish endosymbiotic relationship and build an ecological niche as also seen in the case of plant pathogens (Berg et al. 2005).

4.3.1 Bacterial Endophytes

Organisms flourish in their ecological niche by associating themselves with the neighbouring organisms directly or indirectly benefitting them in the ecological system. Plants by no means deviate from this rule. Plants for their plentiful survival associate with the surrounding microorganisms and provide them shelter in their intracellular spaces. Endophytes that colonize inside plants form a deep relationship with the host plant, thus helping plants to survive and grow under normal as well as stressful environments (Santoyo et al. 2016). Endophytic bacteria do so by initiating allelopathic effects in plants, thereby enhancing plant growth in turn (Mishra et al. 2013). Crops benefitted by endophytic associations include tomato, potato, canola, rice, wheat and several other crops (Mei and Flinn 2010; Ma et al. 2011). Almost all plant tissues like root, stem, leaves, seeds, fruits, tubers, ovules and nodules house bacterial endophytes (Nair and Padmavathy 2014); they are, however, more ubiquitous in the root tissues as compared to other aerial parts (Rosenblueth and Martínez-Romero 2006).

4.3.2 Fungal Endophytes

Fungal endophytes like their bacterial counterparts live in close associations with their host plant that are mutually beneficial, where the fungal partner confers resistance to the plant against biotic and abiotic stresses and the plant partner in turn provides nutrition and shelter to the fungal partner (Suryanarayanan 2017). Plant tissue system such as roots, stems, branches, leaves, flowers and fruits hosts the fungal endophytes in an asymptomatic manner (Saikkonen et al. 2006) and contributes significantly to the huge fungal biodiversity. Fungal endophytes confer resistance to host plants against the negative effects of pests and pathogens by interrupting different developmental processes in the pests such as feeding patterns and reproductive stages (Vega 2018).

Suppression of pest-induced damage due to fungal endophytes has been shown in a number of crops including maize (Cherry et al. 2004), tomato, cotton, and coffee plants, banana, faba bean and common bean plants (Qayyum et al. 2015; Klieber and Reineke 2016). Fungal endophytes caused a reduction in damage to crop productivity by producing secondary mycotoxigenic metabolites that are toxic to the pests (Gurulingappa et al. 2010). Fungal endophytes also produce plant growth-promoting metabolically active compounds (Lahrmann et al. 2013; Jia et al. 2016) such as vital plant hormonal compounds like piperine, GA and IAA/auxins (Jia et al. 2016; Chithra et al. 2017). Fungal endophytes also enable plants to tolerate environmental stresses such as salinity, drought, flood and temperature fluctuations (Leitão and Enguita 2016; Yao et al. 2017).

4.3.3 Endophytes and Bioremediation

Amelioration of agricultural soils involves enhancement in soil fertility, productivity and ecological functioning (Sun et al. 2016). Reports have extensively shown that bioremediation of degraded agricultural land can be very efficiently done by microorganism-associated plants. The microorganisms may be rhizospheric or endophytic. Such practices are collectively known as phytoremediation/bioremediation (Gerhardt et al. 2009; Ma et al. 2011). The speedy interaction or communication between the plants and their microorganisms has made bioaccumulation of contaminants from polluted soils very efficient (Ma et al. 2011). Numerous bacterial and fungal endophytes are known to stimulate the bioavailability of contaminants precipitated in the soil colloids for uptake by plants (Deng and Cao 2017). Different types of contaminants like organic compounds, chlorinated solvents, heavy metals, polycyclic aromatic hydrocarbons (PAHs), insecticides, pesticides, explosives and even radionuclides may be removed from contaminated soils by bioremediation technique (Wang et al. 2002; Oh et al. 2013).

Under stress conditions, the beneficial effects of endophytes may even surpass the effects of rhizospheric bacteria (Barka et al. 2006; Hardoim et al. 2008). Beneficial effects of endophytes under biotic stress are conferred either by ISR in plants against pathogens or by themselves acting as biocontrol agents. Similarly, under abiotic stress, plant growth is stimulated either by production of plant growth hormones (GA/IAA) or making nutrient available to plants via N fixation, P solubilization, siderophore production, etc. (Ryan et al. 2008).

4.4 ROLE OF ENDOPHYTES IN IMPROVING FOOD, FODDER, FUEL AND FIBRE

4.4.1 Improvement of Food Production

Although knowledge is sparse about the mechanisms by which endophytes bring about plant growth, it is expected that endophytes do so through both direct and indirect methods. Though a majority of endophytes begin their journey as rhizospheric microorganisms, and are proposed to hold on to their attributes inside the host plant as well (Kumar et al. 2020), phyllosphere as well as seeds may also serve as the site of origin of these microorganisms (Ryan et al. 2008).

Host plants are benefitted by endophytic partners in a number of direct ways that promote plant growth by increasing nutrient uptake which ultimately increases overall crop yield (Muthukumrasamy et al. 2002). A number of highly established endophytes fix nitrogen in both leguminous and non-leguminous plants (Bhattacharjee et al. 2008). Besides, endophytes also stimulate the production, regulation and activity of plant hormones like auxins, cytokinins, gibberellins and ethylene. Reports have shown that endophytes produce an enzyme ACC deaminase (1-aminocyclopropane-1-carboxylate deaminase) that reduces the effect of stress ethylene under biotic/abiotic stresses (Barnawal et al. 2012). Endophytic strains like *Burkholderia phytofirmans* was reported to enhance the growth and yield of many crops, including potato and tomato, and *Burkholderia kururiensis* is reported to enhance growth in rice plant by IAA (indole-3-acetic acid) hormone production (Mattos et al. 2008). A variety of fungal endophytes promoted height, biomass and tiller number in numerous crops, e.g., *Stagonospora* spp., a seed-borne endophytic fungus, increased *Phragmites australis* crop yield (Ernst et al. 2003). *Penicillium citrinum* strain generated a large amount of physiologically active gibberellins than *Gibberlla fujikuroi* (wild type) which offers to generate a biologically active gibberellic acid (GA3) as a source of income (Khan et al. 2008).

Endophytes improve plant resistance system against pathogens infestation through antipathogenic activity by regulating genetic expression of the host disturbing physiological responses and plant defence-related pathways. Jasmonic acid and salicylic acid play an important role in plant stress responses against phytopathogens (Gunatilaka 2006). The gibberellins produced by endophytes enhance resistance for insect and phytopathogens via salicylic and jasmonic acid pathways (Waqas et al. 2015). Endophyte *Fusarium solani* elicits ISR in tomato plants, foliar endophytic fungi, *Colletotrichum tropicale* inoculated in *Theobroma cacao* enhances tolerance to *Phytophthora* (Mejía et al. 2008). Endophytic *Pseudomonas putida* BP25 associated to black pepper inhibits a range of phytopathogens, viz., *Rhizoctonia solani, Phytophthora capsici, Gibberella moniliformis, Pythium myriotylum, Radopholus similis* and *Colletotrichum gloeosporioides* by secretion of varied compounds (Sheoran et al. 2015).

Various abiotic stress, viz., drought, water logging, salinity, cold, heat, and heavy metal toxicity, causes adverse effects on soil and environmental health (Khare et al. 2018). Phytohormones play a vital role in imparting plant tolerance to environmental stresses (Wani et al. 2016) such as abscisic acid (ABA)-mediated stomatal closure plays an important role in reducing osmotic stress in plants. Recently, Ilangumaran and Smith (2017) reported salt tolerance in wheat modulated by an ABA-signalling surge by halo-tolerant *Dietzia natronolimnaea*. In another study, Pandey et al. (2016) exhibited that inoculation of endophytic *Trichoderma harzianum* upregulated malonialdehyde aquaporin and dehydrin genes in tolerating abiotic stress in rice. Endophytes also colonizes *Arabidopsis* root and shoots and stimulates tolerance against salt stress by synthesizing KMBA (2-keto-4-methylthiobutyric acid). At high soil temperatures *Curvularia protuberate*, an endophytic fungus has been linked with *Dichanthelium lanuginosum* for its survival. Heavy metal-induced oxidative injury can also be reduced by endophytes through diverse mechanisms such as sequestration, intracellular accumulation and extracellular precipitation or alteration of toxic metal ions to a minimum or nontoxic form (Mishra 2017).

During the last decade, endophytes have fascinated us and attracted great attention because of their ability to synthesize a wide array of bioactive secondary metabolites (Gouda et al. 2016). These compounds derived from endophytes belong to the various structural groups such as terpenoids, xanthones, steroids, phenols, benzopyranones, isocoumarins, chinones, cytochalasines, tetralones and enniatines (Schulz et al. 2002). Sometimes they also contribute to the variation in already known structures such as a fungal steroid, ergosterol or plant hormone indole-3-acetic acid (Lu and Fedoroff 2000). Some of the endophytes can also be attributed to provide the protection of plants against pests because of these compounds. Some of the endophytic bacterial and fungal strains are listed in Tables 4.1 and 4.2.

TABLE 4.1
Bacterial Endophytes and Their Host Plants

Bacterial Endophyte	Host Plant	Bioactive Influence
Bradyrhizobium sp. SUTNa-2	*Oryza sativa*	Plant growth promoting
Pantoea dispersa IAC-BECa-132, *Pseudomonas* sp., *Enterobacter* sp.	*Saccharum offcinarum*	Plant growth promoting
Enterobacter cloacae RCA25 *Herbaspirillum huttiense* RCA24	*O. sativa*	Plant growth promoting
Pseudomonas granadensis T6 *Rhizobium larrymoorei* E2	*O. sativa*	Plant growth promoting and pesticide tolerance
Bacillus amyloliquefaciens EPP90 *Bacillus subtilis*; *Bacillus pumilus*	*Pennicetum glaucum*	PGP and abiotic stress Tolerance
Gordonea terrae	*Avicena marina*	Plant growth promoting
Pantoea, *Pseudomonas*, *Enterobacter*	*Eleusine coracana*	Plant growth promoting
B. subtilis LE24 *B. amyloliquefaciens* LE109 *Bacillus tequilensis* PO80	*Citrus* sp.	Biocontrol of pathogens
Curtobacterium sp. SAK 1	*Glycine max*	PGP and salinity stress Tolerance
B. tequilensis (PBE1)	*Solanum lycopersicum*	PGP and biocontrol of pathogens

Source: Omomowo and Babalola (2019). DOI: 10.3390/microorganisms7110481.

TABLE 4.2
Fungal Endophytes and Their Host Plants

Fungal Endophyte	Host Plant	Bioactive Influence
Penicillium aurantiogriseum 581PDA3; *Alternaria alternate* 581PDA5; *Trichoderma harzianum* 582PDA7	*Triticum aestivum*	Plant growth promoting and abiotic stress tolerance
Mucor sp.	*Arabidopsis arenosa*	Metal toxicity tolerance
Fusarium sp.	*Dendrobium moniliforme*	Plant growth promoting
Piriformospora indica	*Cymbidium aloifolium*	Plant growth promoting andabiotic stress tolerance
Porostereum spadiceum AGH786	*Glycine max*	Plant growth promoting and salinity stress tolerance
Aspergillus awamori W11	*Withania somnifera*	Plant growth promoting
Aspergillus fumigatus TS1 *Fusarium proliferatum* BRL1	*Oxalis corniculata*	Plant growth promoting
Yarrowia lipolytica	*Euphorbia milii* L.	Plant growth promoting and salinity stress tolerance
Aspergillus oryzae	*Raphanus sativus*	Plant growth promoting and biocontrol
Paecilomyces variotii, *Penicillium purpurogenum*	*Caralluma acutangula*	Plant growth promoting

Source: Omomowo and Babalola (2019). DOI: 10.3390/microorganisms7110481.

4.4.2 Improvement of Fodder Production

Lolitrem B is an indole diterpene. It causes ryegrass staggers in grassing animals and is known as tremorgenic mycotoxins that are responsible for neurotoxicoses (Gallagher et al. 1984). Clavines are ergot alkaloids that are derived from ergopeptines, dimethylallyl tryptophan and lysergic acid amines (Schardl et al. 2006). Fescue toxicosis is mainly caused by ergovaline, a major ergopeptine product. Fescue toxicosis may have a number of symptoms such as loss of weight, high body temperature, convulsions, infertility, gangrene in the extremities of body parts and in extreme cases may even prove to be fatal (Bacon 1995). Peramine is chemically a pyrrolopyrazine that acts as a feeding deterrent to argentine stem weevil (ASW) (Rowan 1993; Tanaka et al. 2005). Another group of feeding deterrents is lolines that also have potent insecticidal properties but are not toxic to livestock making them highly desirable in agricultural forage (Schardl et al. 2007).

Thus, endophytes that provide bio-protective properties (due to presence of peramine or lolines) to the host plant against insect pests and at the same time do not cause ryegrass staggers (due to lolitrem B or ergovaline) were exploited to produce better quality pastures (Tapper and Latch 1999). One such strain developed was AR1 (Fletcher 1999). AR1 applied pastures when used as cattle feed resulted in 9% increased milk production by the cattle as compared to animals fed on common toxic endophyte-rich grassland (Bluett et al. 2005). Besides the discussed 4 secondary metabolites, at least 35 more putative molecules do exist that may impart positive or negative effects on agriculture.

Fungal growth and mutualism inside the host plant is coordinated and controlled by certain symbiotic signalling processes brought about by molecules of some classes of proteins/peptides. These molecules include epichloenin A, a fungal siderophore, involved in the uptake and storage of iron (Koulman et al. 2012; Johnson et al. 2013), nicotinamide adenine dinucleotide phosphate (NADPH)oxidase of fungal origin that generates reactive oxygen species (Scott and Eaton 2008), and cAMP, a signalling molecule involved in numerous other signalling pathways. Useful endophytic strains utilized to improve pastures are shown in Table 4.3.

4.4.3 Improvement in Fuel Production

Socio-economic development requires energy as a critical input. In view of the energy concerns of the globe, conventional fuels will be of uttermost importance in the decades to come. Non-renewable fuel, also known as conventional fuel, is limited in amount and causes pollution of land, water and air. Moreover, the prices of nonrenewable fuels, especially imported crude oil, are always rising. As such fuels are needed to meet about 80% of the country's domestic need, the problems associated with nonrenewable fuels cannot be ruled out, making it necessary to use it prudently (Anonymous 2008). The threat to the world's energy security will continue until alternative sources to replace/add-on petro-based fuels are developed. Such sources may be based on locally generated feedstock that is renewable. Amongst the various alternatives of renewable resources available, biodiesel that also happens to be environmentally friendly, may prove to be the most promising one. The problem of climate change due to carbon emissions associated with conventional fuel can also be efficiently addressed by the use of biofuels. Biodiesel is a combination of alkyl esters produced by trans-esterification of alcohol and animal fat or vegetable oil mediated by the presence of a catalyst (Meher et al. 2006). Presently plants serve as the major source of biodiesels comprising nearly 90% of the total biodiesel production (Durrett et al. 2008). Biodiesels are increasingly being produced globally in both developed and developing countries, causing a serious clash land for fuel versus food (Li et al. 2008). To overcome these problems, microorganisms with the potential of producing biodiesel must be preferred over plants.

Microorganisms can be multiplied and biodiesel produced in large-scale fermentation tanks and therefore do have some obvious advantages over biodiesels of plant and animal origin. Features like non-requirement of land for cultivation, ability to accumulate high levels of lipids and no competition in food production make them an important candidate for biofuel production. Solar energy trapped by green photosynthetic microorganisms is stored in the form of chemical energy in the biomass, which can later be released via biochemical reactions. The structural and storage carbohydrates in these microorganisms need to be scaled up to produce biodiesel. Microbial fermentation, in absence or limited supply of air or oxygen, is commonly used and may prove to be highly effective method for such biochemical conversion process. Commercially, valuable fuel originating from microorganisms principally depends on obtaining the microorganisms capable of efficiently producing the desired fuel. The vast biodiversity (Bhagobaty and Joshi 2009) and wide adaptability

TABLE 4.3
Endophyte-Based Formulations and Their Properties for Improvement in Fodder Production

Commercial or Common Name	Fungal Species	Notable Alkaloids Produced	Key Traits
Common toxic (wild type)	*Neotyphodium lolii*	Lolitrems Peramine Ergovaline	Ryegrass staggers Negative impacts on animal health Good ASW and black beetle resistance
Common toxic (wild type)	*N. coenophialum*	Peramine Lolines Ergovaline	Fescue toxicosis Broad-spectrum insect resistance
Common type (wild type)	*N. uncinatum*	Lolines	Broad-spectrum insect resistance
Endosafe	*N. lolii*	Peramine	No ryegrass staggers Good ASW resistance
MaxQ	*N. coenophialum* strain AR542 and AR584 (MaxQII)	Lolines Peramine	No fescue toxicosis Broad-spectrum insect resistance
MaxP	*N. coenophialum* strain AR542 and AR584	Lolines Peramine	No fescue toxicosis Broad-spectrum insect resistance
AR1	*N. lolii*	Peramine	No ryegrass staggers and good ASW resistance
Endo5	*N. lolii*	Peramine Ergovaline	Good ASW and black beetle resistance No ryegrass staggers
NEA2	Mix of *N. lolii* strains	Lolitrems Peramine Ergovaline	Good black beetle resistance
AR37	*Neotyphodium* sp.	Epoxy-janthitrems	Broad-spectrum insect pest resistance, excellent animal performance but some ryegrass staggers
Avanex	*N. coenophialum* strain AR601	Ergovaline Lolines	Bird and wildlife deterrent
Avanex	*N. lolii* strains AR94/95	Peramine Ergovaline Lolitrem B (only for AR95)	Bird and wildlife deterrent

Source: Johnson et al. (2013). https://www.researchgate.net/publication/251565969.

of microorganisms, especially endophytic fungi, make them principle forms to be studied and developed for biofuel generation.

Fermentation by yeast utilizes sugar and starch to produce ethanol. The process being expensive makes fuel-related hydrocarbons (mycodiesel) produced by microorganisms stand as an excellent alternative. Endophytic fungi that make molecules

TABLE 4.4
Concentration of Biodiesel Present in Endophytic Fungi

Endophytic Fungi	Concentration of Biodiesel
Xylaria (NICL3)	66.7%
Penicillium (PAOE)	83.1%
Penicillium brasilianum	50.8%
Penicillium griseoroseum	40.5%
Xylaria (NICL5)	91.0%
Trichoderma (T19)	67.8%
Trichoderma (T25)	11.0%
Trichoderma (T27)	40.1%
Trichoderma harvezionum	40.1%
Soy biodiesel	90.7%

Source: Anjum et al. (2016).

like alkane, cyclohexane, cyclopentane, mono-terpenoid, alkyl alcohols and ketone, benzene and polyaromatic hydrocarbon molecules closely related to molecules in diesel have recently been discovered and reported. The volatile compounds thus were named "mycodiesel" and the concept of production of fuel-related substances by endophytic fungi was proposed. The volatile compounds include glycosides, phenolics, terpenoides and complex lipids, in addition to peptides and substituted peptides. Table 4.4 shows a list of such volatile compounds. The cellulosic materials in the plant-based agricultural wastes are converted to hydrocarbons by the biodiesel-producing organisms. These compounds can be efficiently used in fuel-drop technology-oriented engines and are also renewable. Microorganisms directly produce these biofuels by competent and lesser number of biosynthetic steps, thus making them favoured candidates for recent internal combustion engines. Microbial activities also produce some methyl esters like palmitic, stearic, oleic, linoleic and lenolenic acids, which are also present in plant-derived biodiesel.

Examples of fuel-producing endophytic fungi include *Muscodor albus*, *Muscodor crispans*, *Muscodor roseus*, *Muscodor vitigenus*, and *Muscodor sutura*, besides *Colletotrichum* spp. and *Alternaria* spp. Biofuel-producing endophytes are also isolated from some oil-bearing plants like *Jatropha curcas*, *Ricinus communis*, *Helianthes annuus*, *Madhuca longifolia* and *Pongamia pinnata*. Other endophytes known to produce biofuel were isolated from *Myrothecium inundatum*, *Acalypha indica* L., *Ocimum sanctum* and *Brassica juncea* (Bhagobaty 2014). Table 4.5 shows some fungal endophytes with biofuel-producing potential.

4.4.4 Improvement in Fibre Quality and Yield

The backbone of a rural economy especially of dryland areas largely depends on some commercial cash crops besides food crops. One such cash crop is cotton, which is as

TABLE 4.5
Endophytic Fungi with Biofuel-Producing Potential

Native Host Plant	Endophytic Fungus	Biofuel-Related Hydrocarbons/ Precursors Produce
Acalypha indica	*Myrothecium inundatum*	Octane 1,4-cyclohexadiene 1-methyl- and cyclohexane (1-ethylpropyl)
Ocimum sanctum and *Brassica junce*	*Colletotrichum* sp. and *Alternaria* sp.	Oleic acid, linoleic acid, palmitic acid, stearic acid, linolenic acid
Jatropha curcas	*Colletotrichum truncatum* strain EF9, EF10, EF13	Palmitic acid, stearic acid, oleic acid, linoleic acid

Source: Bhagobaty (2014).

old as it is important and thus is righteously called "white gold". Pests and diseases can damage about half of the total cotton produced annually (Sundarmurthy and Basu 1985). Enhancement of disease resistance in the host by ISR may be brought about by endophytic microorganisms, thereby promoting plant growth as well. In experiments carried out by Rajendran et al. (2007), endophytes were demonstrated to degrade complex chitin polymer present as an important component in insect integument. ISR in cotton may be induced due to the elicitation of defence mechanisms involving signal molecules such as peroxidase (POD), polyphenoloxidase (PPO), PAL (phenylalanine ammonia-lyase), phenols and chitinases released by endophytes. The studies further revealed that endophytes also promoted growth in cotton plants (Rajendran et al. 2007). Similar studies by Adams and Kloepper (2002) and Li et al. (2010) showed some endophytic strains that were antagonistic to *Verticillium dahliae* and *Fusarium oxysporum* and also promoted cotton plant growth and yield. These strains were further reported to show positive results in both greenhouse and field conditions (Li et al. 2012). Li et al. (2014) have investigated the role of *Penicillium simplicissimum*, *Leptosphaeria* spp., *Talaromyces flavus* and *Acremonium* spp. as biocontrol agents and further showed that all the four endophytes improved fibre quality in terms of fibre length, strength, micronaire, length uniformity and elongation. Ball weight and lint percentage showed significant improvement under controlled conditions. Yuan et al. (2017) have shown that disease control when taken up at initial stages increased survival rate and yield of cotton balls.

Hemp, an industrially important crop, is grown primarily for its oilseed- and edible seed-derived products, fibre products and drugs with non-psychoactive medicinal properties (Small 2016). Marijuana derived from hemp is useful both medicinally and recreationally. *Cannabis* produces a plethora of secondary metabolites; some of which may improve growth and health of plants and provide protection against different biotic and abiotic disturbances (Gonçalves et al. 2019). A few major components of *Cannabis* include phytocannabinoids like delta-tetrahydrocannabinoids (THC), cannabidiol-carboxylic acid, cannabigerol, cannabinol (CBN), cannabidiol (CBD) and cannabichromene, whose therapeutic effects are currently under research

(Small 2017). Taghinasab and Jabaji (2020) have reported the role of some bacterial endophytes such as *Pseudomonas*, *Pantoea* and *Bacillus* and fungal endophytes such as *Aureobasidium*, *Alterneria* and *Cochliobolus* in promoting growth and health in hemp plants. The ability of these endophytes to produce siderophore, cellulose and solubilize phosphate made them important PGPM. Redman et al. (2002) and Taghinasab (2018) also studied the plant protection and growth promotion abilities of both bacterial and fungal candidates in hemp. ISR and systemic acquired resistance-induced plant protection in hemp was studied by Burketova et al. (2015) and Busby et al. (2016) and by antibiosis and mycoprasitisim by Silva et al. (2019). Some other endophytic friends were responsible for inducing abiotic stress resistance in hemp via antibiotic production, ISR induction, parasitism, competition and quorum sensing (Eljounaidi et al. 2016). *Curvularia* spp. provided protection to hemp plants from high temperatures (Redman et al. 2002). A multispecies consortium made up of *Gluconacetobacter diazotrophicus*, *Burkholderia ambifaria*, *Azospirillum brasilense* and *Herbaspirillum seropedicae*, obtained from roots and shoots of sugarcane, sorghum, corn and Bermuda grass (Botta et al. 2013), have improved hemp biomass. Conant et al. (2017) reported a significant increase in marijuana bud yield and plant height when treated with the microbial biostimulant Mammoth PTM. Mammoth PTM is consortium made up of four bacterial taxa *P. putida*, *Enterobacter clocae*, *Citrobacter freundii* and *Comamonas testosterone* (Baas et al. 2016). Root inoculation of hemp by AMF improved tolerance of hemp to heavy metal contamination of Cd, Ni and Cr (Citterio et al. 2005).

Jute (*Corchorus* spp.) is a major cash crop of South East Asian countries. The best quality fibres are obtained from mainly two species of *Corchorus*, namely, *C. olitorius* and *C. capsularis*. *Macrophomina phaseolina*, a phytopathogen, drastically reduces its productivity. Chaudhry and Patil (2016) conducted an experiment to identify and detect plant growth-promoting endophytes with attributes like N-fixation, siderophore production, phosphate solubilization and IAA production abilities. Antagonistic activities of the selected endophytes were further tested against *M. phaseolina*. The beneficial entities included *Pseudomonas*, *Bacillus*, *Micrococcus*, *Staphylococcus*, *Ralstonia*, *Brevibacteruim* and *Kocuria* that proved to be important biofertilizers and biocontrol agents for not only *C. olitorius* but other crops as well. Vigour index, rate of seed germination, fresh and dry weight, and root length and number were among the growth parameters improved by the endophytic partners. A leaf endophyte from jatropa, *Methylobacterium* has been reported to fix significant amount of nitrogen and thus improve plant growth (Madhaiyan et al. 2015). Saleem et al. (2020) have reported the role of endophytes isolated from flax (*Linum usitatissimum* L.) in phytoremediation and their biological and economic importance.

4.5 CONCLUSION AND FUTURE PERSPECTIVES

The need to develop eco-friendly and sustainable means to increase plant productivity, reduce disease incidences, and win competition imposed by weeds is a call of the hour. The over-exploitation of land for non-sustainable agricultural practices has made soil unfit for further use. Indiscriminate chemical inputs, inappropriate water use techniques accompanied with practices such as deforestation and deplantation have made land degradation a major problem. Endophytic and soil microorganisms could be engaged to enhance productivity of the 4Fs, i.e., food, fodder, fuel and fibre by improving plant health. Growth-promoting attributes of endophytes include reduction in disease incidence, damage induced due to insects and competition with weeds. Endophytes may promote plant growth using lower amounts of fertilizers, fungicides, insecticides or herbicides. Endophytic associations may also prove beneficial in growth and yield improvement under abiotic or environmental stress such as temperature, water, salinity or nutrient stress. In association with endophytic microorganisms, those plants that survive and thrive well in stressful environments may be grown in degraded lands, which may go a long way in replenishing nutrients otherwise lost from degraded land by means of nutrient cycling. Nutrients may be brought back to the soil and in turn to plants by supplementing microbial diversity through microorganism amendments in soil. Our future visions must include studies to focus on relationship optimization between plant, soil and microorganisms, i.e., plant-soil-microorganism interaction needs to be strengthened to make plant growth in harmony with nature.

REFERENCES

Adams, P.D., Kloepper, J.W. 2002. Effect of host genotype on indigenous bacterial endophytes of cotton *Gossypium hirsutum* L. *Plant Soil* 240: 181–189.

Ajilogba, C.F., Babalola, O.O. 2019. GC–MS analysis of volatile organic compounds from Bambara groundnut rhizobacteria and their antibacterial properties. *World J Microbiol Biotechnol* 35: 83.

Anjum, N., Pandey, V.K., Chandra, R. 2016. Exploitation of endophytic fungus as a potential source of biofuel. *Carbon – Sci Technol* 8(2): 55–62.

Anonymous. 2008. National Policy on Biofuels. Ministry of New and Renewable Energy, Government of India. http://mnre.gov. in/file-manager/UserFiles/biofuel_policy.pdf

Aremu, B.R., Alori, E.T., Kutu, R.F., Babalola, O.O. 2017. Potentials of microbial inoculants in soil productivity: An outlook on African legumes. In: Panpatte, D., Jhala, Y., Vyas, R.H.S. (Eds.), "Microorganisms for Green Revolution". Springer, Berlin/Heidelberg, Germany. pp. 53–75.

Arnold, A.E. 2007. Understanding the diversity of foliar endophytic fungi: Progress, challenges, and frontiers. *Fungal Biol Rev* 21(2–3): 51–66.

Arnold, A.E., Lutzoni, F. 2007. Diversity and host range of foliar fungal endophytes: Are tropical leaves biodiversity hotspots. *Ecology* 88(3): 541–549.

Awasthi, A., Singh, M., Soni, S.K., Singh, R., Kalra, A. 2014. Biodiversity acts as insurance of productivity of bacterial communities under abiotic perturbations. *ISME J* 8(12): 2445–2452.

Baas, P., Bell, C., Mancini, L.M., Lee, M.N., Conant, R.T., Wallenstein, M.D. 2016. Phosphorus mobilizing consortium Mammoth P(TM) enhances plant growth. *Peer J* 4, e2121. doi: 10.7717/peerj.2121.

Bacon, C. 1995. Toxic endophyte-infected tall fescue and range grasses: Historic perspectives. *J Anim Sci* 73: 861–870.

Barka, E.A., Nowak, J., Clément, C. 2006. Enhancement of chilling resistance of inoculated grapevine plantlets with a plant growth-promoting rhizobacterium, *Burkholderia phytofirmans* strain PsJN. *Appl Environ Microbiol* 72: 7246–7252.

Barnawal, D., Bharti, N., Maji, D., Chanotiya, C.S., Kalra, A. 2012. 1-Aminocyclopropane-1-carboxylic acid (ACC) deaminase-containing rhizobacteria protect *Ocimum sanctum* plants during waterlogging stress via reduced ethylene generation. *Plant Physiol Biochem* 58: 227–235.

Berg, G., Eberl, L., Hartmann, A. 2005. The rhizosphere as a reservoir for opportunistic human pathogenic bacteria. *Environ Microbiol* 7(11): 1673–1685.

Bhagobaty, R.K. 2014. Endophytic fungi: Prospects in biofuel production. *Proc Natl Acad Sci, India, Sect B Biol Sci.* doi: 10.1007/s40011-013-0294-3 2014.

Bhagobaty, R.K., Joshi, S.R. 2009. Endophytes: A biotechnological goldmine. In: Mishra, C.S.K., Champagne, P. (Eds.), "Biotechnology Applications". IK International Publishing House, India. pp. 300–308.

Bhattacharjee, R.B., Singh, A., Mukhopadhyay, S.N. 2008. Use of nitrogen-fixing bacteria as biofertiliser for non-legumes: Prospects and challenges. *Appl Microbiol Biotechnol* 80(2): 199–209.

Bluett, S., Thom, E.R., Clark, D., MacDonald, K., Minnee, E. 2005. Effects of perennial ryegrass infected with either AR1 or wild endophyte on dairy production in the Waikato. *New Zealand J Agric Res* 48: 197–212.

Botta, A.L., Santacecilia, A., Ercole, C., Cacchio, P., Del Gallo, M. 2013. In vitro and in vivo inoculation of four endophytic bacteria on *Lycopersicon esculentum. New Biotechnol* 30: 666–674. doi: 10.1016/j.nbt.2013.01.001.

Burketova, L., Trda, L., Ott, P.G., Valentova, O. 2015. Bio-based resistance inducers for sustainable plant protection against pathogens. *Biotechnol Adv* 33: 994–1004.

Busby, P.E., Ridout, M., Newcombe, G. 2016. Fungal endophytes: Modifiers of plant disease. *Plant Mol Biol* 90: 645–655.

Card, S.D., Johnson, L.J., Teasdale, S., Caradus, J. 2016. Deciphering microbial behaviour – The link between endophyte biology and efficacious biological control agents. *FEMS Microbiol Ecol* 92 (8): fiw114. doi: 10.1093/femsec/fiw114. Epub 2016 May 23. PMID: 27222223.

Chaudhry, V., Patil, P.B. 2016. Genomic investigation reveals evolution and lifestyle adaptation of endophytic *Staphylococcus epidermidis. Sci Rep* 6: 19263.

Cherry, A.J., Banito, A., Djegui, D., Lomer, C. 2004. Suppression of the stem-borer *Sesamia calamistis* (Lepidoptera; Noctuidae) in maize following seed dressing, topical application and stem injection with African isolates of *Beauveria bassiana. Int J Pest Manage* 50: 67–73.

Chithra, S., Jasim, B., Mathew, J., Radhakrishnan, E.K. 2017. Endophytic *Phomopsis* sp. colonization in *Oryza sativa* was found to result in plant growth promotion and piperine production. *Physiol Plant* 160: 437–446.

Citterio, S., Prato, N., Fumagalli, P., Aina, R., et al. 2005. The arbuscular mycorrhizal fungus *Glomus mosseae* induces growth and metal accumulation changes in *Cannabis sativa* L. *Chemosphere* 59: 21–29. doi: 10.1016/j.chemosphere.2004.10.009.

Conant, R.T., Walsh, R.P., Walsh, M., Bell, C.W., Wallenstein, M.D. 2017. Effects of a microbial biostimulant, Mammoth PTM, on *Cannabis sativa* bud yield. *J Hortic* 4: 5.

Deng, Z., Cao, L. 2017. Fungal endophytes and their interactions with plants in phytoremediation: A review. *Chemosphere* 168: 1100–1106.

Durrett, T.P., Benning, C., Ohlrogge, J. 2008. Plant triacylglycerols as feedstocks for the production of biofuels. *Plant J* 54: 593–607.

Eljounaidi, K., Lee, S.K., Bae, H. 2016. Bacterial endophytes as potential biocontrol agents of vascular wilt diseases – Review and future prospects. *Biol Control* 103: 62–68.

Enebe, M.C., Babalola, O.O. 2018. The influence of plant growth-promoting rhizobacteria in plant tolerance to abiotic stress: A survival strategy. *Appl Microbiol Biotechnol* 102: 7821–7835.

Ernst, M., Menden, K.W., Wirsel, S.G.R. 2003. Endophytic fungal mutualists: Seed-borne *Stagonospora* spp. enhances reed biomass production in axenic microcosmos. *MIPMI* 16: 580–587.

Fletcher, L.R. 1999. "Non-toxic" endophytes in ryegrass and their effect on livestock health and production. *Grassl Res Prac Ser* 7: 133–139.

Gallagher, R., Hawkes, A., Steyn, P., Vleggaar, R. 1984. Tremorgenic neurotoxins from perennial ryegrass causing ryegrass staggers disorder of livestock: Structure elucidation of lolitrem B. *J Chem Soc, Chem Commun* 9: 614–616.

Gerhardt, K.E., Huang, X.D., Glick, B.R., Greenberg, B.M. 2009. Phytoremediation and rhizoremediation of organic soil contaminants: Potential and challenges. *Plant Sci* 176: 20–30.

Gonçalves, J., Rosado, T., Soares, S., et al. 2019. *Cannabis* and its secondary metabolites: Their use as therapeutic drugs, toxicological aspects, and analytical determination. *Medicines (Basel, Switzerland)* 6(31). doi: 10.3390/medicines6010031.

Gouda, S., Das, G., Sen, S.K., Shin, H-S., Patra, J.K. 2016. Endophytes: A treasure house of bioactive compounds of medicinal importance. *Front Microbiol* 7: 1538.

Gunatilaka, A.A.L. 2006. Natural products from plant-associated microorganisms: Distribution, structural diversity, bioactivity, and implications of their occurrence. *J Nat Prod* 69: 509–526. doi: 10.1021/np058128n.

Gurulingappa, P., Sword, G.A, Murdoch, G., McGee, P.A. 2010. Colonization of crop plants by fungal entomopathogens and their effects on two insect pests when in plants. *Biol Control* 55: 34–41.

Hardoim, P.R., van Overbeek, L.S., van Elsas, J.D. 2008. Properties of bacterial endophytes and their proposed role in plant growth. *Trends Microbiol* 16: 463–471.

Hardoim, P.R., Van, O., Verbeek, L.S., Berg, G. 2015. The hidden world within plants: Ecological and evolutionary considerations for defining functioning of microbial endophytes. *Microbiol Mol Biol Rev* 79(3): 293–320.

Hungria, M., Campo, R.J., Souza, E.M., Pedrosa, F.O. 2010. Inoculation with selected strains of *A. brasilense* and *A. lipoferum* improves yields of maize and wheat in Brazil. *Plant Soil* 331(1–2): 413–425.

Igiehon, N.O., Babalola, O.O. 2018. Below-ground-above-ground plant-microbial interactions: Focusing on soybean, rhizobacteria and mycorrhizal fungi. *Open Microbiol J* 12: 261–279.

Igiehon, N.O., Babalola, O.O., Aremu, B.R. 2019. Genomic insights into plant growth promoting rhizobia capable of enhancing soybean germination under drought stress. *BMC Microbiol* 19: 159.

Ilangumaran, G., Smith, D.L. 2017. Plant growth promoting rhizobacteria in amelioration of salinity stress: A systems biology perspective. *Front Plant Sci* 8: 1768.

Jain, R., Bhardwaj, P., Pandey, S.S., Kumar, S. 2021. *Arnebia euchroma*, a plant species of cold desert in the Himalayas, harbors beneficial cultivable endophytes in roots and leaves. *Front Microbiol.* https://doi.org/10.3389/fmicb.2021.696667.

Jia, M., Chen, L., Xin, H.L., et al. 2016. A friendly relationship between endophytic fungi and medicinal plants: A systematic review. *Front Microbiol* 7: 906.

Johnson, L.J., Koulman, A., Christensen, M., et al. 2013. An extracellular siderophore is required to maintain the mutualistic interaction of *Epichloë festucae* with *Lolium perenne*. *PLoS Pathog*. doi: 10.1371/journal.ppat.1003332.

Khan, S.A., Hamayun, M., Yoon, H., et al. 2008. Growth promotion and *Penicillium citrinum*. *BMC Microbiol* 8: 231.

Khare, E., Mishra, J., Arora, N.K. 2018. Multifaceted interactions between endophytes and plant: Developments and prospects. *Front Microbiol.* https://doi.org/10.3389/fmicb.2018.02732

Klieber, J., Reineke, A. 2016. The entomopathogen *Beauveria bassiana* has epiphytic and endophytic activity against the tomato leaf miner *Tuta absoluta*. *J Appl Entomol* 140: 580–589.

Koulman, A., Lee, T.V., Fraser, K., et al. 2012. Lolitrems, peramine and paxilline: Mycotoxins of then ryegrass/endophyte interaction. *Agric Syst Environ* 44: 103–122.

Kumar, A., Droby, S., Singh, V.K., Singh, S.K., White, J.F. 2020. Entry, colonization and distribution of endophytic microorganisms in plants. Microbial Endophytes: Functional Biology and Applications. Elsevier Publication. doi: 10.1016/B978-0-12-819654-0.00001-6.

Lahrmann, U., Ding, Y., Banhara, A., et al. 2013. Host-related metabolic cues affect colonization strategies of a root endophyte. *Proc. Natl. Acad. Sci. U.S.A.* 110: 13965–13970.

Le Cocq, K., Gurr, S.J., Hirsch, P.R., Mauchline, T.H. 2017. Exploitation of endophytes for sustainable agricultural intensification. *Mol Plant Pathol* 18: 469–473.

Leitão, A.L., Enguita, F.J. 2016. Gibberellins in *Penicillium* strains: Challenges for endophyte-plant host interactions under salinity stress. *Microbiol Res* 183: 8–18.

Li, Q., Du, W., Liu, D. 2008. Perspectives of microbial oils for biodiesel production. *Appl Microbiol Biotechnol* 80: 749–756.

Li, C.H., Zhao, M.W., Tang, C.M., Li, S.P. 2010. Population dynamics and identification of endophytic bacteria antagonistic toward plant-pathogenic fungi in cotton root. *Microb Ecol* 59: 344–356.

Li, C.H., Shi, L., Han, Q., et al. 2012. Biocontrol of *Verticillium* wilt and colonization of cotton plants by an endophytic bacterial isolate. *Journal of Applied Microbiology* 113: 641–651.

Li, Z.F., Wang, L.F., Feng, Z.L., Zhao, L.H., Shi, Y.Q., Zhu, H.Q. 2014. Diversity of endophytic fungi from different *Verticillium*-wilt-resistance *Gossypium hirsutum* and evaluation of antifungal activity against *V. dahliae* in vitro. *J Microbiol Biotechnol* 24: 1149–1161. PMID: 24836187.

Lu, C., Fedoroff, N. 2000. A mutation in the arabidopsis HYL1 gene encoding a dsRNA binding protein affects responses to abscisic acid. *Auxin, Cytokinin.* https://doi.org/10.1105/tpc.12.12.2351.

Ma, Y., Prasad, M., Rajkumar, M., Freitas, H. 2011. Plant growth promoting rhizobacteria and endophytes accelerate phytoremediation of metalliferous soils. *Biotechnol Adv* 29: 248–258.

Madhaiyan, M., Alex, T.H., Ngoh, S.T., Prithiviraj, B., Ji, L. 2015. Leaf-residing *Methylobacterium* species fix nitrogen and promote biomass and seed production in *Jatropha curcas*. *Biotechnol Biofuels* 8: 222.

Mattos, K., Padua, V., Romeiro, A., et al. 2008. Endophytic colonization of rice (*Oryza sativa* L.) by the diazotrophic bacterium *Burkholderia kururiensis* and its ability to enhance plant growth. *Ana Acad Bras Ciênc* 80(3). https://doi.org/10.1590/S0001-37652008000300009

Meher, L.C., Vidyasagar, D, Naik, S.N. 2006. Technical aspects of biodiesel production by trans-esterification – A review. *Renew Sustain Energy Rev* 10: 248–268.

Mei, C., Flinn, B.S. 2010. The use of beneficial microbial endophytes for plant biomass and stress tolerance improvement. *Recent Pat. Biotechnol* 4: 81–95.

Mejía, L.C., Rojas, E., Maynard, Z., Herre, E.A. 2008. Endophytic fungi as biocontrol agents of *Theobroma cacao* pathogens. *Biol Control* 46(1): 4–14.

Mishra, G.K. 2017. Microbes in heavy metal remediation: A review on current trends and patents. *Recent Pat Biotechnol* 11(9): 188–196.

Mishra, S., Upadhyay, R.S., Nautiyal, C.S. 2013. Unravelling the beneficial role of microbial contributors in reducing the allelopathic effects of weeds. *Appl Microbiol Biotechnol* 97: 5659–5668.

Muthukumrasamy, R., Revathi, G., Seshadri, S., Lakshminarasimhan, C. 2002. *Gluconacetobacter diazotrophicus* (syn. *Acetobacter diazotrophicus*), a promising diazotrophic endophyte in tropics. *Curr Sci* 83(2):137–145.

Nair, D.N., Padmavathy, S. 2014. Impact of endophytic microorganisms on plants, environment and humans. *Sci World J 2014*(2): 250693.

Nisa, H., Kamili, A.N., Nawchoo, I.A., Shafi, S., Shameem, N., Bandh, S.A. 2015. Fungal endophytes as prolific source of phytochemicals and other bioactive natural products: A review. *Microb Pathog* 82: 50–59.

O'Riordan, T. (Ed.). 2000. "Environmental Science for Environmental Management" (2nd edition). Routledge. https://doi.org/10.4324/9781315839592.

Oh, K., Li, T., Cheng, H., et al. 2013. Development of profitable phytoremediation of contaminated soils with biofuel crops. *J. Environ. Prot* 4(4): 58–67.

Omomowo, O.I., Babalola, O.O. 2019. Bacterial and fungal endophytes: Tiny giants with immense beneficial potential for plant growth and sustainable agricultural productivity. *Microorganisms* 7: 481.

Pandey, V., Ansari, M.W., Tula, S., et al. 2016. Dose-dependent response of *Trichoderma harzianum* in improving drought tolerance in rice genotypes. *Planta* 243(5). doi: 10.1007/s00425-016-2482-x.

Pankaj, U. 2020. Multifarious benefits of biochar application in different soil types. In: Singh, J.S., Singh, C. (Eds.), "Biochar Application in Agricultural and Environmental Management". Springer Nature, Switzerland. pp. 259–272.

Pankaj, U., Singh, D.N., Singh, G., Verma, R.K. 2019. Microbial inoculants assisted growth of *Chrysopogon zizanioides* promotes phytoremediation of salt affected soil. *Indian J. of Microbiology* 59(2): 137–146.

Qayyum, M.A., Wakil, W., Arif, M.J., Sahi, S.T., Dunlap, C.A. 2015. Infection of *Helicoverpa armigera* by endophytic *Beauveria bassiana* colonizing tomato plants. *Biol Control* 90: 200–207.

Raghukumar, C. 2008. Marine fungal biotechnology: An ecological perspective. *Fungal Divers* 31: 5–19.

Rajendran, L., Samiyappan, R., Raguchander, T., Saravanakumar, D. 2007. Endophytic bacteria mediate plant resistance against cotton bollworm. *J Plant Interac* 2(1): 1–10.

Redman, R.S., Sheehan, K.B., Stout, R.G., Rodriguez, R.J., Henson, J.M. 2002. Thermotolerance generated by plant/fungal symbiosis. *Science* 298: 1581.

Rosenblueth, M., Martínez-Romero, E. 2006. Bacterial endophytes and their interactions with hosts. *Mol Plant Microb Interac* 19: 827–837.

Rowan, D. 1993. Identification of extracellular siderophores and a related peptide from the endophytic fungus Epichloë festucae in culture and endophyte-infected *Lolium perenne*. *Phytochemistry* 75: 128–139.

Ryan, R.P., Germaine, K., Franks, A., Ryan, D.J., Dowling, D.N. 2008. Bacterial endophytes: Recent developments and applications. *FEMS Microbiol Lett* 278: 1–9.

Saikkonen, K., Lehtonen, P., Helander, M., Koricheva, J., Faeth, S.H. 2006. Model systems in ecology: Dissecting the endophyte–grass literature. *Trends Plant Sci* 11: 428–433.

Saleem, M.H., Ali, S., Hussain, S., et al. 2020. Flax (*Linum usitatissimum* L.): A potential candidate for phytoremediation? Biological and economical points of view. *Plants* 9: 496. doi: 10.3390/plants9040496.

Santoyo, G., Moreno-Hagelsieb, G., del Carmen Orozco-Mosqueda, M., Glick, B.R. 2016. Plant growth-promoting bacterial endophytes. *Microbiol Res* 183: 92–99.

Schardl, C.L., Panaccione, D.G., Tudzynski, P. 2006. Ergot alkaloids biology and molecular biology. In: Cordell GA (Ed.), "The Alkaloids: Chemistry and Biology". Academic Press, San Diego. pp. 45–86.

Schardl, C., Grossman, R., Nagabhyru, P., Faulkner, J., Mallik, U. 2007. Loline alkaloids: Currencies of mutualism. *Phytochemistry* 68: 980–996.

Schulz, B., Boyle, C., Draeger, S., Römmert, A.K., Krohn, K. 2002. Endophytic fungi: A source of novel biologically active secondary metabolites. *Mycol Res* 106(9): 996–1004.

Scott, B., Eaton, C.J. 2008. Role of reactive oxygen species in fungal cellular differentiations. *Curr Opin Microbiol* 11: 488–493.

Shentu, X., Zhan, X., Ma, Z., Yu, X., Zhang, C. 2014. Antifungal activity of metabolites of the endophytic fungus *Trichoderma brevicompactum* from garlic. *Braz J Microbiol* 45(1): 248–254.

Sheoran, N., Nadakkakath, A.V., Munjal, V., et al. 2015. Genetic analysis of plant endophytic *Pseudomonas putida* BP25 and chemo-profiling of its antimicrobial volatile organic compounds. *Microbiol Res* 173: 66–78.

Silva, N.I., Brooks, S., Lumyong, S., Hyde, K.D. 2019. Use of endophytes as biocontrol agents. *Fungal Biol Rev* 33: 133–148. doi: 10.1016/j.fbr.2018.10.001.

Small, E. 2016. "Cannabis: A Complete Guide" (1st ed.). CRC Press. https://doi.org/10.1201/9781315367583

Stocking, M.A. 2001. Land degradation. "International Encyclopedia of the Social & Behavioral Sciences." Elsevier, Amsterdam. pp. 8242–8247.

Sun, J., Pan, L., Zhan, Y., et al. 2016. Contamination of phthalate esters, organochlorine pesticides and polybrominated diphenyl ethers in agricultural soils from the Yangtze River Delta of China. *Sci Total Environ* 544: 670–676.

Sundarmurthy, V.T., Basu, A.K. 1985. Management of cotton insect pests in polycrop system in India. *Outlook Agric* 14: 79–82.

Suryanarayanan, T. 2017. Fungal endophytes: An eclectic review. *Kavaka* 48: 1–9.

Swarnalatha, Y., Saha, B., Choudary, Y.L. 2015. Bioactive compound analysis and antioxidant activity of endophytic bacterial extract from *Adhathoda beddomei*. *Asian J Pharm Clin Res* 8(1): 70–72.

Taghinasab, M. 2018. The root endophytes *Trametes versicolor* and *Piriformospora indica* increase grain yield and P content in wheat. *Plant Soil* 426: 10–348.

Taghinasab, M., Jabaji, S. 2020. Cannabis microbiome and the role of endophytes in modulating them, production of secondary metabolites: An overview. *Microorganisms* 8: 355.

Tanaka, A., Tapper, B.A., Popay, A., Parker, E.J., Scott, B. 2005. A symbiosis expressed non-ribosomal peptide synthetase from a mutualistic fungal endophyte of perennial ryegrass confers protection to the symbiotum from insect herbivory. *Mol Microbiol* 57(4): 1036–1050.

Tapper, B.A., Latch, G.C.M. 1999. Selection against toxin production in endophyte-infected perennial ryegrass. Ryegrass endophyte: An essential New Zealand symbiosis. *Grassl Res Pract Ser* 7: 107–111.

UN News. 2019-06-16. 24 billion tons of fertile land lost every year, warns UN chief on World Day to Combat Desertification. Retrieved 2021-07-29.

UN Reports. 2017. World population projected to reach 9.8 billion in 2050, and .... Department of Economic and Social Affairs. https://www.un.org.

Uzoh, I.M., Igwe, C.A., Okebalama, C.B., Babalola, O.O. 2019. Legume-maize rotation effect on maize productivity and soil fertility parameters under selected agronomic practices in a sandy loam soil. *Sci Rep* 9: 8539.

Vega, F.E. 2018. The use of fungal entomopathogens as endophytes in biological control: A review. *Mycologia* 110: 4–30.

Verma, S.K., Pankaj, U., Khan, K., Singh, R., Verma, R.K. 2016. Bio-inoculants and vermicompost improve *Ocimum basilicum* yield and soil health in a sustainable production system. *Clean – Soil, Air, Water* 44 (9999): 1–8.

Wang, Y., Inagawa, Y., Saito, T., Kasuya, K.I., Doi, Y., Inoue, Y. 2002. Enzymatic hydrolysis of bacterial poly(3-hydroxybutyrate-*co*-3-hydroxypropionate) by poly(3-hydroxyalkanoate) depolymerase from *Acidovorax* sp. TP4. *Biomacromolecules* 34: 828–834.

Wani, S.H., Kumar, V., Shriram, V., Sah, S.K. 2016. Phytohormones and their metabolic engineering for abiotic stress tolerance in crop plants. *Crop J* 4(3): 162.

Waqas, M., Khan, A.L., Hamayun, M., et al. 2015. Endophytic fungi promote plant growth and mitigate the adverse effects of stem rot: an example of *Penicillium citrinum* and *Aspergillus terreus*. *J Plant Interac* 10(1): 280–287.

Yao, Y.Q., Lan, F., Qiao, Y.M., Wei, J.G., Huang, R.S., Li, L.B. 2017. Endophytic fungi harbored in the root of *Sophora tonkinensis* Gapnep: Diversity and biocontrol potential against phytopathogens. *Microbiol Open* 6: e00437.

Yuan, Y., Feng, H., Wang, L., et al. 2017. Potential of endophytic fungi isolated from cotton roots for biological control against *Verticillium* wilt disease. *PLoS ONE* 12(1): e0170557. doi: 10.1371/journal.pone.0170557.

5 Microbial Strategies for Improving Food Crop Productivity under Salinity and Drought

Savitha T.
Tiruppur Kumaran College for Women, Tiruppur, India

Abhishek Sharma and A. Sankaranarayanan
Uka Tarsadia University, Surat, India

Ashraf Y. Z. Khalifa
King Faisal University, Kingdom of Saudi Arabia
and
University of Beni-Suef, Beni-Suef, Egypt

CONTENTS

5.1 INTRODUCTION

Plant development, advancement, profitability and protection from climatic pressure are presently the significant subjects of concern intended for effective farming and applying biotechnological tools on plants. Abiotic and biotic stresses cause huge misfortunes in farming practices. It is notable to facilitate that drought and salinity are chief abiotic stresses that trim down the worldwide profitability of significant harvests (Singh et al. 2018). Drought or saltiness could essentially decrease plant acquiesce, cause land debasement and impact plant-microorganism relations

DOI: 10.1201/9781003147077-5

(Lesk et al. 2016). In recent times, Santos-Medellin et al. (2017) investigated the rhizobium-dwelling microbiomes that were the forerunner of compositional alterations in the light of the drought.

Drought-resistant plant growth-promoting microorganisms (PGPM) may help the host plants to alleviate different abiotic stresses, because of their commitment to plant resistance to dry season and their capacity to shield plants from disease by microorganisms (Eke et al. 2019; Halo et al. 2020). The isolation, categorization and classification of microorganisms having the capacity to elevate plant resistance to drought may be utilized to alleviate plant misfortunes beneath antagonistic environmental transform circumstances. Similarly, a research report (Xu et al. 2018) exhibited that drought enormously decreased microorganism's network variety in the rhizobiome area and expanded the plenitude of *Actinobacteria* and *Firmicutes*, generally articulated in the endosphere of the root.

Besides, drought stress influenced the reallocation of plant metabolites by the plant root. The discoveries recommend that there are sub-atomic discoursed/connections between microorganisms and host plants for restructuring rhizosphere biota to acclimatize to/adjust to drought pressure. Unwinding the sub-atomic discourse may propel basic information on utilizing beneficial organisms to improve plant pressure resistance.

Crop production climate and their related abiotic factors influenced harvests and thereby forestalled the complete acknowledgement of the hereditary capability of the plant regarding yield and eminence. Depending upon the agro-natural circumstance and cultivation practices, crops experience a wide assortment of abiotic stresses, which may be arranged comprehensively as those of climatic inception and edaphic cause introduced in Figure 5.1.

Recently, farming profitability is declining because of the negative impacts of environmental change, worldwide temperature rise and expanded ecological burdens. Along these lines, to accomplish reasonable improvement in horticulture and to increment farming items for taking care of the entirety of the world's kin, it appears to be important to utilize suitable arrangements and naturally viable and ecologically agreeable methods to diminish the unfavourable impacts of this stress on the plant. Soil salinity is an issue for agrarian efficiency in the region of the globe. As indicated, crops growing in salt-influenced soils are dependent upon osmotic pressure, reduced physical nature of the soil, nutritional problems, poisonousness and decreased harvest yields. Restricting harvest misfortunes because of salinity stress is a significant region of distress to adapt to the background of expanding food prerequisites.

Novel farming advancements are expected to get better food production in salt-influenced soils. Advantageous halotolerant rhizosphere microorganisms related to the plant have been known to build plant resilience to saltiness throughout components, for example, root framework advancement improved soil structure, expanded water and supplement take-up, decreased sodium ingestion, diminished negative impacts of stress ethylene and expanded articulation of qualities engaged with protection from the stress of salinity (Pankaj et al. 2019a, b, 2020). Microbial inoculation to alleviate stress in plants could be a more financially savvy, well-disposed alternative that could be accessible in a more limited period. Such inoculants supply

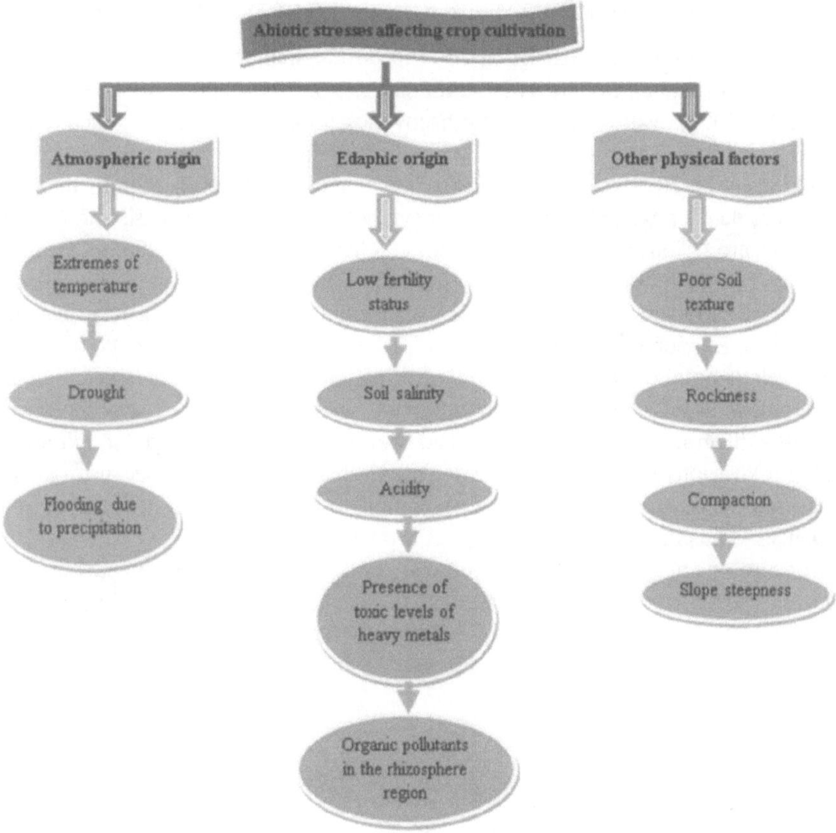

FIGURE 5.1 Various types of abiotic stresses affecting crop growth and yield.

leads to the advancement of feasible horticulture under a salinity-stressed environment (Etesami and Noori 2019).

5.2 MICROBIAL-MEDIATED ALLEVIATION OF SALINITY STRESS IN FOOD CROPS

Soil saltiness considerably reduces plant supplement take-up, particularly phosphorous (P), since phosphate particles accelerate with calcium particles in salt-dominated soil (Grattan and Grieve 1999). Microorganisms involved in phosphate solubilization expand P accessibility to plants (Goldstein 1986; Gyaneshwar et al. 2002), and improved P nourishment can increment biological nitrogen obsession and the accessibility of different supplements because these microorganisms can create plant growth-promoting substances. The improvement of salt-tolerant crop harvests is certainly not a prudent methodology for maintainable agribusiness. In contrast, microbial inoculation to alleviate salt stress is a superior choice since it limits creation expenses and ecological vulnerability (Dixon et al. 1993).

5.2.1 The Mechanisms Involved in Salinity Stress in Food Crops

Salinity is a significant constriction in crop development throughout the universe. It is assessed that salinity stress has influenced more than 800 million hectares of land (FAO 2008). Many examination discoveries imply the reactive oxygen species (ROS) further cause harmful impacts of NaCl on vegetable crops by oxidative pressure (Jungklang et al. 2004) and other vascular plants (Attia et al. 2008). Significant physiological cycles, for example, growth, photosynthesis and ion homeostasis and nitrogen obsession, are seriously harmed by soil saltiness (Bartels and Sunkar 2005). One of the outcomes caused by salt stress leads to irregular ethylene production, and it restrains the root and shoot length as well as overall plant expansion and improvement (Nadeem et al. 2010). Ethylene, a vaporous chemical, is engaged with a wide scope of growth and developmental cycles, for example, germination of seed, root hair stretching and development, abscission of leaf and petal, fruit ripening and organ senescence; yet it is quickly integrated into the light of outside burdens and acknowledged to initiate the outflow of various stress-related genes. The discharge of ethylene because of these circumstances is usually known as pressure ethylene (Saleem et al. 2007). Soil salinity influences plants in three different ways as shown in Figure 5.2.

Concerning the impacts of soil saltiness on the yield of crops, endeavouring to reduce the backward impacts of saltiness on plant development utilizing bacterial inoculation can be incredibly critical. Microorganisms interceding plant stress amelioration has arisen as a significant segment of salt pressure management in plants and

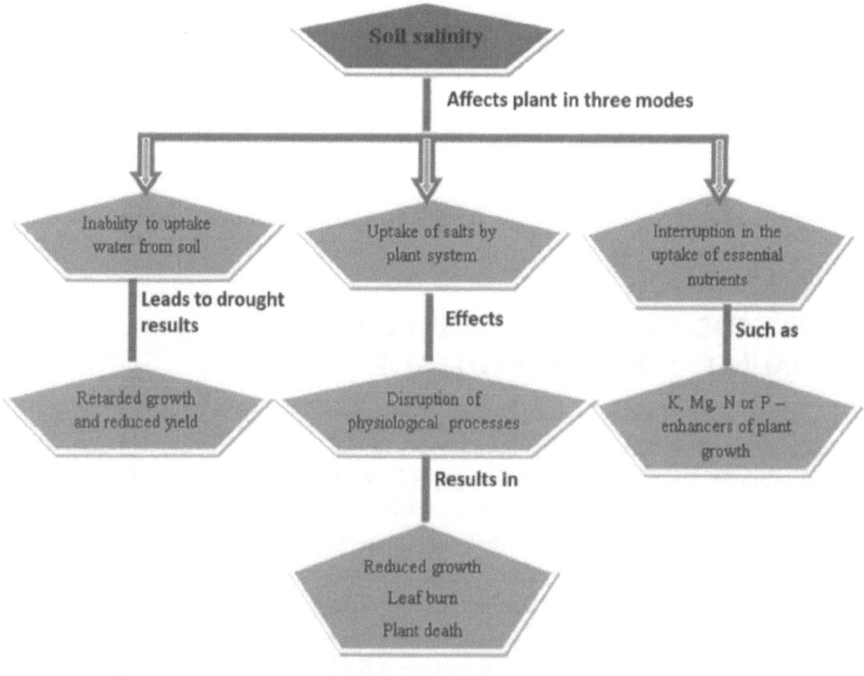

FIGURE 5.2 Effect of soil salinity on crop growth and yield.

their role in convalescing development and profitability has been recognized (Yang et al. 2009). Arbuscular mycorrhizal fungi are known to be related to the underlying foundations of over 80% of terrestrial plant species and have been archived to ease salt-initiated harm in plants (Bharati et al. 2013).

Soil salinity profoundly affects the germination of seeds, which is the main crucial part of fruitful crop production. Under such circumstances, it is fundamental to invigorate seed germination and seedling development. The most suitable arrangement in such circumstances is to utilize salt-tolerant bacterial inoculants that produce auxins, gibberellins and endorse plant development under salinity conditions (Mayak et al. 2004b). Giri and Mukerji (2004) revealed that, in saline soil, higher retention of P in inoculated plants might improve their development rate and salt resistance and smother the antagonistic impact of saltiness stress. As for the impacts of soil saltiness on harvest profitability, endeavouring to alleviate the reverse impacts of saltiness on plant development utilizing bacterial inoculation can be of incredible centrality (Venkateswarlu et al. 2008).

Inoculation of plants with microorganisms that generate indole acidic corrosive and ACC deaminase regularly positively affects the lightening of salt pressure in plants. In an investigation by Barra et al. (2016), halotolerant bacterial consortia were built and formulated from avocado trees to create bio manures to perk up its creation on saline soils.

Different abiotic and biotic stresses are safeguarded by volatile composites from rhizobacteria and they enhanced plant development. An exploration finding of Bhattacharyya et al. (2014) clarified the biological role of volatile compounds from *Alcaligenes faecalis* strain JBCS1294 on the development execution of *Arabidopsis thaliana* under salinity stress. The volatile substances of *Bacillus subtilis* GB03 gave resistance to salinity stress by simultaneously downwarding and updirecting the outflow of an ion transporter, the high-affinity K^+ carrier (HKT1), in *Arabidopsis* roots and shoots, separately, which brought about lower Na^+ accumulation all through the plant (Zhang et al. 2008) and upgraded resilience to osmotic pressure by expanding choline and glycine betaine amalgamation (Zhang et al. 2010). These outcomes accentuate that plant perception of bacterial volatiles may bring about tissue-explicit regulation of genes, which is fundamental for overseeing Na^+ homeostasis in plants stressed by salinity.

Phytohormones are identified to direct plant salinity resilience and Na/K homeostasis. Cytokinin, ethylene and auxin have been accounted for to be mediators of both plant development advancement and the enlistment of fundamental opposition by volatile exhibits from *B. subtilis* GB03 in *Arabidopsis* (Ryu et al. 2004). Signalling mechanisms engaged with salt pressure have been widely concentrated by over articulation or knockdown investigations of conceivably applicable qualities in plants. The tolerance of salinity can be accomplished by the guideline of the statement of key ion channels and carriers, for example, AtHKT1, AtSOS1, AtAVP1 and AtNHX1 in *Arabidopsis*, which bear a positive or negative connection to shooting Na^+ concentrations (Moller et al. 2009; Jha et al. 2010). The proline and malondialdehyde content are additionally firmly identified with salt resistance (Shi et al. 2012).

Plant growth-promoting rhizobacteria (PGPR) instigates physical and chemical changes in plants to bring about improved resilience to abiotic stress and these

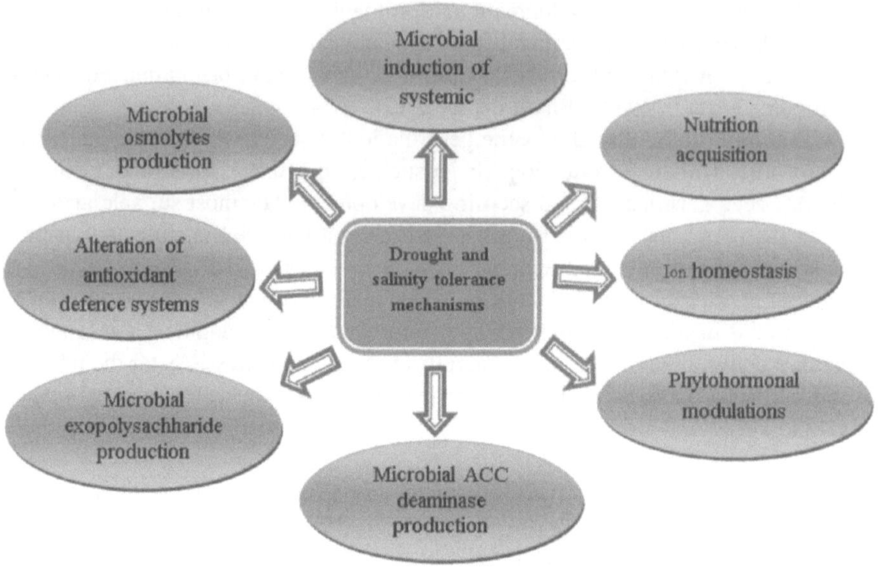

FIGURE 5.3 Mechanisms of plant growth-promoting microorganisms (PGPM) that mediate drought and salinity tolerance in crops.

modifications were termed induced systemic tolerance by Yang et al. (2009). Presently, *Bacillus amyloliquefaciens* NBRISN13 inoculation to rice plants in aqua-farming and soil conditions presented to 200 mM of NaCl expanded plant development and salt resilience (Nautiyal et al. 2013). Figure 5.3 presents the mechanisms of PGPM. In pea plants, the resilience to salinity was surveyed by diminishing stress ethylene levels through ACC deaminase containing rhizobacteria *Arthobacter protophormiae* and advancing plant development through improved colonization of useful microorganisms like *Rhizobium leguminosarum* and *Glomus mosseae* (Barnawal et al. 2014).

5.3 MICROBIAL-MEDIATED ALLEVIATION OF DROUGHT STRESS IN FOOD CROPS

Soil dampness is the expert natural variable since its accessibility coordinates atmosphere and soil conditions, while the dry season is a significant choice of power on organic life forms and can adjust plant network organization and capacity (Holmgren et al. 2006). One of the significant advancements in human agribusiness was the water system which moderated the unconstructive effects of water shortage on yield development (Grover et al. 2010). The deteriorating quality and diminishing quantity of freshwater resources and poor water management system approach were also taken into account during the drought season in an agricultural field. Soil dampness shortage straightforwardly impacts harvest profitability, yet besides, reduces yields through its effect on the accessibility and transport of soil supplements. Drought stress influences plant chemical equilibrium by decreasing the endogenous cytokinin levels and expanding the degrees of abscisic corrosive (ABA) content in the leaves, thereby inspiring stomatal conclusion. The cytokinin ABA threat may be the consequence of metabolic connections since they share typical biosynthetic inceptions (Figueiredo et al. 2008).

Drought reduces the accessibility of CO_2 for photosynthesis, which can prompt the arrangement of ROS, for example, superoxide radicals, hydrogen peroxide and hydroxyl radicals, and cause lipid peroxidation of membranes (Sgherri et al. 2000). Dynamic oxygen species can follow up on unsaturated fats and release the layers, which lastly influence the DNA. Drought stress may harm cells either straightforwardly or by implication, through ROS arrangement, for example, superoxide radicals and H_2O (Mittler and Zilinskas 1994). Drought pressure is likewise known to influence numerous biochemical exercises, for example, nitrate reductase, which catalyses the rate-restricting advance in the nitrate digestion pathway. Under a dry season environment, NR movement diminishes in plants because of the lower take-up of nitrate from the soil by the roots (Caravaca et al. 2005). Drought expands the weakness to supplement misfortunes from the rooting zone to broaden their length, increment their surface territory and modify their architecture to catch less versatile supplements, for example, phosphorus. Drought likewise disturbs root-microorganism relations that assume a significant part in plant supplement attainment.

5.3.1 Drought Stress Alleviating Mechanisms by Microbial Agents

Alleviation of drought stress by rhizobacterial species has earned a lot of consideration previously. Mayak et al. (2004a) revealed that, in transient water pressure, the ACC (1 aminocyclopropane 1 carboxylate) deaminase-producing PGPR *Achromobacter piechaudii* ARV8 fundamentally expanded the new and dry loads of plant seedlings (tomato and pepper) and reduced the production of ethylene under transient drought stress. Dodd et al. (2005) observed that the impacts of inoculation of pea plants with the ACC deaminase delivering bacterium *Variovoras paradoxus* 5C-2 were more articulated and predictable under controlled soil drying. In long-haul tests, plants inoculated with ACC deaminase delivering bacterium gave more seed yield, seed number and seed nitrogen in water pressure conditions, re-established in vaccinated plants. Marulanda et al. (2009) announced that three indigenous bacterial strains, viz., *Pseudomonas putida*, *Pseudomonas* sp. and *Bacillus megaterium*, detached from water stressed on soil could invigorate plant development under dry conditions. Drought stress alleviation by microorganisms in a variety of crops is presented in Table 5.1.

The revelation of plants to dry season pressure, for the most part, brings about a critical decrease in the development and yield of crops, for example, *Hordeum vulgare*, *Oryza sativa*, *Triticum aestivum* and *Zea mays* (Kamara et al. 2003; Samarah 2005; Rampino et al. 2006; Lafitte et al. 2007) because of low dampness in plants, the extreme focus of sunlight, high temperature brought about by dry spells with redesigned breath, photosynthesis and catalyst activity (Fathi and Tari 2016).

In the early phase of drought stress, moderate water exhaustion causes a reduction in shoot improvement, though root advancement is maintained, achieving an overhauled root/shoot extent (Bogeat-Triboulot et al. 2007). Plants presented to

TABLE 5.1
Drought Stress Alleviation by Microbial Origin in a Variety of Crops

Crop Variety	Microbial Source	Mechanism Involved	References
Lettuce	*Bacillus* sp.	Increased AM fungal colonization in roots and enhanced photosynthesis	Vivas et al. (2003)
Cartharanthasus roseus	*Pseudomonas fluorescence*	Improved plant growth	Jaleel et al. (2007)
Bean	*Paenibacillus polymyxa* *Rhizobium tropici*	Altered hormonal balance and stomata conductance	Figueiredo et al. (2008)
Lettuce	*Pseudomonas mendonica*	Increased phosphatase activity in roots and proline accumulation in leaves	Kohler et al. (2008)
Wheat	*Pantoea agglomerans*	Rhizosphere soil aggregation through EPS	Amellal et al. (1998)
Wheat	*Azospirillum* sp.	Improved water relations	Creus et al. (2004)
Pea	*Variovorax paradoxus*	Synthesis of ACC deaminase	Dodd et al. (2004)
Arabidopsis	*Paraphaeosphaeria quadriseptata*	Induction of HSP	McLellan et al. (2007)
Pea	*Pseudomonas* sp.	Decreased ethylene production	Arshad et al. (2008)
Sun flower	*Pesudomonas putida* P45	Improved soil aggregation due to EPS production	Sandhya et al. (2009a)
Trifolium	*Bacillus megaterium*	IAA and proline production	Marulanda et al. (2007)
Tomato	*Azosprillum brasilense*	Nitric oxide as a signalling molecule in the IAA-induced pathway which enhanced lateral root and root hair development	Molina-Favero et al. (2008)
Maize	*Azospirillum lipoferum*	Gibberellins increased ABA levels and alleviated drought stress	Cohen et al. (2009)
Arabidopsis	*Phyllobacterium brassicacearum* STM196	Enhanced ABA content resulted in decreased leaf transpiration	Bresson et al. (2013)
Lavandula dentata	*Bacillus thuringiensis*	IAA resulted in higher K content, proline and decreased the glutathione reductase and ascorbate peroxidase	Armada et al. (2014)
Citrus (*Citrus tangerine*)	*Glomus versiforme*	Higher activities of catalase (CAT), ascorbate peroxidase (APX), superoxide dismutase (SOD)	Wu et al. (2006)
Maize (*Zea mays*)	*Unneliformis mosseae*	Accumulation of amino acids and imino acids, the remarkable increase in trehalose content, and higher trehalose activity	Schellenbaum et al. (1998)

(Continued)

TABLE 5.1 ***(Continued)***
Drought Stress Alleviation by Microbial Origin in a Variety of Crops

Crop Variety	Microbial Source	Mechanism Involved	References
Soybean (*Glycine max*)	*Hizophagus intraradices*	Higher leaf water potential in mycorrhizal plants and mycorrhiza protected the plants against oxidative stress	Meddich et al. (2015)
Sorghum (*Sorghum bicolor*)	*H. intraradices*	Mycorrhiza minimized the adverse effects of drought and increased the grain yield by 17.8%	Alizadeh et al. (2011)
Wheat (*Triticum asetivum*)	*U. mosseae*	Higher biomass and higher grain yields, shoot P and Fe concentration in mycorrhizal plants	Al-karaki et al. (2004)

coordinate dry seasons customarily show a slight change in their improvement, close by only a bit of development in the root mass fraction (RMF) decided as the degree of plant dry mass in roots. The plants seem to maintain top of the ground improvement, and therefore, their force for over the ground resources, to whatever extent, may be achievable under moderate dry season conditions. Poorter et al. (2012) reported that the biomass was decreased by over half differentiated in control plants, reacting with a strong extension in RMF, which can be by and large credited to a decrease in the advancement of the stems. Under moderate or outrageous dry season conditions, the accretion of salts and ions in upper soil layers prompts osmotic pressing factors and particle destruction in plants. With the development in the degree of dry season pressure, the turgor pressing factor of the plant cells decreases, causing the wrinkling of plant cell.

This may lead to diminishing the size and number of the leaves similar to the new weight and water substance of plants (Jaleel et al. 2009). Under gentle or moderate dry spell pressure, the roots may change their plan and dispersion of assets to avoid parchedness (Hasibeder et al. 2015). Regardless, under outrageous dry spell pressure conditions, the roots contract and the photosystem II becomes broken in the leaves (Fathi and Tari 2016). During soil dry spell and climatic aridity (high-fume pressure deficit), the root exudation profiles may also change the rhizosphere soil properties (Song et al. 2012). Drought stress may also decrease supplement scattering and mass movement of water-dissolvable enhancements (Selvakumar et al. 2012).

In terms of drought, stomata close logically alongside an equal decrease in water-use effectiveness and net photosynthetic action. Aside from different boundaries, the modifications in photosynthetic pigments are firmly related to drought resilience. Self-defensive reactions to the pressure at the leaf level should then be set off rapidly to shield the photosynthetic apparatus from being irreversibly harmed. Searching of ROS by enzymatic and non-enzymatic frameworks, cell layer steadiness, articulation of stress-responsive qualities and proteins are basic components of drought resistance. Besides, metabolite changes emphatically add to dry spell variation, especially polyphenols, lipophilic compounds and some tricarboxylic acid (TCA) cycle metabolites are associated with safeguard and security. In contrast, other compounds, for example, sugars, amino acids and polyols, supply osmoregulation.

5.4 MICROBIAL ROLE IN THE GROWTH OF FOOD CROPS IN DEGRADED LAND

Soil ecosystem changes with the degree of disturbances such as physical, chemical, biological, hydrological and geological parts of the earth. The basic micro-counterpart, i.e., microorganisms, plays a significant role in terrestrial ecosystems. Microorganisms act as a source and sink of nutrients and play a basic role in nutrient conservation in a dry tropical environment (Dwivedi and Soni 2011). A list of various factors directly disturbed and diminished the vegetated surface in terrestrial and natural product-producing ability (Cooke and Johnson 2002). To improve world food security and maintain environmental quality, numerous research findings have investigated the consequences of land degradation, needless to mention its role in decreasing soil fertility by loss of soil organic matter and nutrients (Singh 2010;

Singh and Singh 2013), and affected the microbial biomass in soil and its activity (Alqarawi et al. 2014). Land restoration may alter the ecosystem function by changing biological status, i.e., changes in microbial biomass and organic matter deterioration (Potthoff et al. 2006).

5.4.1 Diversity of Microorganisms in Degraded Land Restoration

The status of the microbial ecosystem can be accessed through the hereditary description of the soil microbial commune and it also gives the idea about the eminence of the soil and the advancement of restoration after degradation (Singh 2015). To control the microbial community, soil organic carbon and pH were the main significant factors and soil total nitrogen was a potentially imperative factor for soil microbial composition and function as well as soil moisture, cation exchange capacity and physical structure to a minor extent. Micro fauna recycles organic matter that is trapped in bacteria, fungi and protozoa. They create more surface area for fungi and bacteria to act upon by breaking down organic matter. It makes nutrients in a more stable form and therefore the nutrients are easily available for plant uptake. Degraded land when added with bacterial and fungal feeding nematodes can be restored in fewer periods than the soil without the nematodes. The feeding material for fungal and bacterial are the dead cells from the plant roots as well as sugars, amino acids and organic acids that leak from roots. To keep the plant roots healthy and aid them to grow faster on degraded lands, some of the microorganisms produce antibiotic compounds and hormones to recycle nutrients more rapidly. In legumes, VAM fungi supply the phosphorus required by rhizobium bacteria to fix nitrogen efficiently. Soil microorganisms are one of the most important soil components (Figure 5.4).

Bacteria engage in recreation of an essential role in organic material decomposition at high moisture conditions. Fungi play a predominant role on decomposition at later stages. Rhizobacteria can take nitrogen from the air and renovate it into a form of nitrogen called ammonia which is used by plants (Gil-Sotres et al. 2005). Free-living, as well as symbiotic PGPR, can enhance plant growth directly by providing bioavailable P for plant uptake, fixing N for plant use, sequestering trace elements

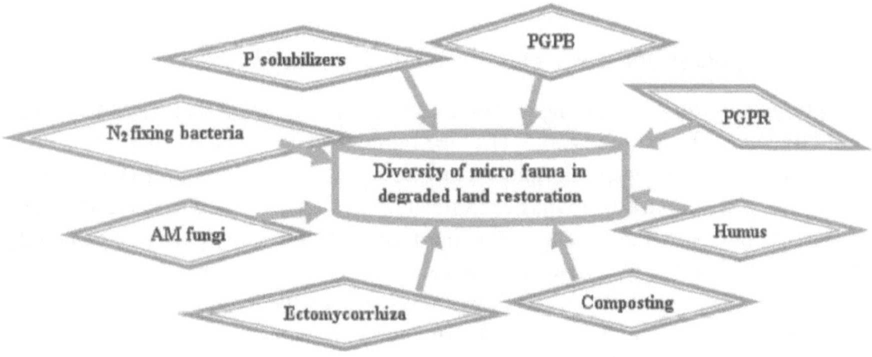

FIGURE 5.4 The diversity of microorganisms has contributed to degraded land restoration.

like iron for plants by siderophores, producing plant hormone-like auxins, cytokinins and gibberellins and lowering of plant ethylene levels (Khan 2004).

5.4.2 Mechanisms and General Adaptations of Microorganisms in Degraded Land in the Growth of Food Crops

In native soil, the soil microbiota incorporates assorted gatherings of organisms that dwell in the soil and complete a wide scope of fundamental functions that are essential for ordinary and healthy soil. The essential component of soil microorganisms is to decay natural matter and deliver supplements into the plant-accessible structures. It additionally directs the creation and annihilation of contaminations like nitrous oxides, methane, nitrates and other naturally harmful mixtures (Doran and Linn 1993). It impacts the enduring and solubilization of minerals and adds to soil design and accumulation. They additionally structure advantageous relationships with roots. All living beings in the biosphere rely upon microbial movement since it prompts the debasement of natural materials and gives food. Numerous anthropogenic exercises like city improvement, agribusiness, mining, utilization of pesticides and contamination can influence soil microbial variety.

Soil organisms particularly plant development promoting microorganisms and AM fungi are worn for appealing plant development and yields of horticulture crops under ordinary and stress conditions. They improve plant development on different physiological boundaries of the plant because of outer upgrades by a few distinct components. Those incorporate plant development controllers, creation of various metabolites and change of climatic nitrogen into alkali and so on, by immediate and roundabout ways. Moreover, it additionally gives resistance against biotic segments through prompted fundamental opposition and foundational procured resistance. Plant-microorganism interaction adding to plant development promotion and infectious prevention under changing climates and empowering more supportable agribusiness without trading off biological system is working. Plant development regulators continue valuable plant-organism cooperation, for example, the association between plant development advancing rhizobacteria and arbuscular fungi. The microbial diversity in rhizospheric soil maintains soil well-being and efficiency.

5.5 CONCLUSION AND FUTURE PERSPECTIVES

Drought and salinity are among the main ecological variables that have hampered rural efficiency around the world. The two burdens can initiate a few morphological, physiological, biochemical and metabolic adjustments through different components, at last, impacting plant development, advancement and profitability. The reactions of plants to these pressure conditions are exceptionally perplexing and rely upon various factors, for example, the species and genotype, plant age and size, the pace of movement just as the power and span of the burdens. These components strongly affect plant reaction and characterize if identified relief measures with acclimation happen. Application of biological approaches decreases dependency on pests and fertilizers. This method is more sustainable, cost-effective and increases crop yield. The application of microbial consortium has reduced the depletion of soil organic material and environmental pollution.

REFERENCES

Alizadeh, A., Alizade, V., Nassery, L., Eivazi, A. 2011. Effect of drought stress on apple dwarf rootstocks. *J Appl Sci Eng Technol.* 13: 86–94.

Al-Karaki, G.N., McMichael, B., Zah, J. 2004. Field response of wheat to arbuscular mycorrhizal fungi and drought stress. *Mycorrrhiza.* 14: 263–269.

Alqarawi, A.A., Abd Allah, E.F., Hashem, A. 2014. Alleviation of salt-induced adverse impact via mycorrhizal fungi in *Ephedra aphylla* forssk. *J Plant Interact.* 9(1): 802–810.

Amellal, N., Burtin, G., Bartoli, F., Heulin, T. 1998. Colonization of wheat roots by EPS-producing *Pantoea agglomerans* and its effect on rhizosphere soil aggregation. *Appl Environ Microbiol.* 64: 3740–3747.

Armada, E., Roldan, A., Azcon, R. 2014. Differential activity of autochthonous bacteria in controlling drought stress in native *Lavandula* and *Salvia* plants species under drought conditions in natural arid soil. *Microb Ecol.* 67: 410–420.

Arshad, M., Shaharoona, B., Mahmood, T. 2008. Inoculation with *Pseudomonas* spp. containing ACC-deaminase partially eliminates the effects of drought stress on growth, yield and ripening of Pea (*Pisum sativum* L.). *Pedosphere.* 18(5): 611–620. https://doi.org/10.1016/S1002-0160(08)60055-7.

Attia, H., Arnaud, N., Karray, N., Lachaal, M. 2008. Long term effects of mild salt stress on growth, ion accumulation and superoxide dismutase expression on *Arabidopsis rosette* leaves. *Physiol Plant.* 132: 293–305.

Barnawal, D., Bharti, N., Maji, D., Chanotiya, C.S., Kalra, A. 2014. ACC deaminase containing *Atrthobacter protophormiae* induces NaCl stress tolerance through reduced ACC oxidase activity and ethylene production resulting in improved nodulation and mycorrhization in *Pisum sativum. J Plant Physiol.* https://dx.doi.org/ 10.1016/j.jplph.2014.03.007.

Barra, P.J., Inostroza, N.G., Acuna, J.J., Mora, M.L., Crowley, D.E., Jorquera, M.A. 2016. Formulation of bacterial consortia from avocado (*Persea americana* Mill) and their effect on growth, biomass and superoxide dismutase activity of wheat seedlings under salt stress. *Appl Soil Ecol.* 102: 80–91. https://dx.doi.org/10.1016/j.apsoil.2016.02.014.

Bartels, D., Sunkar, R. 2005. Drought and salt tolerance in plants. *Crit Rev Plant Sci.* 24: 23–58.

Bharati, N., Baghel, S., Barnawal, D., Yadav, A., Kalra, A. 2013. The greater effectiveness of *Glomus mosseae* and *Glomus intraradices* in improving productivity, oil content and tolerance of salt stressed menthol mint (*Mentha arvensis*). *J Sci Food Agric.* 93: 2154–61.

Bhattacharyya, D., Yu, S.M., Lee, Y.H. 2014. Volatile compounds from *Alcaligenes faecalis* strain JBCS1294 confer salt tolerance in *Arabidopsis thaliana* through the auxin and gibberellins pathways and differential modulation of gene expression in root and shoot tissues. *Plant Growth Regul.* doi: 10.1007/s10725-014-9953-5.

Bogeat-Triboulot, M.B., Brosche, M. et al. 2007. Gradual soil water depletion results in reversible changes of gene expression, protein profiles, ecophysiology and growth performance in *Populus euphratica*, a popular growing in arid regions. *Plant Physiol.* 143: 876–892. doi: 10.1104/pp.106.088708.

Bresson, J., Varoquaux, F., Bontpart, T., Touraine, B., Vile, D. 2013. The PGPR strain *Phyllobacterium brassicacearum* STM196 induces a reproductive delay and physiological changes that result in improved drought tolerance in *Arabidopsis. New Phytol.* 200: 558–569.

Caravaca, F., Alguacil, M.M., Hemandez, J.A., Roldan, A. 2005. Involvement of antioxidant enzyme and nitrate reductase activities during water stress and recovery of mycorrhizal *Myrtus communis* and *Phillyrea angustifolia* plants. *Plant Sci.* 169: 191–197.

Cohen, A.C., Travaglia, C.N., Bottini, R., Piccoli, P.N. 2009. Participation of abscisic acid and gibberellins produced by endophytic *Azospirillum* in the alleviation of drought effects in maize. *Botanique.* 87: 455–462.

Cooke, J.A., Johnson, M.S. 2002. Ecological restoration of land with particular reference to the mining of metals and industrial minerals: A review of theory and practice. *Environ Rev.*10: 41–71.

Creus, C., Sueldo, R.J., Barassi, C.A. 2004. Water relations and yield in *Azospirillum*-induced wheat exposed to drought in the field. *Can J Bot.* 82(2): 273–281. doi: 10.1139/b03-119.

Dixon, R.K., Garg, V.K., Rao, M.V. 1993. Inoculation of *Leucaena* and *Prosopis* seedlings with *Glomus* and *Rhizobium* species in saline soil: Rhizosphere relations and seedlings growth. *Arid Soil Res Rehabil.* 7: 133–144.

Dodd, I.C., Belimov, A.A., Sobeih, W.Y., Safronova, V.I., Grierson, D., Davies, W.J. 2004. Will modifying plant ethylene status improve plant productivity in water-limited environments? In *Proceedings for the 4th International Crop Science Congress*, Brisbane, Australia, 26 September–1 October 2004.

Dodd, A.N., Salathia, N., Hall, A., Kévei, E., Tóth, R., Nagy, F., Hibberd, J.M., Millar, A.J., Webb, A.A. 2005. Plant circadian clocks increase photosynthesis, growth, survival, and competitive advantage. *Science* 309(5734): 630–633. doi: 10.1126/science.1115581.

Doran, J.W., Linn, D.M. 1993. Microbial ecology of conservation and management. In: Hatfield, J.L., Stewart, B.A. (eds.). *Advances in soil science.* Lewis Publishers, Boca Raton.

Dwivedi, V., Soni, P. 2011. A review on the role of soil microbial biomass in eco-restoration of degraded ecosystem with special reference to mining areas. *J Appl Nat Sci.* 3(1): 151–158.

Eke, P., Kumar, A., Sahu, K.P., Wakam, L.N. et al. 2019. Endophytic bacteria of desert cactus (*Euphorbia trigonas* Mill.) confer drought tolerance and induce growth promotion in tomato (*Solanum lycopersicum* L.). *Microbiol Res.* 228: 126302. doi: 10.1016/j.micres.2019.126302.

Etesami, H., Noori, F. 2019. Soil salinity as a challenge for sustainable agriculture and bacterial mediated alleviation of salinity stress in crop plants. In: Kumar, M., Etesami, H., Kumar, V. (eds.). *Saline soil-based agriculture by halotolerant microorganisms.* Springer, Singapore. https://doi.org/10.1007/978-981-13-8335-9_1.

FAO. 2008. FAO land and plant nutrition management service. http://www.fao.org/ag/agl/agll/spush.

Fathi, A., Tari, D.B. 2016. Effect of drought stress and its mechanism in plants. *Int J Life Sci.* 10: 1–6. doi: 10.3126/ijls.v10il.14509.

Figueiredo, M.V.B., Burity, H.A., Martinez, C.R., Chanway, C.P. 2008. Alleviation of drought stress in common bean (*Phaseolus vulgaris* L.) by co-inoculation of *Paenibacillus polymyxa* and *Rhizobium tropici*. *Appl Soil Ecol.* 40: 182–188.

Gil-Sotres, F., Trasr-Cepeda, C., Leiros, M.C., Seoane, S. 2005. Different approaches to evaluating soil quality using biochemical properties. *Soil Biol Biochem.* 37: 877–887.

Giri, B., Mukerji, K.G. 2004. Mycorrhizal inoculants alleviate salt stress in *Sesbania aegyptiaca* and *Sesbania grandiflora* under field conditions: Evidence for reduced sodium and improved magnesium uptake. *Mycorrhiza.* 14: 307–312.

Goldstein, A.H. 1986. Bacterial phosphate solubilisation: Historical perspective and future perspective. *Am J Altern Agric.* 1: 57–65.

Grattan, S.R., Grieve, C.M. 1999. Salinity mineral nutrient relations in horticultural crops. *Sci. Hortic (Amsterdam).* 78: 127–157. doi: 10.1016/S0304-4238(98)00192-7.

Grover, M., Ali, S.K.Z., Sandhya, V., Rasul, A., Venkateswarlu, B. 2010. Role of microorganisms in adaptation of agriculture crops to abiotic stress. *World J Microbiol Biotechnol.* doi: 10.1007/s11127-010-0572-7.

Gyaneshwar, P., Naresh, K.G., Parekh, L.J. 2002. Effect of buffering on the phosphate solubilising ability of microorganisms. *World J Microbiol Biotechnol.* 14: 669–673. doi: 10.1023/A:1008852718733.

Halo, B.A., Al-Yahyai, R.A. Al-Sadi, A.M. 2020. An endophytic *Talaromyces omanensis* enhances reproductive physiological and anatomical characteristics of drought stressed tomato. *J Plant Physiol.* 249: 153163. doi: 10.1016/j.jplph.2020.153163.

Hasibeder, R., Fuchslueger, L., Richter, A., Bahn, M. 2015. Summer drought alters carbon allocation to roots respiration in mountain grassland. *New Phytol.* 205: 1117–1127. doi: 10.1111/nph.13146.

Holmgren, M. et al. 2006. Extreme climatic events shape arid and semiarid ecosystems. *Front Ecol Environ.* 4: 87–95.

Jaleel, C.A., Ragupathi, G. Manivannan, P., Panneerselvam, R. 2007. Response of antioxidant defense system of *Cathranthus roseus* L.G. Don. to paclobutrazol treatment under salinity. *Acta Physiol Plant.* 29(3): 205–209. doi: 10.1007/s11738-007-0025-6.

Jaleel, C.A., Manivannan, P., Wahid, A., Farooq, M. et al. 2009. Drought stress in plants: A review on morphological characteristics and pigments composition. *Int J Agric Biol.* 11: 100–105.

Jha, D., Shirley, N., Tester, M., Roy, S.J. 2010. Variation in salinity tolerance and shoot sodium accumulation in *Arabidopsis* ecotype linked to differences in the natural expression levels of transporters involved in sodium transport. *Plant Cell Environ.* 33: 793–804.

Jungklang, J., Sunohara, Y., Mastsumoto, H. 2004. Antioxidative enzymes to NaCl stress in salt tolerant *Sesbania rostrata*. *Weed Biol Manage.* 4: 81–85.

Kamara, A.Y., Menkir, A., Badu-Apraku, B., Ibikunle, O. 2003. The influence of drought stress on growth yield and yield components of selected maize genotypes. *J Agric Sci.* 141: 43–50. doi: 10.1017/S0021859603003423.

Khan, A.G. 2004. Mycotrophy and its significance in wetland ecology and wetland management. In: Wong, M.H. (ed.). *Developments in ecosystems,* vol 1, 97–114, Elsevier, Northhampton, UK.

Kohler, J., Hernandez, J.A., Caravaca, F., Roldan, A. 2008. Plant growth promoting rhizobacteria and arbuscular mycorrhizal fungi modify alleviation biochemical mechanisms in water stressed plants. *Funct Plant Biol.* 35(2). doi: 10.1071/FP07218.

Lafitte, H.R., Yongsheng, G., Yan, S., Li, Z.K. 2007. Whole plant responses, key processes and adaptation to drought stress: The case of rice. *J Exp Bot.* 58: 169–175. doi: 10.1093/jxb/erl101.

Lesk, C., Rowhani, P., Ramankutty, N. 2016. Influence of extreme weather disasters on global crop production. *Nature.* 529: 84–87. doi: 10.3389/fpls.2013.00272.

Marulanda, A., Porcel, R., Barea, J.M., Azcon, R. 2007. Drought tolerance and antioxidant activities in lavender plants colonized by native drought-tolerant or drought-sensitive *Glomus* species. *Microb Ecol.* 54: 543–552.

Marulanda, A., Barea, J.M., Azcon, R. 2009. Stimulation of plant growth and drought tolerance by native microorganisms (AM fungi and bacteria) from dry environments: Mechanisms related to bacterial effectiveness. *J Plant Growth Regul.* 28: 115–124.

Mayak, S., Tieosh, T., Glick, B.R. 2004a. Plant growth promoting bacteria that confer resistance to water stress in tomato and peppers. *Plant Sci.* 166: 525–530.

Mayak, S., Tieosh, T., Glick, B.R. 2004b. Plant growth promoting bacteria that confer resistance to water stress in tomato plants to salt stress. *Plant Physiol Biochem.* 42: 565–572.

McLellan, C.A., Turbyville, T.J., Wijeratne, K., Kerschen, A., Vierling, E., Queitsch, C., Whiteshell, L., Gunatilaka, A.A.L. 2007. A rhizosphere fungus enhances *Arabidopsis* thermotolerance through production of an HSP90 inhibitor. *Plant Physiol.* 145: 174–182.

Meddich, A., Jaitiv, F., Bourzik, W., El asli, A. 2015. Use of mycorrhizal fungi as a strategy for improving the drought tolerance in date palm (*Phoenix dactylifera*). *Sci Hortic*. 192: 468–474. doi: 10.1016/j.scienta.2015.06.024.

Mittler, R., Zilinskas, B.A. 1994. Regulation of pea cytosolic ascorbate peroxidase and other antioxidant enzymes during the progression of drought stress and following recovery from drought. *Plant J.* 5: 397–405.

Molina-Favero, C., Creus, C.M., Simontacchi, M., Puntarulo, S., Lamattina, L. 2008. Aerobic nitric oxide production by *Azosprillum brasilense* Sp245 and its influence on root architecture in tomato. *Mol Plant Microb Interact.* 2: 1001–1009.

Moller, I.S., Gilliham, M., Jha, D., Mayo, G.M., Roy, S.J., Coates, J.C., Haseloff, J., Tester, M. 2009. Shoot Na^+ exclusion and increased salinity tolerance engineered by cell type-specific alteration of Na^+ transport in *Arabidopsis*. *Plant Cell.* 21: 2163–2178.

Nadeem, S.M., Zahir, Z.A., Naveed, M., Ashraf, M. 2010. Microbial ACC-deaminase: Prospects and applications for inducing salt tolerance in plants. *Crit Rev Plant Sci.* 29: 360–393.

Nautiyal, C.S., Srivastava, S., Chauhan, P.S., Seem, K., Mishra, A., Sopory, S.K. 2013. Plant growth promoting bacteria *Bacillus amyloliquefaciens* NBRISN13 modulates gene expression profile of leaf and Rhizosphere community in rice during salt stress. *Plant Physiol Biochem.* 66: 1–9.

Pankaj, U., Singh, D.N., Singh, G., Verma, R.K. 2019a. Microbial inoculants assisted growth of *Chrysopogon zizanioides* promotes phytoremediation of salt affected soil. *Indian J. Microbiol.* 59(2): 137–146.

Pankaj, U., Verma, R.S., Yadav, A., Verma, R.K. 2019b. Effect of arbuscular mycorrhizae species on essential oil yield and chemical composition of commercially grown palmarosa (*Cymbopogon martinii*) varieties in salinity stress soil. *J Essent Oil Res*. 31(2): 145–153.

Pankaj, U., Singh, D.N., Mishra, P., Gaur, P., Vivekbabu, C.S., Shanker, K., Verma, R.K. 2020. Autochthonous halotolerant plant growth promoting rhizobacteria promote bacoside A yield of *Bacopa monnieri* (L.) Nash and phytoextraction of salt-affected soil. *Pedosphere*. 30(5): 671–683.

Poorter, H., Niklas, K.J., Reich, P.B., Oleksyn, J., Poot, P., Mommer, L. 2012. Biomass allocation to leaves stems and roots: Meta-analyses of interspecific variation and environmental control. *New Phytol.* 193: 30–50. doi: 10.1111/j.1469-8137.2011.03952.x.

Potthoff, M., Steenwerth, K.L., Jackson, L.A., Drenovsky, R.E., Scow, K.M., Joergensen, R.G. 2006. Soil microbial community composition as affected by restoration practices in California grassland. *Soil Biol Biochem.* 38: 1851–1860.

Rampino, P., Pataleo, S., Gerardi, C., Perotta, C. 2006. Drought stress responses in wheat: Physiological and molecular analysis of resistant and sensitive genotypes. *Plant Cell Environ.* 29: 2143–2152. doi: 10.1111/j.1365-3040-2006.01588.x.

Ryu, C.M., Farag, M.A., Hu, C.H., Reddy, M.S., Kloepper, J.W., Pare, P.W. 2004. Bacterial volatiles induce systemic resistance in *Arabidopsis*. *Plant Physiol.* 134: 1017–1026.

Saleem, M., Arshad, M., Hussain, S., Bhatti, A.S. 2007. Perspective of plant growth promoting rhizobacteria (PGPR) containing ACC deaminase in stress agriculture. *J Ind Microbiol Biotechnol.* 34(10):635–648. doi: 10.1007/s10295-007-0240-6.

Samarah, N.H. 2005. Effects of drought stress on growth and yield of barley. *Agron Sustain Dev.* 25: 145–149. doi: 10.1051/agro: 2004064.

Sandhya, V., Ali, S.Z., Grover, M.M., Kishore, N., Venkateswarlu, B. 2009a. *Pseudomonas* sp. strain P45 protects sunflowers seedlings from drought stress through improved soil structure. *J Oilseed Res.* 26: 600–601.

Santos-Medellin, C., Edwards, J., Liechty, Z., Nguyen, B., Sundaresan, V. 2017. Drought stress results in a compartment specific restructuring of the rice root-associated microbiomes. *mBio.* 8: e00764-17. doi:10.1128/mBio.00764-17.

Schellenbaum, L., Muller, J., Boller, Th., Wiemken, A., Schuepp, H. 1998. Effects of drought on non-mycorrhizal maize: Changes in the pools of non-structural carbohydrates, in the activities of invertase and trehalase and in the pools of amino acids and imino acids. *New Physiol.* 138: 59–66.

Selvakumar, G., Panneerselvam, P., Ganeshamurthy, A.N. 2012. In Maheshwari, D.K. (ed.). *Bacterial mediated alleviation of abiotic stress in crops in bacteria in agrobiology stress management*, 205–224. Springer-Verlag, Berlin-Heidelberg.

Sgherri, C.L.M., Maffei, M., Navari-Izzo, F. 2000. Antioxidative enzymes in wheat subjected to increasing water deficit and dewatering. *J Plant Physiol.* 157: 273–279.

Shi, H.T., Li, R.J., Cai, W., Liu, W., Wang, C.L., Lu, Y.T. 2012. Increasing nitric oxide content in *Arabidopsis thaliana* by expressing rat neuronal nitric oxide syntheses resulted in enhanced stress tolerance. *Plant Cell Physiol.* 53: 344–357.

Singh, B.K. 2010. Exploring microbial diversity for biotechnology: The way forward. *Trends Biotechnol.* 28(3): 111–116.

Singh, J.S. 2015. Microbes: The chief ecological engineers in reinstating equilibrium in degraded ecosystems. *Agric Ecosys Environ.* 203: 80–82.

Singh, J., Singh, D.P. 2013. Impact of anthropogenic disturbances on methanotrophs abundance in dry tropical forest ecosystems, India. *Expert Opin Environ Biol.* 2: 1–3.

Singh, V.K., Singh, A.K., Singh, P.P., Kumar, A. 2018. Interaction of plant growth promoting bacteria with tomato under abiotic stress: A review. *Agric Ecosyst Environ.* 267: 129–140. doi: 10.1016/j.agree.2018.08.020.

Song, F., Han, X., Zhu, X., Herbert, S.J. 2012. Response to water stress of soil enzymes and root exudates from drought and non-drought tolerant corn hybrids at different growth stages. *Can J Soil Sci.* 92: 501–507. doi: 10.4141/cjss2010-057.

Venkateswarlu, B., Desai, S., Prasad, Y.G. 2008. Agriculturally important for stressed ecosystems: Challenges in technology development and application. In: Khachatourians, G.G., Arora, D.K., Rajendran, T.P., Srivastava, A.K. (eds). *Agriculturally important microorganisms*, 225–246. Academic World, Bhopal.

Vivas, A., Marulanda, A., Ruiz-Lozano, J.M., Barea, J.M., Azcon, R. 2003. Influence of *Bacillus* spp on physiological activities of two arbuscular mycorrhizal fungi and plant responses to PEG-induced drought stress. *Mycorrhiza.* 13: 249–256.

Wu, Q.S., Xia, R.X., Zou, Y.N. 2006. Reactive oxygen metabolism in mycorrhizal and non-mycorrhizal citrus (*Poncirus trifoliate*) seedlings subjected to water stress. *J Plant Physiol.* 163: 1101–1110.

Xu, L., Dan, N., Zhaobin, D., Tuesday, S., Grady, P., Hixson, K.K. et al. 2018. Drought delays development of the sorghum root microbiome and enriches for monoderm bacteria. *Proc Natl Acad Sci USA.* 115: E4284–E4293. doi: 10.1073/pnas.1717308115.

Yang, J., Kloepper, J., Ryu, C.N. 2009. Rhizosphere bacteria help plants tolerate abiotic stress. *Trends Plant Sci.* 14: 1–4.

Zhang, H., Kim, M.S., Sun, Y., Dowd, S.E., Shi, H., Pare, P.W. 2008. Soil bacteria confer plant salt tolerance by tissue-specific regulation of the sodium transporter HKT1. *Mol Plant Microbe Interact.* 21: 737–744.

Zhang, H., Murzello, C., Sun, Y., Kim, M.S., Xie, X., Jeter, R.M., Zak, J.C, Dowd, S.E., Pare, P.W. 2010. Choline and osmotic-stress tolerance induced in *Arabidopsis* by the soil microbe *Bacillus subtilis* (GB03). *Mol Plant Microb Interact.* 23: 1097–1104.

6 Secondary Metabolites

Stress Busters for the Development of Resilience in Plants to Sustainable Agriculture

Priyanka Sati and Deepa Minakshi
Graphic Era (Deemed to Be) University, Dehradun, India

Eshita Sharma
CSIR–Institute of Himalayan Bioresource Technology, Palampur, India

Ruchi Soni
Regional Centre of Organic Farming, Ghaziabad, India

Praveen Dhyani
Kumaun University, Bhimtal, India

CONTENTS

DOI: 10.1201/9781003147077-6

6.1 INTRODUCTION

In the natural environment, plants face a varied number of stress conditions (biotic and abiotic) such as high anthropogenic activity, habitation in extreme situations of temperature, radiation, and pressure (Taârit et al. 2012; Flora et al. 2013). However, in the agricultural ecosystem, plants are exposed to numerous stress conditions also, such as those arising from the altered levels of acidity-alkalinity, salinity, heavy metal concentrations, and toxic compounds in the soil. Although the duration, intensity, and occurrence of different stress conditions are mostly unpredictable, it is evident that the stress either in combination or alone adversely affects plant growth and agriculture productivity. The prevalent environmental stress factors even in the lowest signal intensity induce varied non-specific, reversible, and non-reversible changes at different physiological and functional levels (such as alterations in respiration, electron transport system, and photosynthesis) leading to the altered quantity and quality of metabolite synthesis (Ksouri et al. 2007), with reproductive stages affected the most. All of these altered conditions affect the inhibition of plant growth, as plant growth and development are dependent on different environmental settings along with nutritional factors.

Amidst prevailing stress/adverse conditions, the survival and growth of plants is a massive task. Plants, being sessile, in nature essentially entails an explicit adaptive mechanism to contest adverse conditions. Therefore, plants develop a wide range of strategies to cope with stressful situations according to the magnitude of the stress signal. When the stress is of short term and low intensity, the plant growth recovery remains facilitated in its course; however, when the stress signal is non-tolerable, the plants' metabolic functions are severely affected leading to the hindrance of phenological stages and eventually to death (Taârit et al. 2010). To combat the adverse effects of different abiotic and biotic stress environment, the plants undergo different physiological and biochemical alterations accompanied with modifications in the structure and function of cell membranes by initiating several molecular, cellular, and physiological changes (Taârit et al. 2010; Pankaj et al. 2020).

However, such alterations in metabolic activities towards channelizing the metabolic flux towards acclimatization and adaptation adversely affect normal growth and development (Krasensky and Jonak 2012). Moreover, this can lead to an extreme loss of active compounds (primary and secondary metabolites [SMs]), especially in the case of medicinal plants, which are important sources of medicinal compounds of pharmaceutical importance (Pankaj et al. 2019a, b, 2020).

SMs, natural products synthesized by plants, bacteria, and fungi, play a variety of roles in numerous adaptive developmental and physiological mechanisms in plants for survival in the changing stress factors. The SMs have been shown to be attributed to wide ecological adaptive functions to defend against pathogens, insects, and herbivore attacks (Hartmann 2004). A varied number of SMs are biosynthesized from primary metabolites synthesized and accumulated in the plant cells, and this production is generally regulated by the tight coordination and exchange of signalling molecules between the root and shoot.

The signalling molecules from the root part are often stimulated directly and indirectly by the soil environment, which is also a hot spot of microbial abundance and activity (Ortíz-Castro et al. 2009; Glick 2012). The microorganisms can directly or indirectly help in activating plants' immunity, growth and morphogenesis, and regulation of plants' strategy to different stress, by producing a diverse range of phytohormones. Owing to microorganisms' ability to cope up with extreme environmental surroundings (Mirete et al. 2016), the plant-microorganism interaction has been of interest, as knowledge of these processes could lead to the development of novel agricultural applications. Apropos, in this chapter, we explore the different SMs primarily responsible for tolerance to biotic and abiotic stress conditions. Further, we explore the role of microorganisms and microbial formulations to elicit the SMs production in plants.

6.2 SECONDARY METABOLITES FOR BIOTIC STRESS TOLERANCE

Diverse physiological changes produce SMs in plants that are derived from primary metabolites. SMs produced from primary metabolites enable the plants to grow and survive under diverse environmental stresses (Kurepin et al. 2017). Based on their biosynthetic pathway, there exist three main groups of SMs in plants, i.e., nitrogenous compounds (cyanogenic glycosides, glucosinolates, and alkaloids), phenolic compounds (flavonoids and phenylpropanoids), and terpenes (isoprenoids) (Ashraf et al. 2018).

6.2.1 Alkaloids

Alkaloids are low-molecular-weight compounds, and their biosynthesis in plants takes place from amino acids, particularly aspartic acid, tryptophan, tyrosine, and lysine (Khalil 2017). Nicotine, tropane alkaloids, purine alkaloids, terpenoid indole alkaloids, and benzylisoquinoline alkaloids are all examples of alkaloids. There are 3000 compounds (approximately) which represent the terpenoid indole alkaloids class, e.g., camptothecin and vinblastine (antineoplastic agents), strychnine (poison for rats), and quinine (antimalarial drug). Most of the alkaloids, including the pyrrolizidine alkaloids, are toxic and used primarily in protection against microbial infection and herbivore attack. Tryptophan is a prime amino acid that is involved in terpene indole alkaloids biosynthesis and contributor to protecting the plants from biotic stresses (Facchini 2001). Medicinally important terpenoid indole alkaloids include ajmaline and vinblastine which are employed to treat cardiovascular and cancer diseases, respectively (Akhgari et al. 2017). Terpenoid indole alkaloids bearing medicinal values are abundant in plants, i.e., *Rauwolfia serpentina* and *Catharanthus roseus*. Benzylisoquinoline alkaloids are formed by condensation of tyrosine amino acids and include pharmacologically important compounds like analgesics (morphine), antimicrobial agents (sanguinarine and berberine), cough suppressants (codeine), and muscle relaxants (papaverine). These alkaloids are generally from plant families, i.e., Berberidaceae, Menispermaceae, Papaveraceae, and Ranunculaceae. Tropane alkaloids such as scopolamine and hyoscyamine are derived from *Atropa*, *Hyoscyamus*, and *Datura*. Commonly used glucosidase inhibitors, calystegines (widespread tropane alkaloids) belong to the Convolvulaceae and Solanaceae families. Ornithine involved in the biosynthesis of nicotine results in addiction by influencing the receptor for nicotinic acetylcholine (Ashraf et al. 2018).

6.2.2 Cyanogenic Glycosides

Other than alkaloids, cyanogenic glycosides and glucosinolates are nitrogenous protective chemicals found in plants (Patra et al. 2013). Hydrogen cyanide (HCN), a deadly gas produced by Graminae, Rosaceae, and Leguminosae families of plants, is because of cyanogenic glycosides found in them (Ning et al. 2013). In plants, HCN can generate cyanogenic glycosides along with enzymes responsible for sugar hydrolysis. However, HCN is a toxic gas that inhibits the activity of certain key metalloproteins, viz., cytochrome oxidase. The significant amounts of cyanogenic glycosides present in tubers of *Manihot esculenta* can cause limb paralysis. The presence of cyanogenic glucosides in cassava makes it suitable for long time storage without being attacked by pests. Glucosinolates are composed of another important class of glycosides and also known as mustard oil glycosides, which upon hydrolysis produce volatile substances. The presence of glucosinolates in vegetables like radish, broccoli, and cabbage gives them a particular taste and smell (Ashraf et al. 2018).

6.2.3 Phenolics

Phenolics is an important family of aromatic plant SMs, containing one or more acidic hydroxyl groups attached to a phenyl ring (Ali et al. 2013). They constitute an important part of plants' defence system against pests and diseases including root parasitic nematodes. Phenolics in plants are composed of approximately 10,000 compounds, which are soluble in both water and organic solvents (Ashraf et al. 2018). The shikimic and malonic acid routes are involved in phenolic biosynthesis. The malonic acid pathway is widespread in microorganisms such as fungi and bacteria; however, the shikimic pathway is more frequent in plants, fungi, and bacteria (Butkhup et al. 2016). The shikimic acid pathway employs precursors obtained from glycolysis and pentose phosphate pathway producing a diverse range of aromatic amino acids via intermediate phenylalanine (Taiz and Zeiger 2006). Coumarins, flavonoids, lignins and lignans, arylpyrones and styrylpyrones, stilbenes, Phenylpropanoids (cinnamic acid, *p*-coumaric acid, and their derivatives), and tannins are among the many compounds that make up phenolics (Fang et al. 2011; Ashraf et al. 2018). They are an active set of molecules possessing a varied range of anti-microbial activity against fungi and bacteria (Brooker et al. 2008). These cyclic molecules are natural pesticidal defenders for plants and their derivatives possess enhanced antifungal activity and stability against a range of soilborne plant pathogenic fungi.

Another type of coumarin is furanocoumarins present abundantly in members of the Umbelliferae family which includes celery parsnip and parsley. Under normal conditions, these compounds are not toxic, but on activation by light (UV-A), which excites furanocoumarins to higher energy states, can insert and bind themselves into the double helix of DNA and pyrimidine bases, respectively, thereby blocking transcription and repair followed by cell death (Mazid et al. 2011). A basic linear furanocoumarin, psoralin is used in the treatment of fungal defence (Ali et al. 2008). Lignin is a complex family of phenolics that is generated by the phenylpropanoid groups of prominently branched polymers and is often found in plants after cellulose. The lignin is formed by simultaneous and random condensation of three different monomeric alcohols, viz., coniferyl, coumaryl, and sinapyl which are oxidized to free radicals by a ubiquitous plant enzyme-peroxidase. Its physical toughness and chemical durability make it relatively indigestible to herbivores. Lignifications block the growth of pathogens and are a frequent response to infection or wounds (Taiz and Zeiger 2006).

Phenolic compounds, i.e., anthocyanins, isoflavones, flavonols, and flavones, are the main contributors to the defence response against insects, microorganisms, and regulate plant growth. Isoflavonoids are critical for plant development and defence response and are generally secreted by the legumes and promote the formation of nitrogen-fixing nodules by symbiotic rhizobia (Sreevidya et al. 2006). Flavonoids have excellent antioxidant potential against reactive oxygen species (ROS). Phenolic compounds act as a substrate for different peroxidases as the first line of defence against a variety of environmental stresses like metal stress (Posmyk et al. 2009).

Phenolics have a number of key roles in plants, including growth, reproduction, and tolerance to a variety of abiotic and biotic challenges (Giménez et al. 2014). Plants containing a high amount of phenolics bear allelopathic activity which suppresses the growth of surrounding plants (Ashraf et al. 2018). By detoxifying singlet oxygen, phenolic compounds have a high redox potential, making them strong antioxidants (Huang and Bie 2010). Various reports have suggested that phenolics significantly act as anticancer agents in humans (Verma and Shukla 2015). The phenolics serve essential tasks in plants, such as food intake, photosynthesis, protein synthesis, and enzyme activity. Higher production of phenolic compounds in plants is a sign of stress accumulation (Verma and Shukla 2015). Some phenolics in plants give them a distinct aroma and flavour, preventing insects and animals from eating them (Dey and Harborne 1997). Some phenolics in plants impart them a distinct aroma and flavour, preventing insects and animals from eating them (Dey and Harborne 1997). Pollinators are attracted to beautiful petal colours which are produced by phenolic compounds, i.e., anthocyanins (Ashraf et al. 2018).

6.2.4 Terpenes

In plants, the methylerythritol 4-phosphate and mevalonate pathways produce terpenes in the plastids and cytoplasm, respectively. The methylerythritol 4-phosphate pathway synthesizes the side chain of chlorophyll, carotenoids, plant hormones (abscisic acid [ABA], gibberellins), phytol, isoprene, mono- and diterpenes, tocopherols, phylloquinones, and plastoquinones. However, isopentenyl diphosphate produced in the mevalonate pathway is involved in the biosynthesis of brassinosteroids, sterols, sesquiterpenes, and polyphenols (Verma and Shukla 2015).

Isoprenoids, also known as terpenoids, are the most common type of plant SMs, with acetyl coenzyme A serving as a precursor in their formation. Both MEP and MVA pathways use isopentenyl diphosphate for the synthesis of an intermediate compound resulting in terpenes formation (Thimmappa et al. 2014). Based upon isoprene unit (five-carbon chain), terpenes are classified as monoterpenes (two C_5 units), sesquiterpenes (three C_5 units), diterpenes (four C_5 units), triterpenes (six C_5 units), tetraterpenes (eight C_5 units), and polyterpenoids $[(C_5)n]$ (Verma and Shukla 2015). Plant hormones, ABA (sesquiterpenes), gibberellins (diterpenes), carotenoids (tetraterpenes), and sterol (triterpenes), play a critical role in plant growth. Terpenes produced by stressed plants are often poisonous to mammals and insects. Moreover, plants also contain saponins and essential oils, which are classified as terpene glycosides and volatile terpenes, respectively, are toxic because they disrupt cell membranes and interfere with sterol absorption (Tomar et al. 2014).

Under plant stress, terpenes formed are generally toxic to mammals and insects. Besides terpenes, plants contain saponins and essential oils as terpene glycosides and volatile terpenes, respectively. Saponins are considered poisonous as they cause damage to cell membranes and interfere with sterol uptake (Tomar et al. 2014).

6.2.5 Other Secondary Plant Metabolites

Plants are capable of adjusting their developmental programs in combination with external abiotic signals. In order to prevent adverse environmental conditions, plants work against stress by producing various biochemical, physiological, and molecular processes. Endogenous levels of SMs, ion homeostasis, gene expression, and protein activity regulation are all affected by these mechanisms. In the biological organization, gene regulation at transcription levels is the major controlling factor which is referred to by different regulators and transcription factors. Small quantities of growth regulators are plant hormones which include ABA, auxins, cytokinin, ethylene, gibberellic acid, jasmonic acid, and salicylic acid (Wani et al. 2016).

In plants, phytohormones regulate plant responses under environmental stresses and thus aid to the growth and developmental response (Wolters and Jurgens 2009). Plant responses to biotic and abiotic stresses are mediated by plant hormones, i.e., ethylene, jasmonic acid, and ABA. However, auxins, cytokinins, and gibberellic acids are key pointers in both plant development and regulation of plant responses to stress. Changes in transcription levels have been reported when plants are supplemented with plant hormones (Chapman and Estelle 2009). A short exposure of plants to plant hormone induced significant alteration in gene expression in *Arabidopsis* reporting overexpression and repression of 10–300 genes (Paponov et al. 2008), although a longer treatment period of plant hormones showed larger changes in gene expression (Eyidogan et al. 2012). Auxin-treated plant tissues showed alterations, viz., cell differentiation, division, cell expansion, transcriptional level modifications, and electrophysiology. The early active integration of various genes mediated through auxin signals enables the plants to respond to environmental stresses, thereby maintaining plant growth and development (Ashraf et al. 2018). Gibberellic acids are actively involved in various plant developmental processes such as pollen maturation, stem elongation, and seed germination (Olszewski et al. 2002). Gibberellic acid constituents-mediated signalling pathways are vital acknowledged as growth determinants as well as a defence response. Gibberellic acid plays an important function in plant defence responses by interfering in the signalling pathways of other plant hormones in plants under environmental restraints which leads to changes in the expression (Achard et al. 2006).

6.3 SECONDARY METABOLITES FOR ABIOTIC STRESS TOLERANCE

During various ontogenic phases, plants are frequently exposed to abiotic stresses such as temperature stress, light intensity, chemical fertilizers, soil type and composition, water availability, and salinity. An appropriate quantity of abiotic components is needed by plants and alteration in the concentration of any of these components affects the biosynthesis of SMs (Figure 6.1), which in turn have a profound consequence on the productivity and growth of the plant (Ashraf et al. 2018; Pankaj et al. 2019b). Environmental stresses, i.e., temperature, wounds, UV radiation, and nutrient deficiency alter SM concentration.

6.3.1 Chemical Stress

Plants require a range of chemicals, i.e., plant SMs, nutrients, growth regulators, and elicitors for sustaining plant growth and development, and their deficit causes chemical stress in plants. A number of chemicals like elicitors, growth regulators,

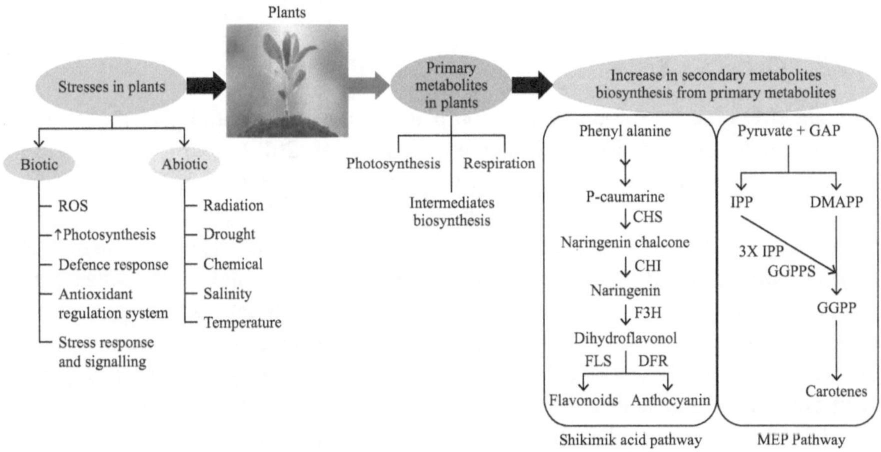

FIGURE 6.1 Plants undergo a variety of stresses and biotic disturbances in their natural growth conditions resulting in the stress initiation and defence responses interceded by signalling processes as well as pathways involving several molecules to perform cellular functions. The physiological processes in plants have a profound impact on primary metabolism that provides biosynthetic intermediates for secondary metabolism which in turn affect bioactive compounds' biosynthesis. This figure depicts secondary metabolic pathways induced during abiotic stress conditions in plants. Upregulation of genes, i.e., GGPS (geranylgeranyl pyrophosphate synthase), DFR (dihydroflavonol4-reductase), (CHS) chalcone synthase, (CHI) chalcone isomerase, (F3H) flavanone 3-hydroxylase, and (FLS) flavonol synthase, involved in flavonoid biosynthesis is depicted in the box.

minerals, ozone, pesticides, pollutants, and trace metals are contributors to chemical stress in plants. In *Cassia angustifolia*, when supplemented with micronutrients caused an increase in the concentration of primary SMs (Verma and Shukla 2015). Endogenous levels of phenols, proteins, and chlorophyll are altered by compounds such as $CuSO_4$, $FeSO_4$, and $ZnSO_4$ (Shitole and Dhumal 2012). The application of phosphorus and nitrogen, which are significant factors of plant growth and development, caused variations in flavonoid concentration in St. John's wort. In *Arabidopsis* and other plant species, nitrogen fertilizers stimulate the expression of certain genes. Phosphorous influences the biosynthesis of SMs in plants and consequently improves the growth and development of the plant and increases in active constituents. However, phosphorus application did not change the concentration of essential oils in *Salvia officinalis*, but it did result in a significant rise in rosmarinic acid, total phenolics, and leaf biomass (Nell et al. 2009). The varying levels of nitrogen cause alterations in plants at gene levels which consequently cause developmental, physiological, and metabolic changes. In order to understand the plant's response to different nitrogen levels and metabolic pathways, different in silico approaches like bioinformatics and genetics are being employed (Vidal and Gutiérrez 2008). Gases like carbon dioxide and oxygen also synthesize SMs in plants and complete vital biological mechanisms. For example in *Taxus baccata*, *Echinacea purpure*, and *Hypericum perforatum*, the variance in SMs is connected to CO_2 (Save et al. 2007). Hence, chemicals are directly related to plants for their growth and development which in turn influence the biosynthesis and alteration in the amount of SMs (Verma and Shukla 2015).

6.3.2 Drought Stress

Drought stress has a profound impact on plant photosynthesis and growth, resulting in biochemical and physiological changes (Lisar et al. 2012). Due to a decrease in photosynthesis, physiological processes like enzyme activities, membrane integrity, and stomatal closure are interrupted in drought stress. The drought stress causes a rise in SM concentration in several medicinal plants, i.e., *Artemisia annua*, *C. roseus*, and *H. perforatum* (Ashraf et al. 2018). A rise in total phenolics and photosynthetic pigments along with a decrease in a plant's fresh and dry biomass in *Trachyspermum ammi* was recorded because of drought stress (Azhar et al. 2011). On the other hand, exposure of St. John's wort plants to drought stress caused a significant fall in photosynthesis but simultaneous enhancement in SM levels like hyperforin, hypericin, and pseudohypericin (Zobayed et al. 2007). The drought stress caused the improved quality of artemisinin in *Artemisia* and rutin, quercetin in *Hypericum brasiliense* (Verma and Shukla 2015). In plants like *Ocimum americanum* and *Ocimum basilicum*, water-stress conditions resulted in alteration in levels of essential oils, macronutrients, proline, and carbohydrates (Khalid 2006). When *Glechoma longituba* was grown under drought stress conditions, a rise in total flavonoids was recorded (Zhang et al. 2012).

6.3.3 Radiation Stress

Light is a crucial abiotic component determining the quality and concentration of SMs in plants which is highly dependent on the intensity of light necessary for growth and photosynthesis. Sunlight supports the SM accumulation (coumarins) in *Mikania glomerata*. The exposure of different parts of *M. glomerata* to shorter or longer light duration resulted in alterations at the endogenous level of coumarins. The leaves and stems of plants, when exposed to a shorter period of light, decrease the coumarin's concentration, while a longer light period caused a significant increase in the same. Hence, photoperiod in leaves and stems significantly influences the coumarin content (de Castro et al. 2006). Previous research has found that photoperiod and light intensity affect the biosynthesis and storage of SMs in plants (Verma and Shukla 2015). Therefore, light intensity along with photoperiod have a major effect on the accrual of plant SMs.

6.3.4 Salinity Stress

Plant SMs endure an increase or decrease in their content due to salinity-induced osmotic stress or specific ion toxicity which exhibits downfall in growth, photosynthesis as well as nutrients uptake in plants (Ashraf et al. 2015). The availability of nutrients affects the levels of primary and SMs (Verma and Shukla 2015). A considerable rise in alkaloid content was observed because of salinity stress in *Achillea fragrantissima* and *C. roseus*. *C. roseus* and *Rauvolfia tetraphylla* on exposure to salinity stress in the growth medium resulted in a considerable increase of vincristine alkaloids and reserpine (Said-Al Ahl and Omer 2011). *A. fragrantissima* and *Matricaria chamomilla* under salinity stress caused enhanced content of different phenolics, i.e., chlorogenic, caffeic, and protocatechuic acids (Verma and Shukla 2015). A marked increase in phenols was reported by *Mentha pulegium* and *Nigella sativa* under salinity stress (Bourgou et al. 2010; Oueslati et al. 2010). Essential oils in *Origanum vulgare*, *Mentha suaveolens*, *T. ammi*, and *Thymus maroccanus* decrease as a result of salinity stress; however, an increase in essential oil content in *Matricaria recutita* is observed (Said-Al Ahl and Omer 2011). Moreover, in *Plantago ovata* roots, salinity stress increases the amounts of saponins, flavonoids, proline, ajmalicine (an alkaloid), and antioxidant properties (Haghighi et al. 2012).

The salinity stress resulted in a decrease in essential oils in *O. vulgare*, *M. suaveolens*, *T. ammi*, and *T. maroccanus*. On the contrary, salinity stress triggered the raise in contents of essential oil in *M. recutita* (Said-Al Ahl and Omer 2011). The salinity stress resulted in an increase in the levels of saponins, flavonoids, proline, ajmalicine (an alkaloid), and antioxidant property in *P. ovata* roots (Haghighi et al. 2012).

6.3.5 Temperature Stress

An appropriate temperature range is very necessary for the proper growth and development of the plants. The wide variation in temperature range influences the growth and productivity of the plant (Yadav 2010) via an altered biosynthesis of SMs (Verma and Shukla 2015). The heat stress causes the decrease in growth and productivity in plants due to lesser stomatal conductance and, hence, declined net CO_2 fixation and photochemical efficiency of photosystem II. High-temperature stress may increase or decrease the production of SMs. The increase in ginsenosides was reported in *Panax quinquefolius* under high-temperature stress (Jochum et al. 2007). Low-temperature stress causes a noticeable decline in chlorophyll content, total soluble protein, biosynthesis, and storage of SMs (Verma and Shukla 2015). This stress inhibits the water uptake and metabolic reactions in plants which hampers plant growth and productivity. Plants undergo alteration in physicochemical and molecular processes which help them endure low-temperature stress conditions (cold acclimation) (Chinnusamy et al. 2007). In case of *Capsicum annuum* grown at low temperatures, the photosynthesis, growth, and yield are suppressed.

6.4 MICROORGANISMS AS ELICITORS TO THE PRODUCTION OF SECONDARY METABOLITES

The total of mankind is dependent on plants for various sources of carbohydrates, proteins, vitamins, food, and shelter. Plants have been crucial components and nutritional factors over the decades. Apart from primary metabolites, plants can also produce a large number of low-molecular-weight components. A variety of organic compounds that are produced by plants allows the interaction with the biotic environment as well as the establishment of a defence mechanism known as plant SMs (Naik and Al-Khyari 2016).

Microbial SMs are valuable organic components that are synthesized during the idiophase stage of microbial growth and are dispensable for the organism's vital functions (Ruiz et al. 2010). Under unfavourable conditions, many microbial regulatory mechanisms regulate the production of SMs, which are known to give protection to the producing bacteria (Singh et al. 2017). Despite their little level of production, SMs boost an organism's ability to withstand harsh environmental circumstances such as biotic and abiotic stresses, temperature changes, salinity, osmotic, and drought stresses; metallic toxicities; and pathogen attack (Vimal et al. 2017, 2019).

Various kinds of organisms, including filamentous actinobacteria, eukaryotic fungus, and unicellular bacteria, synthesize SMs in a flexible and consistent manner. The most important producers of diverse functional SMs include endophytic fungal species, imperfect fungi, basidiomycetes, and ascomycetes. Accordingly, *Aspergillus niger*, *Trichoderma reesei*, *Fusarium niger*, *Penicillium niger*, and *Aureobasidium niger* have demonstrated high efficacy as pharmacological, anticancer, chemopreventive, and immunomodulatory agents (Kelly et al. 2015). The filamentous actinomycetes are the greatest producers of SM, accounting for 45% of all known microbial bioactive chemicals.

6.4.1 Plant-Microorganism Interaction for the Alleviation of Abiotic Stress

Plant-microorganism interactions are the main factors that alleviate and survive under abiotic stress to plants. SMs from microbial origins benefit the plants to adapt under stress conditions (Figure 6.2) by regulating the plant's intrinsic metabolic pathways as well as inducing systemic resistance.

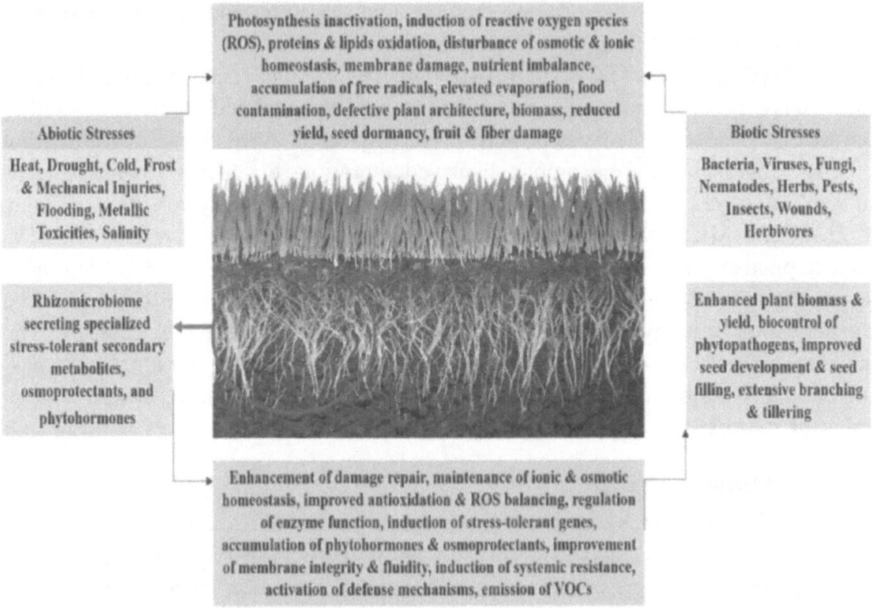

FIGURE 6.2 Plant-microorganism interactions alleviate abiotic and biotic stress conditions in plants. (Reprinted from "Microbial Secondary Metabolites: Effectual Armors to Improve Stress Survivability in Crop Plants". In *Microbial Services in Restoration Ecology* (Shahid and Mehnaz 2020), with permission from Elsevier.)

6.4.1.1 Salinity Stress

The presence of large concentrations of salts (K^+, Ca^{2+}, Mg^{2+}, Na^+, and Cl^-) in the soil has a substantial impact on plant productivity. Plant metabolism is modulated by microbial SMs, allowing them to endure severe salinity environments. Salinity stress, for example, is frequently associated with the production of significant levels of ethylene, which is toxic to plants. The enzyme 1-aminocyclopropane-1-carboxylate (ACC) deaminase is retained in plant growth-promoting rhizobacteria (PGPR), and it reduces ethylene concentration by converting it to ketobutyrate and ammonia (Siddikee et al. 2015).

Microorganisms that induce salinity tolerance include *Bacillus*, *Pseudomonas*, *Rhizobium*, *Azotobacter*, *Enterobacter* spp., arbuscular mycorrhiza fungi (AMF), etc. (Vimal et al. 2019).

6.4.1.2 Drought Stress

Drought stress is typically regarded as the most prevalent stress that has a significant impact on plants and hence results in significant plant mortality (Vimal et al. 2016). Drought causes osmotic stress in plants, which is linked to the accumulation of phytohormones and ABA, which leads to an increase in ROS (Mauch-Mani and Mauch 2005; Pankaj et al. 2020). Plants withstand drought circumstances due to a variety of natural processes; nonetheless, they are unable to withstand the deterioration produced by drought. Several plants, including dune grass, tomato, soybean, and rice, have been researched for their ability to consume less water and produce more after being inoculated with drought-tolerant PGPB.

The capacity of these PGPB and AMF to increase solute concentration, SM accumulation, and transcriptional regulation of genes in inoculated plants compared to uninoculated control plants is primarily recognized as a possible drought-tolerant process. Cacao plants, for example, displayed alterations in gene expression patterns in response to *Trichoderma* inoculation in order to withstand dry conditions (Bailey et al. 2006). Microbial volatile organic compound (VOC)-induced drought tolerance in plants is responsible for the accumulation of nitric oxide and H_2O_2. Drought-stressed plants also accumulate choline and other osmoprotectants, which promote the activation of many innate stress-tolerance mechanisms (Zhang et al. 2010). The accumulation of osmoprotective SMs in plant tissues prevents water loss by increasing cellular osmotic pressure and lowering the free water potential, as well as contributing to the plant's membrane integrity (Zhang et al. 2010).

6.4.1.3 Osmotic Stress

Osmotic stress is another abiotic stress that is harmful to plants. The chlorophyll contents and leaf gaseous exchangeability are significantly reduced by oxidative stress. Plants also have a variety of metabolic, cellular, and molecular responses to osmotic stress, which can be triggered by a variety of primary and secondary stress signals such as ROS, phytohormones (e.g., ABA, ethylene), and intracellular second messengers (e.g., phospholipids). Using microbial SMs, oxidative stress damage to plants can be reduced. Seedlings inoculated with osmotolerant *Pseudomonas simiae*, for example, revealed that photosynthetic machinery was maintained in plants exposed to oxidative stress (Kwon et al. 2010). Proline- and glycine betaine-producing bacteria, particularly PGPR, have a synergistic impact by enhancing plant osmotic stress tolerance among other microbial SMs.

6.4.1.4 Temperature Stress

Unpredictable fluctuations in global climatic circumstances have made it difficult for plants to cope with temperature instability. Temperature fluctuations have a negative impact on the plant's internal metabolic operations and disrupt the homeostatic balance, resulting in a decrease in productivity. Plants' innate tolerance mechanisms also respond to frost injury and excessive temperatures, Although more recently, the facilitation of temperature and freezing shocks on plants by inhabiting the plant rhizome-microbiome has been investigated in detail, the expression of anti-freeze and cold shock proteins, pH, and compatible solute production pathways, genomic and metagenomic analyses of various cultivable and non-cultivable plant microorganisms have revealed the processes of cold and high-temperature adaptations (Theocharis et al. 2012).

Inoculating with temperature-resistant PGPR and AMF has also resulted in improved nutritional absorption, osmolyte aggregation, and photosynthetic aptitude and efficiency. *Dichanthelium lanuginosum* (grass) in Yellowstone National Park, for example, can withstand temperatures of 38–65°C, thanks to the root-colonizing fungus *Curvularia protuberata* (Lata et al. 2018). Some of the abiotic stress-tolerance microorganisms and their mode of stress amendment in different plants are listed in Table 6.1.

TABLE 6.1
Some of the Abiotic Stresses, Microbial Strains, and Mechanisms in Plants

Organisms	Crop	Type of Stress	Mechanisms	References
Pseudomonas frederiksbergensis OS261	*Capsicum* sp., tomato	Salt, cold	Synthesis of ACC deaminase	Chatterjee et al. (2017), Subramanian et al. (2016)
Bacillus amyloliquefaciens	*Oryza sativa*, wheat	Salt, temperature	Abscisic acid and salicylic acid	Shahzad et al. (2017), Timmusk et al. (2014)
Bacillus subtilis GB03	*Arabidopsis thaliana*, white clover	Salt	VOCs, regulation of Na^+ transporter HKT1	Zhang et al. (2008)
Pseudomonas simiae AU	Soybean	Salt	Proline, 4-nitroguaiacol AND quinolone	Vaishnav et al. (2015)
Pseudomonas koreensis AK-1	*Glycine max* L. Merrill	Salt	Improve proline content	Kasotia et al. (2015)
Pseudomonas vancouverensis OB155-gfp	*Solanum lycopersicum*	Cold	Reduction in ROS	Subramanian et al. (2015)
Burkholderia phytofirmans PsJN	Grapevine	Cold	Proline, starch, phenolic accumulation	Sheibani-Tezerji et al. (2015)
Serratia sp. XY21, *B. subtilis* SM21, *Bacillus cereus* AR156	*Cucumis sativa*	Drought	Proline, antioxidants, monodehydro ascorbate	Wang et al. (2012)
Trichoderma harzianum	Tomato	Drought	Proline, VOCs, antioxidants	Mona et al. (2017)
Bacillus megaterium	*Zea mays*	Osmotic	Increased root expression of osmoprotectants	Aroca et al. (2007)
Pseudomonas sp.	*Nicotiana tobaccum*, *Arabidopsis*, *Epacrids*	Osmotic	Glucan, water dikinase, starch degradation	Sarma et al. (2011)
Azospirillum barsilense	Wheat	Temperature	Secondary metabolites production	Timmusk et al. (2014)
Pseudomonas putidam, *Enterobacter cloacae*, *Achromobacter piechaudii*	Tomato	Flooding, salt, drought	Synthesis of ACC deaminase	Mayak et al. (2004)

(Continued)

TABLE 6.1 *(Continued)*
Some of the Abiotic Stresses, Microbial Strains, and Mechanisms in Plants

Organisms	Crop	Type of Stress	Mechanisms	References
Azospirillum sp.	Wheat	Drought	Improve water relation	Creus et al. (2004)
Burkholderia phytotirmans PSJN	Grapevine	Temperature	Synthesis of ACC deaminase	Barka et al. (2006)
Pseudomonas fluorescens	Groundnut	Salt	Synthesis of ACC deaminase	Saravanakumar and Samiyappan (2007)
Pseudomonas sp.	Pea	Drought	Decreased ethylene production	Arshad et al. (2008)
Pseudomonas putida P45	Sunflower	Drought	Improved soil aggregation due to EPS production	Sandhya et al. (2009)
Bacillus subtilis SU47 and *Arthobacter* sp. SU18	Wheat	Salt	Increase in dry biomass, total soluble sugars, and proline content	Upadhyay et al. (2012)

6.4.2 Plant-Microorganism Interaction for the Alleviation of Biotic Stress

Plants are generally susceptible to various bacteria, fungus, viruses, and soil-dwelling phytopathogens, which hinder plant growth and result in substantial productivity loss. Rhizo-microbiome is responsible for providing proper plant health by numerous direct and indirect processes. By competitive antagonism or by triggering plant defence mechanisms such as induced systemic resistance, bioactive microbial SMs limit the growth of invading pathogens and parasites (Pierson and Pierson 2010). Bacterial SMs have been widely acknowledged in the development of biofertilizers and biofungicides over the last few decades. *Pseudomonas* spp. bioactive SMs have a tremendously varied range of chemical structures, as well as various biological roles and varying degrees of fungitoxicity.

Antimicrobials are classified into different groups based on structural similarities and variances. Phenazines, rhizoxin, cyclic lipopeptides (CLPs), acetaminophen, polyketides, diketopiperazines, and volatile antimicrobial substances are some of the primary antagonistic metabolites of *Pseudomonas* spp. (Mehnaz et al. 2013). *In vitro* tests on vine plantlets revealed that pyrrolnitrin-producing *Pseudomonas fluorescens* Pf1TZ competitively inhibited the plant disease *Botrytis cinerea*, and it was also found to be herbicidal against a variety of field herbs (Kilani-Feki et al. 2010).

Phylogenomic and antiglycation studies of 2,4-DAPG-producing *Pseudomonas* spp. endophytes were evaluated for their inhibitory actions by 2,4-DAPG and its by-products (Gutiérrez-García et al. 2017). Additionally, Loper et al. (2008) investigated the origins of orfamides in biocontrol strain *Pseudomonas protegens* Pf-5 and found that they have potent antifungal properties against plant diseases. Three families of CLPs, known as surfactin, iturin, and fengycin, have been extensively recognized for their bioactivities among the different active categories of bioactive substances produced by the *Bacillus*. Furthermore, these CLPs are required for *Bacillus subtilis* root colonization due to their antagonistic and cytotoxic properties (Bionda et al. 2013).

Similarly, endophytic and root-colonizing fungi aid in the reduction of biotic stressors in plants by providing a new reservoir of bioactive SMs. Fungal endophytes produce peptides, alkaloids, terpenoids, lignans, steroids, flavonoids, and peptides that boost the persistence and fitness of plants against invading diseases, accounting for approximately 51% of total bioactive chemicals (Kusari et al. 2012).

6.5 MICROBIAL FORMULATIONS FOR STRESS RESILIENCE: RECENT ADVANCES

Various environmental factors including abiotic and biotic play a substantial role in plant growth and crop productivity. Plant growth, development, and reproduction are regulated by major environmental factors like light, water, carbon, and minerals. Major abiotic stresses responsible for affecting plant growth are drought, salinity, water-logging conditions, heavy metal toxicity, nutrient deficiency, and temperature variation, which are further reflected in crop yield and agricultural productivity. However, plants endure various physiological and morphological modifications to sustain in unexpected and unusual changes of extreme environmental conditions.

Shrivastava and Kumar (2015) stated that abiotic stress can result in 20–50% reduction in crop yield which depends on the crop type and cultivation place like arid and semiarid regions and saline conditions, resulting in less biomass and lower water uptake affecting the photosynthesis in plants.

Microorganisms exert a significant role in ecosystems as they perform nitrogen fixation, carbon and nitrogen biogeochemical cycles, plant nutrient acquisition, and further soil rejuvenation. Several soil microorganisms are biofertilizers and they provide complementary limiting nutrients to the plants by different mechanisms including cellulolytic activity (*Aspergillus*, *Trichoderma*, *Penicillium* spp., *Bacillus amyloliquefaciens*), phosphate solubilization (*Bacillus*, *Pseudomonas* spp., arbuscular mycorrhiza), nitrogen fixation (*Azospirillum*, *Azotobacter*, *Rhizobium*), siderophore production (*Acinetobacter* and *Pseudomonas* sp.), soil acidification (*B. subtilis*), and plant growth hormone secretion (Bhattacharyya and Jha 2012). AMF may exert beneficial effects on host plants like enhancing nutritional deficiency which result in the protection from biotic and abiotic stresses (Szczałba et al. 2019). Beneficial yeasts have also been reported to be found in the phyllosphere and rhizosphere and be helpful in controlling many foliar pathogens through direct antagonism and elicitation of systemic defences (Preininger et al. 2018). Soil yeasts decompose organic matter, solubilize phosphate, and promote root growth, control root pathogens, aggregate soil particles, and support plant growth promotion (Sarabia et al. 2018).

Efficient bacteria are renowned PGPR, which reside in close vicinity of plant roots colonizing the rhizosphere including *Burkholderia*, *Bacillus*, *Pseudomonas*, *Serratia*, and *Streptomyces* which are the most studied genera (Bonaldi et al. 2015). The plant-microorganism interactions activate an induced systemic tolerance against biotic-abiotic stresses and further exhibit microorganism-induced plant responses to stress (Pieterse et al. 2014). It is worth communicating that many commercial products based on efficient microorganisms or their consortia acting as stress busters possess multiple attributes including antagonism towards pathogens, induction of plant defences, stimulation of plant growth hormones, and nutritional exchange with synergistic and additive effects.

Efficient bacteria have also been linked to the mineralization of organic contaminants in soil. *Achromobacter*, *Azospirillum*, *Bacillus*, *Burkholderia*, *Enterobacter*, *Methylobacterium*, *Microbacterium*, *Paenibacillus*, *Pantoea*, *Pseudomonas*, *Rhizobium*, *Variovorax*, and other microbial genera have been reported as stress busters and build host plant tolerance under various abiotic-biotic stress environments (Grover et al. 2011). Phytohormones like auxins, cytokinins, and gibberellins are also produced by various efficient microorganisms that possess significant effect on the root length, surface area, and root hair number, thus facilitate nutrient uptake. Moreover, there is a correlation between higher activities of enzymes performed during antioxidant action and tolerance to oxidative stress in cells (Štajner et al. 1997).

However, the significant deployment of these microorganisms in stressed agriculture depends on their ability to withstand and proliferate under stressful environments, viz., temperature, salinity, mineral deficiency, heavy metal toxicity, and so forth. In the current scenario of fast-evolving global climate change, microorganisms offer an ecological approach to support plants coping with a range of biotic-abiotic stress. Table 6.2 depicts different microorganisms prospecting the development of

TABLE 6.2
Effect of Microbial Formulations on the Growth of Different Crops under Biotic-Abiotic Stress Conditions

Name of PGPR	Source of Isolation	Characteristic of PGPR	Crop	Significant Effect on Crop Growth	Stress Condition	References
Bacillus subtilis SU47 and *Arthrobacter* sp. SU18	Wheat rhizosphere	PGP traits and salt tolerant	Wheat	Significant increase in dry biomass, total soluble sugars, and proline content	Salt stress	Upadhyay et al. (2012)
Klebsiella sp. D5A	*Testuca arundinacea* L. rhizospheres	IAA production, phosphate solubilization, siderophores synthesis, saline-alkaline tolerant, and pH (4–10)	*T. arundinacea*	Promotion of host plant growth and phyto-remediation efficiency enhancement	Petroleum contaminated saline-alkaline soil	Liu et al. (2014)
Burkholderia phytofirmans strain PsJN, *Enterobacter* sp. strain FD17	*Zea mays* rhizosphere	PGP traits and drought tolerant	*Z. mays*	Improved water status, photosynthetic activity, relative water content, membrane permeability, shoot fresh and dry weight	Drought	Naveed et al. (2014)
Azospirillum brasilense sp. 245 strain	Arabidopsis *Thaliana* rhizosphere	PGP traits and drought tolerant	Arabidopsis *Thaliana*	Significant effect on abscisic acid content, water loss, relative water content enhancement, stomata conductance, lipid peroxidation, proline concentration, photosynthetic efficiency	Drought	Cohen et al. (2015)
Rhizobacteria	Velvet bean (*Mucuna pruriens*) rhizosphere	1-Aminocyclopropane-1-carboxylate (ACC) deaminase-producing rhizobacteria	Velvet bean (*M. pruriens*)	Significant growth of roots-shoots in normal and water-stress condition, a significant increase in root-shoot length, root-shoot dry weight	Water stress due to the high concentration of ethylene in the rhizosphere	Saleem et al. (2015)

(Continued)

TABLE 6.2 *(Continued)*
Effect of Microbial Formulations on the Growth of Different Crops under Biotic-Abiotic Stress Conditions

Name of PGPR	Source of Isolation	Characteristic of PGPR	Crop	Significant Effect on Crop Growth	Stress Condition	References
Pseudomonas stutzeri, *B. subtilis*, *Stenotrophomonas maltophilia*, and *Bacillus amyloliquefaciens*	Cucumber rhizosphere	Plant growth-promoting (PGP) attributes and antagonism towards Phytophthora	Cucumber	Significantly higher levels of germination, seedling vigour, growth, and N content in root-shoot tissue, also suppress crown rot	Crown rot caused by *Phytophthora capsici*	Islam et al. (2016)
Rhizophagus irregularis *Variovorax paradoxus* 5C-2	–	PGP traits and drought tolerant	Tomato	Enhanced photosynthetic rate, reduced lipid oxidation, and increased root water conductivity and oxidative phosphorylation in the plant	Drought	Calvo-Polanco et al. (2016)
Klebsiella, *Pseudomonas*, *Agrobacterium*, and *Ochrobactrum*	*Arthrocnemum indicum* rhizosphere	Phosphate solubilization, ACC deaminase activity, salt tolerant, and capable of reducing acetylene	Peanut	Significant increase in total N content (up to 76%), tolerance to NaCl ranging from 4% to 8%	Salt stress	Sharma et al. (2016)
Pseudomonas putida MTCC5279 (RA)	Desert regions of Rajasthan, India	PGP traits and drought tolerant	Chickpea	Reduced/Controlled the expression of the stress response gene, maintained water content, osmolyte, membrane structure, and germination rate of the plant	Drought	Tiwari et al. (2016)

(Continued)

TABLE 6.2 *(Continued)*
Effect of Microbial Formulations on the Growth of Different Crops under Biotic-Abiotic Stress Conditions

Name of PGPR	Source of Isolation	Characteristic of PGPR	Crop	Significant Effect on Crop Growth	Stress Condition	References
Azospirillum sp. (Az19)	Wheat rhizosphere	PGP traits and osmotic and drought tolerant	Maize	Improve the growth and productivity of the plant under water stress compared to the control	Drought	García et al. (2017)
Enterobacter ludwigii	Cynodon dactylon rhizosphere	PGP traits and saline tolerant	*Festuca arundinacea*	Membrane transport protein in the microorganism that controls sodium and hydrogen ion movement across bacteria cell and the production of plant hormone, phosphate solubilization, nitrogen fixation contributes towards the growth, tolerance, and plant productivity	Salinity	Kapoor et al. (2017)
Bacillus sp.	Rhizosphere in different agro-climatic zones of Uttar Pradesh, India	ACC-deaminase production and salt tolerant	Rice	Aided the alleviation of salt stress by increasing the biomass and growth of rice seedling via the production of indole acetic acid and deaminase enzyme	Salinity	Misra et al. (2017)
Bacillus sp. *Alcaligenes* sp. *Proteus* sp.	*Commiphora wightii* rhizosphere	PGP traits and drought and saline tolerant	Chilli	Significantly increased root and shoot length more than the control	Salinity	Patel et al. (2017)

(Continued)

TABLE 6.2 *(Continued)*
Effect of Microbial Formulations on the Growth of Different Crops under Biotic-Abiotic Stress Conditions

Name of PGPR	Source of Isolation	Characteristic of PGPR	Crop	Significant Effect on Crop Growth	Stress Condition	References
Pseudomonas sp., *Bacillus cereus*, *Bacillus pumilus*, *Proteus* sp.	Maize and rice rhizosphere	Phytohormone production and drought tolerant	Maize	An increased concentration of phytohormones, viz., abscisic acid, 3-acetic acid, and gibberellic acid in soil and leaves of maize	Drought	Yasmin et al. (2017)
Enterobacter sp.	Rice rhizosphere	ACC-deaminase production and halotolerant	Rice	Promoted the growth of rice seedling and reduced ethylene production and antioxidant enzyme activities in the plant	Salinity	Sarkar et al. (2018)
Pseudomonas fluorescens, *Enterobacter hormaechei*, and *Pseudomonas migulae*	Foxtail millet (*Setaria italica* L.) rhizosphere	ACC-deaminase production and drought tolerant	Foxtail millet (*S. italica* L.)	Exopolysaccharide production and seed germination and seedling stimulation	Drought	Niu et al. (2018)
Acinetobacter calcoaceticus X128	Walnut rhizosphere	PGP traits and drought tolerant	*Sambucus williamsii*	Improved photosynthetic rate, stomatal conductance, intracellular CO_2 concentration, and total chlorophyll content (a + b)	Drought	Liu et al. (2019)
Cupriavidus necator 1C2 (B1) and *P. fluorescens* S3X (B2)	Metal-polluted area	PGP traits, water, metal, and osmotic deficit stress tolerant	*Zea mays* L.	Improved shoot biomass under moderate water deficit	Water deficit stress	Pereira et al. (2020)

(Continued)

TABLE 6.2 ***(Continued)***
Effect of Microbial Formulations on the Growth of Different Crops under Biotic-Abiotic Stress Conditions

Name of PGPR	Source of Isolation	Characteristic of PGPR	Crop	Significant Effect on Crop Growth	Stress Condition	References
B. subtilis HAS31	–	PGP traits and drought tolerant	Potato	Improved chlorophyll concentration, photosynthesis process, relative water content, osmolytes, antioxidants, enzymes and oxidative stress, relative growth rate, tuber, and above-ground biomass production	Drought	Batool et al. (2020)
Alcaligenes sp. *Bacillus* sp.	Alkaline soil	PGP traits and alkaline stress tolerant	*Z. mays* L.	Significantly improved the photosynthetic pigments, soluble sugar content, decreased proline level, significantly enhanced soil enzymes such as dehydrogenase, alkaline phosphatase, and betaglucosidase	Alkaline stress	Dixit et al. (2020)
Burkholderia phytofirmans PsJNT	Onion rhizosphere	PGP traits and efficiency of phytoextraction of Zn, Pb, and Cd	*Z. mays*, *Brassica juncea*, and *Medicago sativa*	Increased the yield of dry biomass and the survival rate of plants grown on soil contaminated with metal like Zn, Pb, and Cd	Metal contamination	Konkolewska et al. (2020)

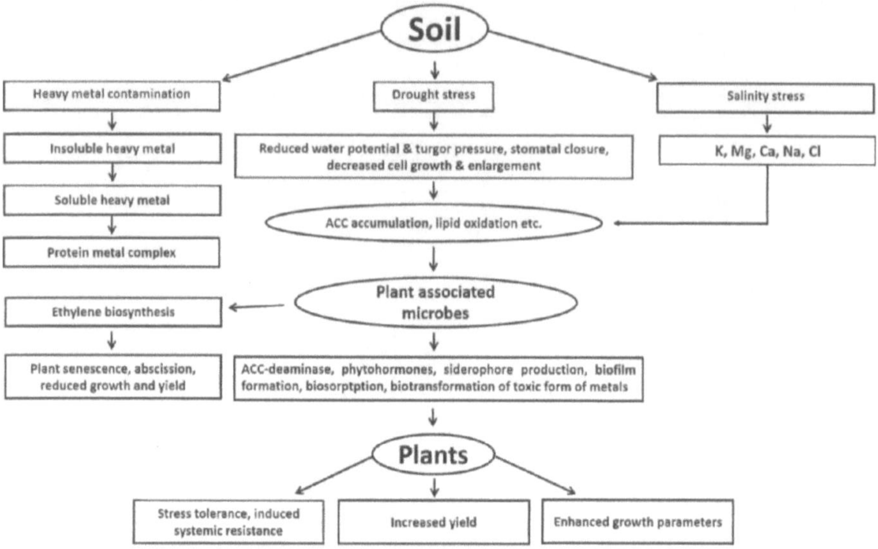

FIGURE 6.3 Schematic representation of the influence of microorganisms in stimulating plant growth and yield under stressful conditions of heavy metals, drought, and salinity.

microbial formulation and further deployed as stress busters in agriculture under chronically unfavourable climatic conditions. The microorganisms confer plant resistance for biotic-abiotic stress, and their application in sustaining nutrient deficiency, nodule inducers, phytohormones, and osmo-protecting molecules, to achieve better adaptation and performance in the field under unfavourable climatic conditions are also explained in Figure 6.3.

6.6 CONCLUSIONS AND FUTURE PERSPECTIVES

Microorganism-mediated stress resistance to increase tolerance levels in plants through increased biosynthesis of SMs has been studied for the past few decades. There are large numbers of bacterial and fungal strains with abundant SMs that have been studied in terms of their role as phyto-stimulators, disease suppressors, and nutrient mobilizers. They are associated with various positive factors including long-term survivability, eco-friendliness, biodegradability, and affordability and hence have attracted the world's attention in the form of chemical fertilizers and fungicide substitutes. Various previous investigations have focussed on the commercialization of several stress-tolerant microbial strains in the form of biofungicides and biofertilizers. Furthermore, broad-spectrum bacterial consortium with functions against harmful pathogens and stress conditions need to design regarding their applications to combat different prevalent biotic and abiotic stress conditions.

REFERENCES

Achard, P., Cheng, H., De Grauwe, L., Decat, J., Schoutteten, H., Moritz, T., Van Der Straeten, D., Peng, J., Harberd, N.P. 2006. Integration of plant responses to environmentally activated phytohormonal signals. *Science* 311(5757):91–94.

Akhgari, A., Oksman-Caldentey, K. M., Rischer, H. 2017. Biotechnology of the medicinal plant *Rhazya stricta*: A little investigated member of the Apocynaceae family. *Biotechnology Letters* 6(39):829–840.

Ali, M., Abbasi, B.H., Ihsan-ul-haq. 2013. Production of commercially important secondary metabolites and antioxidant activity in cell suspension cultures of *Artemisia absinthium* L. *Industrial Crops and Products* 49:400–406.

Ali, S.T., Mahmooduzzafar-Abdin, M.Z., Iqbal, M. 2008. Ontogenetic changes in foliar features and psoralen content of *Psoralea corylifolia* Linn. exposed to SO_2 stress. *Journal of Environmental Biology* 29(5):661–668.

Aroca, R., Porcel, R., Ruiz-Lozano, J.M. 2007. How does arbuscular mycorrhizal symbiosis regulate root hydraulic properties and plasma membrane aquaporins in *Phaseolus vulgaris* under drought, cold or salinity stresses? *New Phytologist* 173(4):808–816.

Arshad, M., Shaharoona, B., Mahmood, T. 2008. Inoculation with *Pseudomonas* spp. containing ACC-deaminase partially eliminates the effects of drought stress on growth, yield, and ripening of pea (*Pisum sativum* L.). *Pedosphere* 18(5):611–620.

Ashraf, M.A., Iqbal, M., Hussain, I., Rasheed, R. 2015. Physiological and biochemical approaches for salinity tolerance. In: Wani, S.H., Hossain, M.A. (Eds.), *Managing Salt Tolerance in Plants: Molecular and Genomic Perspectives*. New York: CRC Press, p. 79.

Ashraf, M.A., Iqbal, M., Hussain, I., Rasheed, R., Hussain, I., Riaz, M., Arif, M.S. 2018. Chapter 8 – Environmental stress and secondary metabolites in plants: An overview. In: Ahmad, P. et al. (Eds.), *Plant Metabolites and Regulation under Environmental Stress*. Academic Press, pp. 153–167.

Azhar, N., Hussain, B., Ashraf, M.Y., Abbasi, K.Y. 2011. Water stress mediated changes in growth, physiology and secondary metabolites of desi ajwain (*Trachyspermum ammi* L.). *Pakistan Journal of Botany* 43(1):15–19.

Bailey, B.A., Bae, H., Strem, M.D., Roberts, D.P., Thomas, S.E., Crozier, J., Samuels, G.J., Choi, I.-Y., Holmes, K.A. 2006. Fungal and plant gene expression during the colonization of cacao seedlings by endophytic isolates of four *Trichoderma* species. *Planta* 224(6):1449–1464.

Barka, E.A., Nowak, J., Clément, C. 2006. Enhancement of chilling resistance of inoculated grapevine plantlets with a plant growth-promoting rhizobacterium, *Burkholderia phytofirmans* strain PsJN. *Applied and Environmental Microbiology* 72(11):7246–7252.

Batool, T., Ali, S., Seleiman, M.F., Naveed, N.H., Ali, A., Ahmed, K., Abid, M., Rizwan, M., Shahid, M.R., Alotaibi, M., Al-Ashkar, I. 2020. Plant growth promoting rhizobacteria alleviates drought stress in potato in response to suppressive oxidative stress and antioxidant enzymes activities. *Scientific Reports* 10(1):1–19.

Bhattacharyya, P.N., Jha, D.K. 2012. Plant growth-promoting rhizobacteria (PGPR): Emergence in agriculture. *World Journal of Microbiology and Biotechnology* 28(4):1327–1350.

Bionda, N., Fleeman, R.M., Shaw, L.N., Cudic, P. 2013. Effect of ester to amide or N-methylamide substitution on bacterial membrane depolarization and antibacterial activity of novel cyclic lipopeptides. *ChemMedChem* 8(8):1394–1402.

Bonaldi, M., Chen, X., Kunova, A., Pizzatti, C., Saracchi, M., Cortesi, P. 2015. Colonization of lettuce rhizosphere and roots by tagged *Streptomyces*. *Frontiers in Microbiology* 6:25.

Bourgou, S., Kchouk, M., Bellila, A., Marzouk, B. 2010. Effect of salinity on phenolic composition and biological activity of *Nigella sativa*. *Acta Horticulturae* 853:57–60.

Brooker, N., Windorski, J., Blumi, E. 2008. Halogenated coumarins derivatives as novel seed protectants. *Communication in Agriculture and Applied Biological Sciences* 73(2):81–89.

Butkhup, L., Chowtivannakul, S., Gaensakoo, R., Prathepha, P., Samappito, S. 2016. Study of the phenolic composition of Shiraz red grape cultivar (*Vitis vinifera* L.) cultivated in north-eastern Thailand and its antioxidant and antimicrobial activity. *South African Journal of Enology and Viticulture* 31(2):89–98.

Calvo-Polanco, M., Sánchez-Romera, B., Aroca, R., Asins, M.J., Declerck, S., Dodd, I.C., Martínez-Andújar, C., Albacete, A., Ruiz-Lozano, J.M. 2016. Exploring the use of recombinant inbred lines in combination with beneficial microbial inoculants (AM fungus and PGPR) to improve drought stress tolerance in tomato. *Environmental and Experimental Botany* 131:47–57.

Chapman, E.J., Estelle, M. 2009. Mechanism of auxin-regulated gene expression in plants. *Annual Review of Genetics* 43:265–285.

Chatterjee, P., Samaddar, S., Anandham, R., Kang, Y., Kim, K., Selvakumar, G., Sa, T. 2017. Beneficial soil bacterium *Pseudomonas frederiksbergensis* OS261 augments salt tolerance and promotes red pepper plant growth. *Frontiers in Plant Science* 8:705.

Chinnusamy, V., Zhu, J., Zhu, J.K. 2007. Cold stress regulation of gene expression in plants. *Trends in Plant Science* 12(10):444–451.

Cohen, A.C, Bottini, R., Pontin, M., Berli, F.J., Moreno, D., Boccanlandro, H., Travaglia, C.N., Piccoli, P.N. 2015. *Azospirillum brasilense* ameliorates the response of *Arabidopsis thaliana* to drought mainly via enhancement of ABA levels. *Physiologia Plantarum* 153(1):79–90.

Creus, C.M., Sueldo, R.J., Barassi, C.A. 2004. Water relations and yield in *Azospirillum*-inoculated wheat exposed to drought in the field. *Canadian Journal of Botany* 82(2):273–281.

de Castro, E.M., Pinto, J.E.B.P., Bertolucci, S.K.V., Malta, M.R., Cardoso, M.G., Silva, F.A.M. 2006. Coumarin contents in young *Mikania glomerata* plants (guaco) under different radiation levels and photoperiod. *Acta Farmacéutica Bonaerense* 25(3):387–392.

Dey, P.M., Harborne, J.B. 1997. *Plant Biochemistry*. London: Academic Press.

Dixit, V.K., Misra, S., Mishra, S.K., Tewari, S.K., Joshi, N., Chauhan, P.S. 2020. Characterization of plant growth-promoting alkalotolerant *Alcaligenes* and *Bacillus* strains for mitigating the alkaline stress in *Zea mays*. *Antonie Van Leeuwenhoek* 113: 889–905.

Eyidogan, F., Oz, M.T., Yucel, M., Oktem, H.A. 2012. Signal transduction of phytohormones under abiotic stresses. In: Khan, N.A. (Ed.), *Phytohormones and Abiotic Stress Tolerance in Plants*. Berlin: Springer, pp. 1–48.

Facchini, P.J. 2001. Alkaloid biosynthesis in plants: Biochemistry, cell biology, molecular regulation, and metabolic engineering applications. *Annual Review of Plant Biology* 52(1):29–66.

Fang, X., Yang, C.-Q., Wei, Y.-K., Ma, Q.-X., Yang, L., Chen, X.Y. 2011. Genomics grand for diversified plant secondary metabolites. *Research in Plant Disease* 33(1):53–64.

Flora, S.J.S., Shrivastava, R., Mittal, M. 2013. Chemistry and pharmacological properties of some natural and synthetic antioxidants for heavy metal toxicity. *Current Medicinal Chemistry* 20(36):4540–4574.

García, J.E., Maroniche, G., Creus, C., Suárez-Rodríguez, R., Ramirez-Trujillo, J.A., Groppa, M.D. 2017. *In vitro* PGPR properties and osmotic tolerance of different *Azospirillum* native strains and their effects on growth of maize under drought stress. *Microbiological Research* 202:21–29.

Giménez, M.J., Valverde, J.M., Valero, D., Guillén, F., Martínez-Romero, D., Serrano, M., Castillo, S. 2014. Quality and antioxidant properties on sweet cherries as affected by preharvest salicylic and acetylsalicylic acids treatments. *Food Chemistry* 160:226–232.

Glick, B.R. 2012. Plant growth-promoting bacteria: Mechanisms and applications. *Scientifica* 2012: 1–15.

Grover, M., Ali, S.Z., Sandhya, V., Rasul, A., Venkateswarlu, B. 2011. Role of microorganisms in adaptation of agriculture crops to abiotic stresses. *World Journal of Microbiology and Biotechnology* 27(5):1231–1240.

Gutiérrez-García, K., Neira-González, A., Pérez-Gutiérrez, R.M. et al. 2017. Phylogenomics of 2,4-diacetylphloroglucinol-producing *Pseudomonas* and novel antiglycation endophytes from *Piper auritum*. *Journal of Natural Products* 80(7):1955–1963.

Haghighi, Z., Modarresi, M., Mollayi, S. 2012. Enhancement of compatible solute and secondary metabolites production in *Plantago ovata* Forsk. by salinity stress. *Journal of Medicinal Plant Research* 6(18):3495–3500.

Hartmann, T. 2004. Plant-derived secondary metabolites as defensive chemicals in herbivorous insects: A case study in chemical ecology. *Planta* 219:1–4.

Huang, X., Bie, Z. 2010. Cinnamic acid-inhibited ribulose-1,5-bisphosphate carboxylase activity is mediated through decreased spermine and changes in the ratio of polyamines in cowpea. *Journal of Plant Physiology* 167(1):47–53.

Islam, S., Akanda, A.M., Prova, A., Islam, M.T., Hossain, M.M. 2016. Isolation and identification of plant growth promoting rhizobacteria from cucumber rhizosphere and their effect on plant growth promotion and disease suppression. *Frontiers in Microbiology* 6:1360.

Jochum, G.M., Mudge, K.W., Thomas, R.B. 2007. Elevated temperatures increase leaf senescence and root secondary metabolite concentrations in the understory herb *Panax quinquefolius* (Araliaceae). *American Journal of Botany* 94(5):819–826.

Kapoor, R., Gupta, M.K., Kumar, N., Kanwar, S.S. 2017. Analysis of nhaA gene from salt tolerant and plant growth promoting *Enterobacter ludwigii*. *Rhizosphere* 4:62–69.

Kasotia, A., Varma, A., Choudhary, D.K. 2015. *Pseudomonas*-mediated mitigation of salt stress and growth promotion in *Glycine max*. *Agricultural Research* 4(1):31–41.

Kelly, J.R., Kennedy, P.J., Cryan, J.F., Dinan, T.G., Clarke, G., Hyland, N.P. 2015. Breaking down the barriers: The gut microbiome, intestinal permeability and stress-related psychiatric disorders. *Frontiers in Cellular Neuroscience* 9:392.

Khalid, K.A. 2006. Influence of water stress on growth, essential oil, and chemical composition of herbs (*Ocimum* sp.). *International Agrophysics* 20(4):289–296.

Khalil, A. 2017. Role of biotechnology in alkaloids production. In: Naeem, M. et al. (Eds.), *Catharanthus Roseus: Current Research and Future Prospects*. Cham: Springer, pp. 59–70.

Kilani-Feki, O., Khiari, O., Culioli, G., Ortalo-Magné, A., Zouari, N., Blache, Y., Jaoua, S. 2010. Antifungal activities of an endophytic *Pseudomonas fluorescens* strain Pf1TZ harbouring genes from pyoluteorin and phenazine clusters. *Biotechnology Letters* *32*(9):1279–1285.

Konkolewska, A., Piechalak, A., Ciszewska, L., Antos-Krzemińska, N., Skrzypczak, T., Hanć, A., Sitko, K., Małkowski, E., Barałkiewicz, D., Małecka, A. 2020. Combined use of companion planting and PGPR for the assisted phytoextraction of trace metals (Zn, Pb, Cd). *Environmental Science and Pollution Research 27(12)*:1–17.

Krasensky, J., Jonak, C. 2012. Drought, salt, and temperature stress-induced metabolic rearrangements and regulatory networks. *Journal of Experimental Botany* 63(4):1593–1608.

Ksouri, R., Megdiche, W., Debez, A., Falleh, H., Grignon, C., Abdelly, C. 2007. Salinity effects on polyphenol content and antioxidant activities in leaves of the halophyte *Cakile maritima*. *Plant Physiology and Biochemistry* 45(3–4):244–249.

Kurepin, L.V., Ivanov, A.G., Zaman, M., Pharis, R.P., Hurry, V., Hüner, N.P.A. 2017. Interaction of glycine betaine and plant hormones: Protection of the photosynthetic apparatus during abiotic stress. In: Hou, H.J.M. et al. (Eds.), *Photosynthesis: Structures, Mechanisms, and Applications*. Cham: Springer, pp. 185–202.

Kusari, S., Hertweck, C., Spiteller, M. 2012. Chemical ecology of endophytic fungi: Origins of secondary metabolites. *Chemistry & Biology* 19(7):792–798.

Kwon, Y.S., Ryu, C.M., Lee, S., Park, H.B., Han, K.S., Lee, J.H., Lee, K., Chung, W.S., Jeong, M.J., Kim, H.K., Bae, D.W. 2010. Proteome analysis of *Arabidopsis* seedlings exposed to bacterial volatiles. *Planta 232*(6):1355–1370.

Lata, R., Chowdhury, S., Gond, S.K., White Jr., J.F. 2018. Induction of abiotic stress tolerance in plants by endophytic microbes. *Letters in Applied Microbiology* 66(4):268–276.

Lisar, S.Y.S., Motafakkerazad, R., Hossain, M.M., Rahman, I.M.M. 2012. Water stress in plants: Causes, effects and responses In: Ismail, Md. Rahman, M., Hasegawa, H. (Eds.), *Water Stress*. United Kingdom: IntechOpen, p. 27305.

Liu, W., Hou, J., Wang, Q., Ding, L., Luo, Y. 2014. Isolation and characterization of plant growth-promoting rhizobacteria and their effects on phytoremediation of petroleum-contaminated saline-alkali soil. *Chemosphere* 117:303–308.

Liu, F., Ma, H., Peng, L., Du, Z., Ma, B., Liu, X. 2019. Effect of the inoculation of plant growth-promoting rhizobacteria on the photosynthetic characteristics of *Sambucus williamsii* Hance container seedlings under drought stress. *AMB Express* 9(1):1–9.

Loper, J.E., Henkels, M.D., Shaffer, B.T., Valeriote, F.A., Gross, H. 2008. Isolation and identification of rhizoxin analogs from *Pseudomonas fluorescens* Pf-5 by using a genomic mining strategy. *Applied and Environmental Microbiology* 74(10):3085–3093.

Mauch-Mani, B., Mauch, F. 2005. The role of abscisic acid in plant–pathogen interactions. *Current Opinion in Plant Biology* 8(4):409–414.

Mayak, S., Tirosh, T., Glick, B.R. 2004. Plant growth-promoting bacteria confer resistance in tomato plants to salt stress. *Plant Physiology and Biochemistry* 42(6):565–572.

Mazid, M., Khan, T.A., Mohammad, F. 2011. Role of secondary metabolites in defense mechanisms of plants. *Biology and Medicine* 3(2):232–249.

Mehnaz, S., Saleem, R.S.Z., Yameen, B., Pianet, I., Schnakenburg, G., Pietraszkiewicz, H., Valeriote, F., Josten, M., Sahl, H.G., Franzblau, S.G., Gross, H. 2013. Lahorenoic acids A–C, ortho-dialkyl-substituted aromatic acids from the biocontrol strain *Pseudomonas aurantiaca* PB-St2. *Journal of Natural Products* 76(2):135–141.

Mirete, S., Morgante, V., González-Pastor, J.E. 2016. Functional metagenomics of extreme environments. *Current Opinion in Biotechnology* 38:143–149.

Misra, S., Dixit, V.K., Khan, M.H., Mishra, S.K., Dviwedi, G., Yadav, S., Lehri, A., Chauhan, P.S. 2017. Exploitation of agro-climatic environment for selection of 1-aminocyclopropane-1-carboxylic acid (ACC) deaminase producing salt tolerant indigenous plant growth promoting rhizobacteria. *Microbiological Research* 205:25–34.

Mona, S.A., Hashem, A., Abd-Allah, E.F., Alqarawi, A.A., Soliman, D.W.K., Wirth, S., Egamberdieva, D. 2017. Increased resistance of drought by *Trichoderma harzianum* fungal treatment correlates with increased secondary metabolites and proline content. *Journal of Integrative Agriculture* 16(8):1751–1757.

Naik, P.M., Al-Khayri, J.M. 2016. Abiotic and biotic elicitors–role in secondary metabolites production through in vitro culture of medicinal plants. In: Shanker, A.K., Shanker, C. (Eds.), *Abiotic and Biotic Stress in Plants-recent Advances and Future Perspectives*. Rijeka: IntechOpen, pp. 247–277.

Naveed, M., Mitter, B., Reichenauer, T.G., Wieczorek, K., Sessitsch, A. 2014. Increased drought stress resilience of maize through endophytic colonization by *Burkholderia phytofirmans* PsJN and *Enterobacter* sp. FD17. *Environmental and Experimental Botany* 97:30–39.

Nell, M., Vötsch, M., Vierheilig, H., Steinkellner, S., Zitterl-Eglseer, K., Franz, C., Novak, J. 2009. Effect of phosphorus uptake on growth and secondary metabolites of garden sage (*Salvia officinalis* L.). *Journal of the Science of Food and Agriculture* 89(6):1090–1096.

Ning, P., Qiu, J., Wang, X., Liu, W., Chen, W. 2013. Metal loaded zeolite adsorbents for hydrogen cyanide removal. *Journal of Environmental Sciences* 25(4):808–814.

Niu, X., Song, L., Xiao, Y., Ge, W. 2018. Drought-tolerant plant growth-promoting rhizobacteria associated with foxtail millet in a semi-arid agroecosystem and their potential in alleviating drought stress. *Frontiers in Microbiology* 8:2580.

Olszewski, N., Sun, T.P., Gubler, F. 2002. Gibberellin signaling biosynthesis, catabolism, and response pathways. *Plant Cell* 14:S61–S80.

Ortíz-Castro, R., Contreras-Cornejo, H.A., Macías-Rodríguez, L., López-Bucio, J. 2009. The role of microbial signals in plant growth and development. *Plant Signaling & Behavior* 4(8):701–712.

Oueslati, S., Karray-Bouraoui, N., Attia, H., Rabhi, M., Ksouri, R., Lachaal, M. 2010. Physiological and antioxidant responses of *Mentha pulegium* (Pennyroyal) to salt stress. *Acta Physiologiae Plantarum* 32(2):289–296.

Pankaj, U., Singh, D.N., Mishra, P., Gaur, P., Vivekbabu, C.S., Shanker, K., Verma, R.K. 2020. Autochthonous halotolerant plant growth promoting rhizobacteria promote bacoside A yield of *Bacopa monnieri* (L) Nash and phytoextraction of salt-affected soil. *Pedosphere* 30(5):671–683.

Pankaj, U., Singh, D.N., Singh, G., Verma, R.K. 2019a. Microbial inoculants assisted growth of *Chrysopogon zizanioides* promotes phytoremediation of salt affected soil. *Indian Journal of Microbiology* 59(2):137–146.

Pankaj, U., Verma, R.S., Yadav, A., Verma, R.K. 2019b. Effect of arbuscular mycorrhizae species on essential oil yield and chemical composition of commercially grown palmarosa (*Cymbopogon martinii*) varieties in salinity stress soil. *Journal of Essential Oil Research* 31(2):145–153.

Paponov, I.A., Paponov, M., Teale, W., Menges, M., Chakrabortee, S., Murray, J.A.H., Palme, K. 2008. Comprehensive transcriptome analysis of auxin responses in *Arabidopsis*. *Molecular Plant* 1(2):321–337.

Patel, S., Jinal, H.N., Amaresan, N. 2017. Isolation and characterization of drought resistance bacteria for plant growth promoting properties and their effect on chilli (*Capsicum annuum*) seedling under salt stress. *Biocatalysis and Agricultural Biotechnology* 12:85–89.

Patra, B., Schluttenhofer, C., Wu, Y., Pattanaik, S., Yuan, L. 2013. Transcriptional regulation of secondary metabolite biosynthesis in plants. *Biochimica et Biophysica Acta* 1829(11):1236–1247.

Pereira, S.I.A., Abreu, D., Moreira, H., Vega, A., Castro, P.M.L. 2020. Plant growth-promoting rhizobacteria (PGPR) improve the growth and nutrient use efficiency in maize (*Zea mays* L.) under water deficit conditions. *Heliyon* 6(10):e05106.

Pierson, L.S., Pierson, E.A. 2010. Metabolism and function of phenazines in bacteria: Impacts on the behavior of bacteria in the environment and biotechnological processes. *Applied Microbiology and Biotechnology* 86(6):1659–1670.

Pieterse, C.M., Zamioudis, C., Berendsen, R.L., Weller, D.M., Van Wees, S.C., Bakker, P.A. 2014. Induced systemic resistance by beneficial microbes. *Annual Review of Phytopathology* 52:347–375.

Posmyk, M.M., Kontek, R., Janas, K.M. 2009. Antioxidant enzymes activity and phenolic compounds content in red cabbage seedlings exposed to copper stress. *Ecotoxicology and Environmental Safety* 72(2):596–602.

Preininger, C., Sauer, U., Bejarano, A., Berninger, T. 2018. Concepts and applications of foliar spray for microbial inoculants. *Applied Microbiology and Biotechnology* 102(17):7265–7282.

Ruiz, B., Chávez, A., Forero, A., García-Huante, Y., Romero, A., Sánchez, M., Rocha, D., Sánchez, B., Rodríguez-Sanoja, R., Sánchez, S., Langley, E. 2010. Production of microbial secondary metabolites: Regulation by the carbon source. *Critical Reviews in Microbiology* 36(2):146–167.

Said-Al Ahl, H., Omer, E. 2011. Medicinal and aromatic plants production under salt stress: A review. *Herba Polonica* 57(2):72–87.

Saleem, A.R., Bangash, N., Mahmood, T., Khalid, A., Centritto, M., Siddique, M.T. 2015. Rhizobacteria capable of producing ACC deaminase promote growth of velvet bean (*Mucuna pruriens*) under water stress condition. *International Journal of Agriculture and Biology* 17(3):663–667.

Sandhya, V.Z.A.S., Grover, M., Reddy, G., Venkateswarlu, B. 2009. Alleviation of drought stress effects in sunflower seedlings by the exopolysaccharides producing *Pseudomonas putida* strain GAP-P45. *Biology and Fertility of Soils* 46(1):17–26.

Sarabia, M., Cazares, S., González-Rodríguez, A., Mora, F., Carreón-Abud, Y., Larsen, J. 2018. Plant growth promotion traits of rhizosphere yeasts and their response to soil characteristics and crop cycle in maize agroecosystems. *Rhizosphere* 6:67–73.

Saravanakumar, D., Samiyappan, R. 2007. ACC deaminase from *Pseudomonas fluorescens* mediated saline resistance in groundnut (*Arachis hypogea*) plants. *Journal of Applied Microbiology* 102(5):1283–1292.

Sarkar, A., Ghosh, P.K., Pramanik, K., Mitra, S., Soren, T., Pandey, S., Mondal, M.H., Maiti, T.K. 2018. A halotolerant *Enterobacter* sp. displaying ACC deaminase activity promotes rice seedling growth under salt stress. *Research in Microbiology* 169(1):20–32.

Sarma, M.V.R.K., Kumar, V., Saharan, K., Srivastava, R., Sharma, A.K., Prakash, A., Sahai, V. Bisaria, V.S. 2011. Application of inorganic carrier-based formulations of fluorescent pseudomonads and *Piriformospora indica* on tomato plants and evaluation of their efficacy. *Journal of Applied Microbiology* 111(2):456–466.

Save, R., de Herralde, F., Codina, C., Sanchez, X., Biel, C. 2007. Effects of atmospheric carbon dioxide fertilization on biomass and secondary metabolites of some plant species with pharmacological interest under greenhouse conditions. *Afinidad* 64(528):237–241.

Shahid, I., Mehnaz, S. 2020. Microbial secondary metabolites: Effectual armors to improve stress survivability in crop plants. In: Singh, J.S., Vimal, S.R. (Eds.), *Microbial Services in Restoration Ecology*. Elsevier, pp. 47–70.

Shahzad, R., Khan, A.L., Bilal, S., Waqas, M., Kang, S.M., Lee, I.J. 2017. Inoculation of abscisic acid-producing endophytic bacteria enhances salinity stress tolerance in *Oryza sativa*. *Environmental and Experimental Botany* 136:68–77.

Sharma, S., Kulkarni, J., Jha, B. 2016. Halotolerant rhizobacteria promote growth and enhance salinity tolerance in peanut. *Frontiers in Microbiology* 7:1600.

Sheibani-Tezerji, R., Rattei, T., Sessitsch, A., Trognitz, F., Mitter, B. 2015. Transcriptome profiling of the endophyte *Burkholderia phytofirmans* PsJN indicates sensing of the plant environment and drought stress. *MBio* 6(5):e00621-15.

Shitole, S., Dhumal, K. 2012. Influence of foliar application of micronutrients on photosynthetic pigments and organic constituents of medicinal plant *Cassia agustifolia* Vahl. *Annals of Biological Research* 3(1):520–526.

Shrivastava, P., Kumar, R. 2015. Soil salinity: A serious environmental issue and plant growth promoting bacteria as one of the tools for its alleviation. *Saudi Journal of Biological Sciences* 22(2):123–131.

Siddikee, M.A., Sundaram, S., Chandrasekaran, M., Kim, K., Selvakumar, G., Sa, T. 2015. Halotolerant bacteria with ACC deaminase activity alleviate salt stress effect in canola seed germination. *Journal of the Korean Society for Applied Biological Chemistry* 58(2):237–241.

Singh, R., Kumar, M., Mittal, A., Mehta, P.K. 2017. Microbial metabolites in nutrition, healthcare and agriculture. *3 Biotech* 7(1):15.

Sreevidya, V.S., Srinivasa, R.C., Rao, C., Sullia, S.B., Ladha, J.K., Reddy, P.M. 2006. Metabolic engineering of rice with soyabean isoflavone synthase for promoting nodulation gene expression in rhizobia. *Journal of Experimental Botany* 57(9):957–1969.

Štajner, D., Kevrešan, S., Gašić, O., Mimica-Dukić, N., Zongli, H. 1997. Nitrogen and *Azotobacter chroococcum* enhance oxidative stress tolerance in sugar beet. *Biologia Plantarum* 39(3):441.

Subramanian, P., Kim, K., Krishnamoorthy, R., Mageswari, A., Selvakumar, G., Sa, T. 2016. Cold stress tolerance in psychrotolerant soil bacteria and their conferred chilling resistance in tomato (*Solanum lycopersicum* Mill.) under low temperatures. *PLoS ONE* 11(8):e0161592.

Subramanian, P., Mageswari, A., Kim, K., Lee, Y., Sa, T. 2015. Psychrotolerant endophytic *Pseudomonas* sp. strains OB155 and OS261 induced chilling resistance in tomato plants (*Solanum lycopersicum* Mill.) by activation of their antioxidant capacity. *Molecular Plant-Microbe Interactions* 28(10):1073–1081.

Szczałba, M., Kopta, T., Gąstoł, M., Sękara, A. 2019. Comprehensive insight into arbuscular mycorrhizal fungi, *Trichoderma* spp. and plant multilevel interactions with emphasis on biostimulation of horticultural crops. *Journal of Applied Microbiology* 127(3):630–647.

Taârit, M.B., Msaada, K., Hosni, K., Marzouk, B. 2012. Physiological changes, phenolic content and antioxidant activity of *Salvia officinalis* L. grown under saline conditions. *Journal of the Science of Food and Agriculture* 92(8):1614–1619.

Taârit, M.B., Msaada, K., Hosni, K., Marzouk, B. 2010. Changes in fatty acid and essential oil composition of sage (*Salvia officinalis* L.) leaves under NaCl stress. *Food Chemistry*, 119(3):951–956.

Taiz, L., Zeiger, E. 2006. Secondary metabolites and plant defense. In: Taiz, L., Zeiger, E. (Eds.), *Plant Physiology*. Sunderland: Sinauer Associates Inc., pp. 283–308.

Theocharis, A., Clément, C., Barka, E.A. 2012. Physiological and molecular changes in plants grown at low temperatures. *Planta* 235(6):1091–1105.

Thimmappa, R., Geisler, K., Louveau, T., O'Maille, P., Osbourn, A. 2014. Triterpene biosynthesis in plants. *Annual Review of Plant Biology* 65:225–257.

Timmusk, S., Abd El-Daim, I.A., Copolovici, L., Tanilas, T., Kännaste, A., Behers, L., Nevo, E., Seisenbaeva, G., Stenström, E. Niinemets, Ü. 2014. Drought-tolerance of wheat improved by rhizosphere bacteria from harsh environments: Enhanced biomass production and reduced emissions of stress volatiles. *PLoS ONE* 9(5):e96086.

Tiwari, S., Lata, C., Chauhan, P.S., Nautiyal, C.S. 2016. *Pseudomonas putida* attunes morphophysiological, biochemical and molecular responses in *Cicer arietinum* L. during drought stress and recovery. *Plant Physiology and Biochemistry* 99:108–117.

Tomar, N.S., Ahanger, M.A., Agarwal, R.M. 2014. *Jatropha curcas*: An overview. In: Ahmad, P., Wani, M.R. (Eds.), *Physiological Mechanisms and Adaptation Strategies in Plants under Changing Environment*, Vol. 2. New York: Springer, pp. 361–383.

Upadhyay, S.K., Singh, J.S., Saxena, A.K., Singh, D.P. 2012. Impact of PGPR inoculation on growth and antioxidant status of wheat under saline conditions. *Plant Biology* 14(4):605–611.

Vaishnav, A., Kumari, S., Jain, S., Varma, A., Choudhary, D.K. 2015. Putative bacterial volatile-mediated growth in soybean (*Glycine max* L. Merrill) and expression of induced proteins under salt stress. *Journal of Applied Microbiology* 119(2):539–551.

Verma, N., Shukla, S. 2015. Impact of various factors responsible for fluctuation in plant secondary metabolites. *Journal of Applied Research on Medicinal and Aromatic Plants* 2(4):105–113.

Vimal, S.R., Patel, V.K., Singh, J.S. 2019. Plant growth promoting *Curtobacterium albidum* strain SRV4: An agriculturally important microbe to alleviate salinity stress in paddy plants. *Ecological Indicators* 105:553–562.

Vimal, S.R., Singh, J.S., Arora, N., Singh, D.P. 2016. PGPR: An effective bio-agent in stress agricultural management. In: Sharma, B.K., Jain, A. (Eds.), *Microbial Empowerment in Agriculture*. New Delhi: Biotech Books, pp. 81–108.

Vimal, S.R., Singh, J.S., Arora, N.K., Singh, S. 2017. Soil-plant-microbe interactions in stressed agriculture management: A review. *Pedosphere* 27(2):177–192.

Wang, C.J., Yang, W., Wang, C., Gu, C., Niu, D.D., Liu, H.X., Wang, Y.P., Guo, J.H. 2012. Induction of drought tolerance in cucumber plants by a consortium of three plant growth-promoting rhizobacterium strains. *PLoS One* 7(12):e52565.

Wani, S.H., Kumar, V., Shriram, V., Sah, S.K. 2016. Phytohormones and their metabolic engineering for abiotic stress tolerance in crop plants. *Crop Journal* 4(3):162–176.

Wolters, H., Jurgens, G. 2009. Survival of the flexible: Hormonal growth control and adaptation in plant development. *Nature Reviews Genetics* 10(5):305–317.

Yadav, S.K. 2010. Cold stress tolerance mechanisms in plants: A review. *Agronomy for Sustainable Development* 30(3):515–527.

Yasmin, H., Nosheen, A., Naz, R., Bano, A., Keyani, R. 2017. L-tryptophan-assisted PGPR-mediated induction of drought tolerance in maize (*Zea mays* L.). *Journal of Plant Interactions* 12(1):567–578.

Zhang, H., Kim, M.S., Sun, Y., Dowd, S.E., Shi, H., Paré, P.W. 2008. Soil bacteria confer plant salt tolerance by tissue-specific regulation of the sodium transporter HKT1. *Molecular Plant-Microbe Interactions* 21(6):737–744.

Zhang, H., Murzello, C., Sun, Y., Kim, M.S., Xie, X., Jeter, R.M., Zak, J.C., Dowd, S.E., Paré, P.W. 2010. Choline and osmotic-stress tolerance induced in *Arabidopsis* by the soil microbe *Bacillus subtilis* (GB03). *Molecular Plant-microbe Interactions* 23(8):1097–1104.

Zhang, L., Wang, Q., Guo, Q., Chang, Q., Zhu, Z., Liu, L., Xu, H. 2012. Growth, physiological characteristics and total flavonoid content of *Glechoma longituba* in response to water stress. *Journal of Medicinal Plant Research* 6(6):1015–1024.

Zobayed, S.M.A., Afreen, F., Kozai, T. 2007. Phytochemical and physiological changes in the leaves of St. John's wort plants under a water stress condition. *Environmental and Experimental Botany* 59(2):109–116.

7 *Bacillus*-Based Biocontrol

Practical Applications for Sustainable Agriculture

Shweta Singh
University of Minho, Braga, Portugal

Lav Sharma
Universidade de Trás-os-Montes e Alto Douro
Vila Real, Portugal
and
Battelle UK Ltd., Havant, United Kingdom

Rupesh Kumar Singh
Estrada de Gil Vaz, Elvas, Portugal
and
University of Minho, Braga, Portugal

Vishnu D. Rajput, Svetlana Sushkova, and Tatiana Minkina
Southern Federal University, Rostov-on-Don, Russia

Krishan K. Verma
Guangxi Academy of Agricultural Sciences, Guangxi, China

CONTENTS

7.1 INTRODUCTION

Plant diseases from microorganisms result in a significant loss in crop yield (Savary et al. 2012). The use of synthetic pesticides to reduce these diseases has been an effective strategy in the past (Keswani et al. 2014). However, the increased use of

DOI: 10.1201/9781003147077-7

these synthetic pesticides has an adverse effect on the environment (Aktar et al. 2009). To achieve agricultural sustainability, the integrated crop management system (ICMS) is recommended (Souza et al. 2015). Biocontrol of plant diseases is a key component of ICMS, which implies the use of living organisms to control the population of plant pathogens with the most negligible negative impact on the natural ecosystem (Heimpel and Mills 2017; Sharma et al. 2018).

Among all the biocontrol agents, *Bacillus* sp. is the most commonly used and well-investigated biocontrol agent. They directly antagonize the plant pathogens by competing with them for nutrients and niche, by producing antimicrobial compounds (lipopeptides, antibiotics, and lytic enzymes) (Beneduzi et al. 2012; Mnif and Ghribi 2015). *Bacillus* sp. can also act as a biofertilizer by enabling the uptake of some nutrients from the environment through nitrogen fixation, and phosphorous solubilization, etc., or by secreting the plant growth regulator in the rhizosphere (Beneduzi et al. 2012). Hence, *Bacillus* sp. presents a substitute to synthetic pesticides and fertilizers. The favorable effect of *Bacillus* sp. has been demonstrated in several crops such as maize, wheat, soybean, tomato, sunflower, cucumber, and potato (Aloo et al. 2019).

Genus *Bacillus* constitutes over 200 species and subspecies from the Firmicutes phylum. They are gram-positive, rod-shaped, and aerobic or facultative anaerobic bacteria (Logan et al. 2009). They are omnipresent, including soil and rhizosphere, where they comprise approximately up to 95% of the gram-positive bacterial population (Prashar et al. 2013). Bacilli are a large group of bacteria. Many species of bacilli and their derived products are safer to use in the natural environment (Bhattacharyya et al. 2016). This chapter mainly focuses on bacilli and their role as a plant growth-promoting bacteria, the bioactives they show as a biocontrol agent, development of strategies for large-scale fermentation, and the strategies for their formulation as a biofertilizer.

7.2 FERMENTATION OPTIMIZATION

The fermentation process has achieved the cultivation of bacterial cells in controlled conditions in a culture medium. Fermentation is the most basic and essential step in developing a potential biocontrol or biostimulant agent in a cost-effective manner for commercial or large-scale purposes in modern agriculture (Figure 7.1). Crafting a standard operation procedure for optimized CFU value at a reasonable cost is a grueling and time-consuming process. Step-by-step procedure with an intense investigation in different parameters during the growth process may facilitate and ensure the optimized methodology development. *Bacillus* species such as *B. amyloliquefaciens*, *B. subtilis*, and *B. licheniformis* are among the best bacteria for producing secreted proteins, and their yields can go up to 20–25 g L^{-1} (van Dijl et al. 2013). In addition, the lack of toxic by-products and ease of genetic engineering have made *B. subtilis* a bacterium of choice for industrial applications (Earl et al. 2008). The presence of high-quality genomes and genome-wide gene function analysis has also made *B. subtilis* an easy choice for fermentation (Harwood 1992; Kunst et al. 1997; Barbe et al. 2009).

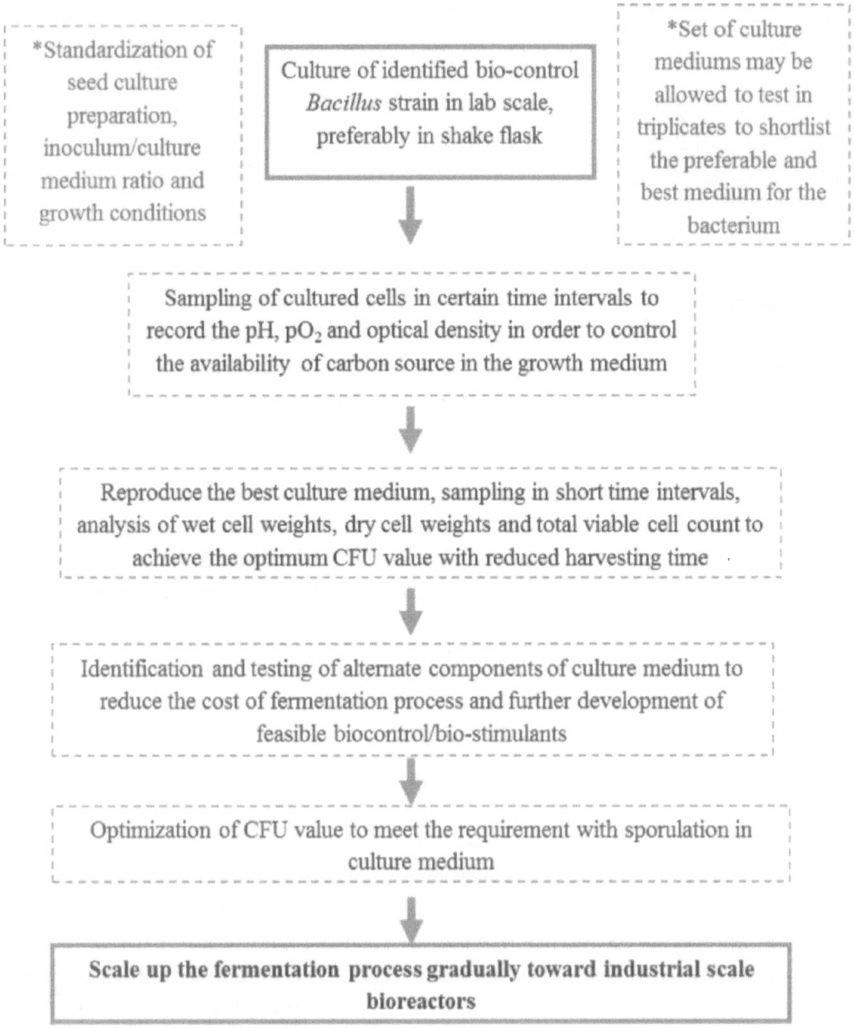

FIGURE 7.1 Diagrammatic step-by-step representation to develop a standard operating procedure for a fermentation method of biocontrol/biostimulant bacterial cells.

B. subtilis degrade numerous substrates that eventually assist the bacterium in surviving in changing environment. Its highly adoptive metabolism makes it easier to grow on cheaper substrates (Su et al. 2020). Compared to *Saccharomyces cerevisiae*, the fermentation cycle of the former is only 48 h in comparison to 180 h of the latter. It also exhibits advantages over *Escherichia coli* as the single-cell membrane of *B. subtilis* facilitates protein secretion and simplifies downstream processing that eventually reduces production cost. Most importantly, *B. subtilis* is generally regarded as safe (Su et al. 2020).

In terms of strategies for the large-scale fermentation of *Bacillus*, Yeh et al. (2006) designed a relatively innovative bioreactor for surfactin, a lipopeptide produced by

B. subtilis ATCC 21332 exhibiting inhibitory effects against a few plant pathogens. Surfactin concentration could reach up to 6.45 g L^{-1}. *B. subtilis* BS-37 could also be used for surfactin production through batch fermentation in a 5-L stirred tank bioreactor, when it was immobilized onto a piece of a cotton towel (Yi et al. 2016). Matar et al. (2009) studied the use of *B. subtilis* strain G-GANA7 as a biological control agent against phytopathogens. To enhance the biomass production and the biocontrol activity of the supernatant, they used batch fermentation where the yields could go up to 3.2 g L^{-1}.

Ahmad et al. (2019) performed the batch fermentation of *B. subtilis* strain B4 as a biocontrol agent. During the exponential phase, a fixed specific growth rate of 0.3 h^{-1} was achieved and a yield of 0.69 g cell g^{-1} glucose was achieved, with maximum biomass yield reaching up to 4.53 g L^{-1} at 11 h. During the exponential growth phase, the gradual depletion of glucose, and the continuous butyric acid production as a by-product, the batch pH decreases. Wei et al. (2003) suggested that the acidic metabolites production during fermentation of surfactin is the responsible factor for the decrease in pH. Later, after consuming glucose, a sporulation stage occurs where the secreted acids are oxidized. The bacterial oxygen demand is low, increasing the levels of dissolved oxygen and an eventual decrease in the biomass. Matar et al. (2009) also observed a growth rate of 0.3 h^{-1} during the exponential phase in the batch fermentation of *B. subtilis* strain G-GANA7. However, the final biomass yield was 0.45 g cell g^{-1} glucose.

7.3 BIOCONTROL FORMULATION

Formulation of biocontrol agent preserves and may enhance the shelf life and ease the application in the field with improved bioactivity. The successful formulation is one of the significant steps during potential biocontrol agent development after a screening of microbial cells and optimizing large-scale quality production of cells by the fermentation process. Formulation of *Bacillus* cell cultures for biocontrol and biostimulant commercial use includes a variety of additives in various states of matter, such as liquids (inverse emulsions by using vegetable oils and water), dusty and granulated (kaolinite clay, diatomaceous earth, grain flours) or encapsulated in beads/microcapsules form. Several other materials also have been considered to meet the requirement of biocontrol agents, for example, sodium benzoate lactose has been used as stabilizers and peptones for nutrients. Arabic gums may also be used as binders, silica gels, anhydrous salts as desiccants, xanthan gums as thickeners, pre-gelatinized corn flour as stickers, microcrystalline cellulose as dispersants, and Tween 80 as surfactants. *Bacillus* strains produce spores and allow scientists to use a vast variety of formulating materials due to high durability and resistance in nature (Emmert and Handelsman 1999; Driks 2004).

Invert emulsion consists of two liquid phases where water may be the dispersed phase and oil serves as a continuous phase. This method includes an emulsifier in the ingredient; the food-grade emulsifiers are generally recommended to avoid toxicity in the final crop products. This method is widely used to form biocontrol agents for plant diseases, insect pests, and noxious weed control. Incorporating water by invert emulsion allows this formulation for application in hot and dry conditions, as water is retained inside the oil film and does not evaporate during application in the field (Batta 2016). Chemical products are being challenged worldwide to reduce the impact of pesticide use, industry pressure to address health and safety issues with their workers, and a growing consumers need for chemical residue-free food. The development of nonchemical disease control is essential for the food industry to remain sustainable and ensure long-term healthy food security.

Encapsulation of microbial cells has been reported as one of the most efficient formulation methods (Rathore et al. 2013). The encapsulation may be in microbeads or microcapsules; beads represent a matrix material where cells are dispersal entrapped while microcapsules represent a core of cells surrounded by the wall material. The size of the beads or capsules may vary according to the need; however, the smaller size of beads/capsules is recommended for efficient diffusion of oxygen, nutrients, and metabolites to ensure the viable cell density and efficacy of biocontrol agents. *Bacillus velezensis* has been encapsulated in different wall materials (sodium alginate, chitosan, and cassava-modified starch) and proved as an efficient formulation method (Luo et al. 2019). Different concentrations of sodium alginate (sterile solution in water) have been used to prepare the mixture and cross-linked by injecting the alginate cells mix into 2% $CaCl_2$. Recent advances in spray drying technology may facilitate the polymerization and drying of beads/capsules in a single-step process. The encapsulation formulation allows the controlled release of biocontrol agents during application in the field while maintaining the cell population during the crop cycle to ensure higher efficacy. Alginate is a polysaccharide obtained from marine algae which form a biodegradable wall material by crosslinking in the presence of $CaCl_2$ (Islam et al. 2010). Alginate (nontoxic, biological origin, biodegradable and biocompatible, non-allergic) properties make it an excellent material for its broad range of use in crops and the food industry. Further, chitosan can be used for value addition due to favorable properties (biological origin, biodegradable, non-toxic, eco-friendly, low cost) to improve the stability and efficacy of formulated beads/capsules. The combination of alginate and chitosan enforce the electrostatic interaction to form alginate-chitosan-alginate microcapsules for higher mechanical stability and structural strength (Li et al. 2005).

Various formulation methods with the inclusion of different matrix materials have been evolved to meet the diverse objectives in the agriculture industry (Cassidy et al. 1996; Bashan 1998; Schisler et al. 2004; John et al. 2011; Vemmer and Patel 2013). Alginate gained high recommendation worldwide among these formulations (Yabur et al. 2007) and it has been proved to develop the plant growth enhancers by encapsulation of microbial cells (Bashan 1998; Young et al. 2006; Yabur et al. 2007; Minaxi 2011), for treatment of wastewater (Cassidy et al. 1996), encapsulation of microbial cells for the food industry (Krasaekoopt et al. 2003; Onwulata 2012; Tripathi and Giri 2014), and pharmaceutical interests (Hunt and Grover 2010; Cook et al. 2012). A higher survival rate of encapsulated *Pseudomonas* fluorescent cells into alginate matrix was recorded (Trevor et al. 1992); additionally shelf life improvement of *Bifidobacterium bifidum* and *Lactobacillus acidophilus* was observed during storage (Özer et al. 2008).

7.4 GENUS *BACILLUS* AS BIOCONTROL AGENTS

The genus *Bacillus* represents a diverse group of pathogenic and non-pathogenic bacteria, which are omnipresent, including soil. These bacteria produce endospores that can survive in adverse environmental conditions (Sharma et al. 2019). *Bacillus* sp. shows predominance in the rhizosphere and the most widespread endophytic bacteria (Silva et al. 2019). These bacteria are preferred for commercial purposes because of their rapid growth in different culture media, their capability to form endospores, and secretion of various bioactive substances (Wu et al. 2015). Moreover, these bacteria grow and flourish in the rhizosphere without disturbing other bacterial populations (Radhakrishnan et al. 2017). *Bacillus* sp. that is used as a biocontrol agent are *B. subtilis*, *B. amyloliquefaciens*, *B. pumilus*, *B. licheniformis*, *B. megaterium*, *B. velezensis*, *B. cereus*, *B. thuringiensis*, etc.

The ability of *Bacillus* sp. to antagonize the effect of phytopathogens is attributed to its property of producing secondary metabolites with antibiotic properties. These are either ribosomally generated low-molecular weight peptide (bacteriocins) or non-ribosomally generated high-molecular weight peptides (polyketides, peptides, and lipopeptides) (Miljaković et al. 2020). Various bacteriocins and bacteriocin-like substances (BLS) (amysin, subtilosin, subtilin, thuricin, etc.) are isolated from *B. amyloliquefaciens*, *B. cereus*, *B. coagulans*, *B. subtilis*, and *B. thuringiensis* (Abriouel et al. 2011). These bacteriocins and BLS have shown antagonistic activities against pathogenic bacteria. *Bacillus* sp. synthesizing the lipopeptides and peptides has shown a wide range of antimicrobial activities (Fira et al. 2018). It has been found that the two most common *Bacillus* species, *B. subtilis* and *B. amyloliquefaciens*, use 4–5% and 8.5% of their respective genetic capacity for synthesizing diverse antimicrobial compounds (Stein 2005; Chen et al. 2009).

Hydrolytic enzymes produced by *Bacillus* sp., such as cellulases, chitinases, chitosanases, glucanases, lipases, and proteases, are also involved in the antimicrobial activity as they effectively hydrolyze the main constituents of bacterial and fungal cell walls. For successful degradation of the microbial cell wall, the mixture of more these hydrolytic enzymes is required (Mardanova et al. 2017). Biocontrol of *Clavibacter michiganesis* by *B. amyloliquefaciens* is due to the lytic enzymes' cellulose, lipase, protease, and chitinase (Gautam et al. 2019). It has been reported that the strains of *Bacillus* sp. which produce hydrolases are more efficient biocontrol agents.

Siderophores, which are also non-ribosomal peptides, also play an important role as a biocontrol of pathogens. Siderophores are metal-chelating peptides synthesized under metal starvation (especially iron) conditions (Khan et al. 2018). Iron is an essential element for various biological processes. Siderophores allow the solubilization and extraction from compounds and reduce their availability for pathogens (Beneduzi et al. 2012).

Bacillus sp. produces a wide variety of siderophores that are associated with plant disease suppression. *B. subtilis* strain-producing siderophore was capable of reducing the incidence of Fusarium wilt (Yu et al. 2011). *B. subtilis* is a potential biocontrol agent against *Bipolaris sorokiniana* for its production of cellulose, chitinase, and siderophores (Villa-Rodríguez et al. 2019). The antimicrobial effect of these compounds on various phytopathogens is summarized in Table 7.1.

TABLE 7.1
Study of Bioactive Compounds Secreted from Various Strains of *Bacillus* sp. and Their Target Pathogens

***Bacillus* Species**	**Plant Disease**	**Target Pathogen**	**Bioactive Compounds**	**References**
B. subtilis	Damping-off of beans	*Pythium ultimum*	Iturin, fengycin	Ongena et al. (2005)
B. licheniformis	Blast disease of rice	*Magnaporthe grisea*	Surfactin	Tendulkar et al. (2007)
B. subtilis	Soilborne diseases in cucumber and pepper	*Fusarium oxysporum* and *Phytophthora capsici*	Bacilysin, iturin	Chung et al. (2008)
B. subtilis	Head blight of wheat	*Fusarium* sp.	Lipopeptides, iturine, fengycin	Dunlap et al. (2011)
B. subtilis	Wilt of pepper	*F. oxysporum*	Siderophores	Yu et al. (2011)
B. subtilis	Bacterial diseases of cucurbits	*Xanthomonas campestris*, *Pectobacterium carotovorum*	Surfactin, iturin, fengycin	Zeriouh et al. (2011)
B. amyloliquefaciens	Head blight of wheat	*Fusarium* sp.	Iturin	Crane et al. (2012)
B. amyloliquefaciens	Wilt of cotton	*Verticillium dahlia*	Iturin	Han et al. (2015)
B. subtilis	Stem canker and black scurf of potato	*Rhizoctonia solani*	Chitinase	Saber et al. (2015)
B. amyloliquefaciens	Wilt disease of tomato	*Fusarium oxysporium*	Protease	Guleria et al. (2016)
B. amyloliquefaciens/ *B. subtilis*	Head blight of wheat	*Fusarium graminearum*	Iturin, surfactin, fengycin	Zalila-Kolsi et al. (2016)
B. amyloliquefaciens	Fungal diseases in field and vegetable crops	*Fusarium oxysporum, F. avenacium, Mucor*	Surfactant, fengycin	Salazar et al. (2017)
B. subtilis	Clove root of garlic	*Fusarium* sp.	Surfactin, lytic enzymes	Bjelić et al. (2018)
B. sp.	Stalk and ear rot of maize	*Fusarium verticillioides*	Protease, glucanase, chitinase, siderophores	Douriet-Gámez et al. (2018)
B. velezensis	Gray mold disease of pepper	*Botrytis cinerea*	Protease, glucanase, chitinase, cellulase	Jiang et al. (2018)
Bacillus sp.	White mold of common bean	*Sclerotinia sclerotiorum*	Surfactin, iturin, fengycin, siderophore	Sabaté et al. (2018)

(Continued)

TABLE 7.1 *(Continued)*
Study of Bioactive Compounds Secreted from Various Strains of *Bacillus* sp. and Their Target Pathogens

***Bacillus* Species**	**Plant Disease**	**Target Pathogen**	**Bioactive Compounds**	**References**
B. subtilis	Take-all of wheat	*Gaeumannomyces graminis*	Iturin, surfactin, plipastatin, bacillomycin, difficidin	Yang et al. (2018)
B. amyloliquefaciens	Bacterial canker of tomato	*Clavibacter michiganensis*	Siderophores, cellulase, lipase, protease, chitinase	Gautam et al. (2019)
B. pumilus and *B. amyloliquefaciens*	Leaf spot disease of sugar beet	*Pseudomonas syringae*	Lipopeptide	Nikolić et al. (2019)
B. subtilis	Spot blotch of wheat	*Bipolaris sorokiniana*	Siderophores, lytic enzymes	Villa-Rodríguez et al. (2019)
B. megaterium	Fungal disease in broccoli	*Alternaria japonica*	Undefined	Vásconez et al. (2020)
B. subtilis	Wilt of watermelon	*Fusarium* sp.	Iturin, bacillomycin, fengycin, surfactin, bacilysin, TasA, and mersacidin	Zhu et al. (2020)

TABLE 7.2
Effect of *Bacillus* sp. on Plant Growth and Development

***Bacillus* sp.**	**Mechanism**	**Plant**	**Effect**	**References**
B. subtilis	CKs	Lettuce	Increased shoot and root weight	Arkhipova et al. (2005)
B. megaterium	CKs	Common bean	Enhanced growth of seedlings	Ortíz-Castro et al. (2008)
Bacillus sp.	IAA, GAs, CKs, ABA	Soybean	Enhanced growth and proline content	Naz et al. (2009)
B. subtilis	IAA, GA	Tomato	Enhanced shoot and root growth, seedling vigor	Chowdappa et al. (2013)
Bacillus sp.	N2 fixation	Maize	Increased germination and root volume	Szilagyi-Zecchin et al. (2014)
B. subtilis	IAA, ACC deaminase	Tomato	Increases shoot and root biomass and chlorophyll content	Xu et al. (2014)
B. subtilis	P-solubilization	Cucumber	Increased growth, increased P accumulation, intake	Garcia-Lopez and Delgado (2016)
B. pumilus	N2 fixation	Maize	Increase in biomass and N content	Kuan et al. (2016)
B. methylotrophicus	GAs	Lettuce	Increased vegetative growth	Radhakrishnan and Lee (2016)
B. aryabhattai	IAA, GAs, ABA, CKs	Soybean	Elongated root and shoot length, enhanced stress tolerance	Park et al. (2017)
B. amyloliquefaciens	ABA	Rice	Increased growth and stress tolerance	Shahzad et al. (2017)
B. megaterium, *B. subtilis*, and *Bacillus simplex*	P solubilization	Eggplant, tomato, pepper	Increased germination and vegetative growth	Bahadir et al. (2018)

Source: Adapted from Miljaković et al. (2020).

7.5 *BACILLUS* AS PLANT GROWTH PROMOTERS

Bacillus sp. can promote plant growth and yield as it produces various metabolites that can increase the nitrogen and phosphorous availability in the soil. Several groups of microorganisms are involved in biological fixation of atmospheric nitrogen. These microorganisms produce nitrogenase enzyme, which helps convert molecular nitrogen into a biologically available form of ammonia. Rhizobacteria, including *Bacillus* sp., contribute a large percentage of total nitrogen fixation by biological agents. Several *Bacillus* sp. including *B. brevis*, *B. cereus*, *B. circulans*, *B. firmus*, *B. licheniformis*, *B. megaterium*, *B. pumilus*, and *B. subtilis* have been reported to show nitrogenase activity (Xie et al. 1998).

In addition to nitrogen, phosphorous is also essential for plant growth. More than 80% of phosphorous present in the soil is not available for plants. The microorganisms that convert insoluble phosphate into soluble form are known as phosphate solubilizing microorganisms (PSM). These microorganisms convert insoluble phosphorous into inorganic and organic acids, carbon dioxide, hydroxyl ions, and siderophores which chelate cations and reduce the pH and release phosphorous (Sharma et al. 2013; Szilagyi-Zecchin et al. 2014). Enzymes such as phosphatase, phytase, and phospholipase are involved in the mineralization of organic phosphate, which is produced by these PSM.

Bacillus stimulates the plant growth and health by enhancing the availability of phosphorus in soil. Isolates of *B. megaterium*, *B. subtilis*, and *B. simplex* exhibited P-solubilizing ability by producing acetic, caproic, isobutyric, isocaproic, propionic, and heptanoic acids and stimulated the growth and germination of seeds of pepper, eggplant, and tomato (Bahadir et al. 2018). Inoculation with *B. subtilis* increased phosphorous uptake and accumulation by cucumber plants (Garcia-Lopez and Delgado 2016).

Bacillus sp. may directly affect the plant growth and yield by imparting the production of phytohormones or plant growth regulators, such as auxins, cytokinins, gibberellins, ethylene, and abscisic acid. Plant hormone biosynthesis by *Bacillus* sp. has been directly related to subsequent growth-promoting traits in different plants (Table 7.2).

7.6 CONCLUSIONS AND FUTURE PERSPECTIVES

Bacillus strains have been tested to control various diseases in economically important crops; additionally, they have been proved to stimulate plant growth and development. Chemical-based disease control in crops has threatened the environment and human health, and new regulations are becoming stricter to decrease chemical use in agriculture practices. There is a possibility of complete prohibition of chemical-based disease control in plants shortly and suggestion of investing more in biocontrol agent identification. For example, wet bubble and dry bubble diseases are widespread in mushrooms and causes significant loss in mushroom production. Sporgon was the only treatment for the diseases which has been prohibited from using commercially since 2020. *Bacillus*-based biocontrol agents with potential formulation may provide better disease control and improve sustainable agriculture practices.

ACKNOWLEDGMENTS

The study was supported by a grant from the Russian Science Foundation (19-74-10046).

REFERENCES

Abriouel, H., Franz, C.M., Ben Omar, N., Gálvez, A. 2011. Diversity and applications of *Bacillus bacteriocins*. *FEMS Microbiology Reviews* 35(1):201–232.

Ahmad, A.G.M., Attia, A.Z.G., Mohamed, M.S., Elsayed, H.E. 2019. Fermentation, formulation and evaluation of PGPR *Bacillus subtilis* isolate as a bioagent for reducing occurrence of peanut soil borne diseases. *Journal of Integrative Agriculture* 18(9):2080–2092.

Aktar, W., Sengupta, D., Chowdhury, A. 2009. Impact of pesticides use in agriculture: Their benefits and hazards. *Interdisciplinary Toxicology* 2:1–12.

Aloo, B.N., Makumba, B.A., Mbega, E.R. 2019. The potential of bacilli rhizobacteria for sustainable crop production and environmental sustainability. *Microbiological Research* 219:26–39.

Arkhipova, T.N., Veselov, S.U., Melentiev, A.I., Martynenko, E.V., Kudoyarova, G.R. 2005. Ability of bacterium *Bacillus subtilis* to produce cytokinins and to influence the growth and endogenous hormone content of lettuce plants. *Plant Soil* 272:201–209.

Bahadir, P.S., Liaqat, F., Eltem, R. 2018. Plant growth promoting properties of phosphate solubilizing *Bacillus* species isolated from the Aegean region of Turkey. *Turkish Journal of Botany* 42:1–14.

Barbe, V., Cruveiller, S., Kunst, F., Lenoble, P., Meurice, G., Sekowska, A., Vallenet, D., Wang, T., Moszer, I., Médigue, C., Danchin, A. 2009. From a consortium sequence to a unified sequence: The *Bacillus subtilis* 168 reference genome a decade later. *Microbiology* 155:1758–1775.

Bashan, Y. 1998. Inoculants of plant growth-promoting bacteria for use in agriculture. *Biotechnology Advances* 16:729–770.

Batta, Y.A. 2016. Invert emulsion: Method of preparation and application as proper formulation of entomopathogenic fungi. *MethodX* 3:119–127.

Beneduzi, A., Ambrosini, A., Passaglia, L.M. 2012. Plant growth-promoting rhizobacteria (PGPR): Their potential as antagonists and biocontrol agents. *Genetics and Molecular Biology* 35:1044–1051.

Bhattacharyya, P.N., Goswami, M.P., Bhattacharyya, L.H. 2016. Perspective of beneficial microbes in agriculture under changing climatic scenario: A review. *Journal of Phytology* 8:26–41.

Bjelić, D., Ignjatov, M., Marinković, J., Milošević, D., Nikolić, Z., Gvozdanović-Varga, J., Karaman, M. 2018. *Bacillus* isolates as potential biocontrol agents of *Fusarium* clove rot of garlic. *Zemdirbyste-Agriculture* 105:369–376.

Cassidy, M.B., Lee, H., Trevors, J.T. 1996. Environmental applications of immobilized microbial cells: A review. *Journal of Industrial Microbiology* 16:79–101.

Chen, X.H., Koumoutsi, A., Scholz, R., Schneider, K., Vater, J., Süssmuth, R., Piel, J., Borriss, R. 2009. Genome analysis of *Bacillus amyloliquefaciens* FZB42 reveals its potential for biocontrol of plant pathogens. *Journal of Biotechnology* 140(1–2):27–37.

Chowdappa, P., Kumar, S.M., Lakshmi, M.J., Mohan, S.P., Upreti, K.K. 2013. Growth stimulation and induction of systemic resistance in tomato against early and late blight by *Bacillus subtilis* OTPB1 or *Trichoderma harzianum* OTPB3. *Biological Control* 65:109–117.

Chung, S., Kong, H., Buyer, J.S., Lakshman, D.K., Lydon, J., Kim, S.D., Roberts, D.P. 2008. Isolation and partial characterization of *Bacillus subtilis* ME488 for suppression of soilborne pathogens of cucumber and pepper. *Applied Microbiology and Biotechnology* 80(1):115–123.

Cook, M.T., Tzortzis, G., Charalampopoulos, D., Khutoryanskiy, V.V. 2012. Microencapsulation of probiotics for gastrointestinal delivery. *Journal of Controlled Release* 162(1):56–67.

Crane, J.M., Gibson, D.M., Richard, H.V., Bergstrom, G.C. 2012. Iturin levels on wheat spikes linked to biological control of fusarium head blight by *Bacillus amyloliquefaciens*. *Phytopathology* 103(2):146–155.

Douriet-Gámez, N.R., Maldonado-Mendoza, I.E., Ibarra-Laclette, E., Blom, J., Calderón-Vázquez, C.L. 2018. Genomic analysis of *Bacillus* sp. strain B25, a biocontrol agent of maize pathogen *Fusarium verticillioides*. *Current Microbiology* 75:247–255.

Driks, A. 2004. The *Bacillus* spore coat. *Phytopathology* 94:1249–1251.

Dunlap, C.A., Schisler, D.A., Price, N.P., Vaughn, S.F. 2011. Cyclic lipopeptide profile of three *Bacillus subtilis* strains: Antagonists of *Fusarium* head blight. *The Journal of Microbiology* 49(4):603–609.

Earl, A.M., Losick, R., Kolter, R. 2008. Ecology and genomics of *Bacillus subtilis*. *Trends in Microbiology* 16:269–275.

Emmert, E.A., Handelsman, J. 1999. Biocontrol of plant disease: A (Gram-) positive perspective. *FEMS Microbiology Letters* 171:1–9.

Fira, D., Dimkić, I., Berić, T., Lozo, J., Stanković, S. 2018. Biological control of plant pathogens by *Bacillus* species. *Journal of Biotechnology* 10(285):44–55.

Garcia-Lopez, A.M., Delgado, A. 2016. Effect of *Bacillus subtilis* on phosphorus uptake by cucumber as affected by iron oxides and the solubility of the phosphorus source. *Agricultural and Food Science* 25:216–224.

Gautam, S., Chauhan, A., Sharma, R., Sehgal, R., Shirkot, C.K. 2019. Potential of *Bacillus amyloliquefaciens* for biocontrol of bacterial canker of tomato incited by *Clavibacter michiganensis* ssp. *michiganensis*. *Microbial Pathogenesis* 130:196–203.

Guleria, S., Walia, A., Chauhan, A., Shirkot, C.K. 2016. Molecular characterization of alkaline protease of *Bacillus amyloliquefaciens* SP1 involved in biocontrol of *Fusarium oxysporum*. *International Journal of Food Microbiology* 232:134–143.

Han, Q., Wu, F., Wang, X., Qi, H., Shi, L., Ren, A., Liu, Q., Zhao, M., Tang, C. 2015. The bacterial lipopeptide iturins induce *Verticillium dahliae* cell death by affecting fungal signalling pathways and mediate plant defence responses involved in pathogen-associated molecular pattern-triggered immunity. *Environmental Microbiology* 17(4):1166–1188.

Harwood, C.R. 1992. *Bacillus subtilis* and its relatives: Molecular biological and industrial workhorses. *Trends in Biotechnology* 10:247–256.

Heimpel, G.E., Mills, N. 2017. *Biological control – Ecology and applications*, Cambridge: Cambridge University Press, p. 386.

Hunt, N.C., Grover, L.M. 2010. Cell encapsulation using biopolymer gels for regenerative medicine. *Biotechnology Letters* 32:733–742.

Islam, M.A., Yun, C.H., Choi, Y.J., Cho, C.S. 2010. Microencapsulation of live probiotic bacteria. *Journal of Microbiology and Biotechnology* 20:1367–1377.

Jiang, C.H., Liao, M.J., Wang, H.K., Zheng, M.Z., Xu, J.J., Guo, J.H. 2018. *Bacillus velezensis* a potential and efficient biocontrol agent in control of pepper gray mold caused by *Botrytis cinerea*. *Biological Control* 126:147–157.

John, R.P., Tyagi, R.D., Brar, S.K., Surampalli, R.Y., Prévost, D. 2011. Bio-encapsulation of microbial cells for targeted agricultural delivery. *Critical Reviews in Biotechnology* 31(3):211–226.

Keswani, C., Mishra, S., Sarma, B.K., Singh, S.P., Singh, H.B. 2014. Unraveling the efficient applications of secondary metabolites of various *Trichoderma* spp. *Applied Microbiology and Biotechnology* 98(2):533–544.

Khan, A., Singh, P., Srivastava, A. 2018. Synthesis, nature and utility of universal iron chelator – Siderophore: A review. *Microbiological Research* 212–213:103–111.

Krasaekoopt, W., Bhandari, B., Deeth, H. 2003. Evaluation of encapsulation techniques of probiotics for yoghurt. *International Dairy Journal* 13(1):3–13.

Kuan, K.B., Othman, R., Abdul Rahim, K., Shamsuddin, Z.H. 2016. Plant growth-promoting rhizobacteria inoculation to enhance vegetative growth, nitrogen fixation and nitrogen remobilisation of maize under greenhouse conditions. *PLoS ONE* 11(3):e0152478.

Kunst, F., Ogasawara, N., Moszer, I., Albertini, A.M., Alloni, G., Azevedo, V. et al. 1997. The complete genome sequence of the gram-positive bacterium *Bacillus subtilis*. *Nature* 390:249–256.

Li, Z., Ramay, H.R., Hauch, K.D., Xiao, D., Zhang, M. 2005. Chitosan-alginate hybrid scaffolds for bone tissue engineering. *Biomaterials* 26:3919–3928.

Logan, N.A., De Vos, P., Garrity, G.M., Jones, D., Krieg, N.R., Ludwig, W., Rainey, F.A., Schleifer, K.H., Whitman, W.B. 2009. Genus *Bacillus*. *Bergey's manual of systematic bacteriology*, 2nd ed. Switzerland: Springer, pp. 21–128.

Luo, W., Liu, L., Qi, G., Yang, F., Shi, X., Zhao, X. 2019. Embedding *Bacillus velezensis* NH-1 in microcapsules for biocontrol of cucumber *Fusarium* wilt. *Applied and Environmental Microbiology* 85:3128–3118.

Mardanova, A.M., Hadieva, G.F., Lutfullin, M.T., Khilyas, I.V., Minnullina, L.F., Gilyazeva, A.G., Bogomolnaya, L.M., Sharipova, M.R. 2017. *Bacillus subtilis* strains with antifungal activity against the phytopathogenic fungi. *Agricultural Science* 8:1–20.

Matar, S.M., El-Kazzaz, S.A., Wagih, E.E., El-Diwany, A.I., Hafez, E.E., Moustafa, H.E., Abo-Zaid, G.A, Serour, E.A. 2009. Molecular characterization and batch fermentation of *Bacillus subtilis* as biocontrol agent, II. *Biotechnology* 8:35–43.

Miljaković, D., Marinković, J., Balešević-Tubić, S. 2020. The significance of *Bacillus* spp. in disease suppression and growth promotion of field and vegetable crops. *Microorganisms* 8(7):1037.

Minaxi, S.J. 2011. Efficacy of rhizobacterial strains encapsulated in nontoxic biodegradable gel matrices to promote growth and yield of wheat plants. *Applied Soil Ecology* 48:301–308.

Mnif, I., Ghribi, D. 2015. Potential of bacterial derived biopesticides in pest management. *Crop Protection* 77:52–64.

Naz, I., Bano, A., Ul-Hassan, T. 2009. Isolation of phytohormones producing plant growth promoting rhizobacteria from weeds growing in Khewra salt range, Pakistan and their implication in providing salt tolerance to *Glycine max* L. *African Journal of Biotechnology* 8:5762–5766.

Nikolić, I., Berić, T., Dimkić, I., Popović, T., Lozo, J., Fira, D., Stanković, S. 2019. Biological control of *Pseudomonas syringae* pv. *aptata* on sugar beet with *Bacillus pumilus* SS-10.7 and *Bacillus amyloliquefaciens* (SS-12.6 and SS-38.4) strains. *Journal of Applied Microbiology* 126(1):165–176.

Ongena, M., Jacques, P., Touré, Y., Destain, J., Jabrane, A., Thonart, P. 2005. Involvement of fengycin-type lipopeptides in the multifaceted biocontrol potential of *Bacillus subtilis*. *Applied Microbiology and Biotechnology* 69(1):29–38.

Onwulata, C.I. 2012. Encapsulation of new active ingredients. *Annual Review of Food Science and Technology* 3:183–202.

Ortíz-Castro, R., Valencia-Cantero, E., López-Bucio, J. 2008. Plant growth promotion by *Bacillus megaterium* involves cytokinin signaling. *Plant Signaling and Behavior* 3:263–265.

Özer, B., Uzun, Y.S., Kirmaci, H.A. 2008. Effect of microencapsulation on viability of *Lactobacillus acidophilus* LA-5 and *Bifidobacterium bifidum* BB-12 during Kasar cheese ripening. *International Journal of Dairy Technology* 61(3):237–244.

Park, Y.G., Mun, B.G., Kang, S.M., Hussain, A., Shahzad, R., Seo, C.W., Kim, A.Y., Lee, S.U., Oh, K.Y., Lee, D.Y. 2017. *Bacillus aryabhattai* SRB02 tolerates oxidative and nitrosative stress and promotes the growth of soybean by modulating the production of phytohormones. *PLoS ONE* 12(3):e0173203.

Prashar, P., Kapoor, N., Sachdeva, S. 2013. Rhizosphere: Its structure, bacterial diversity and significance. *Reviews in Environmental Science and Bio/Technology* 13:63–77.

Radhakrishnan, R., Lee, I.J. 2016. Gibberellins producing *Bacillus methylotrophicus* KE2 supports plant growth and enhances nutritional metabolites and food values of lettuce. *Plant Physiology and Biochemistry* 109:181–189.

Radhakrishnan, R., Hashem, A., Abd Allah, E.F. 2017. *Bacillus*: A biological tool for crop improvement through bio-molecular changes in adverse environments. *Frontiers in Physiology* 8:667.

Rathore, S., Desai, P.M., Liew, C.V., Chan, L.W., Sia Heng, P.W. 2013. Microencapsulation of microbial cells. *Journal of Food Engineering* 116:369–381.

Sabaté, D.C., Brandan, C.P., Petroselli, G., Erra-Balsells, R., Audisio, M.C. 2018. Biocontrol of *Sclerotinia sclerotiorum* (Lib.) de Bary on common bean by native lipopeptide-producer *Bacillus* strains. *Microbiological Research* 211:21–30.

Saber, W.I., Ghoneem, K.M., Al-Askar, A.A., Rashad, Y.M., Ali, A.A., Rashad, E.M. 2015. Chitinase production by *Bacillus subtilis* ATCC 11774 and its effect on biocontrol of *Rhizoctonia* diseases of potato. *Acta Biologica Hungarica* 66:436–448.

Salazar, F., Ortiz, A., Sansinenea, E. 2017. Characterisation of two novel bacteriocin-like substances produced by *Bacillus amyloliquefaciens* ELI149 with broad-spectrum antimicrobial activity. *Journal of Global Antimicrobial Resistance* 11:177–182.

Savary, S., Ficke, A., Aubertot, J., Hollier, C. 2012. Crop losses due to diseases and their implications for global food production losses and food security. *Food Security* 4:519–537.

Schisler, D.A., Sliniger, P.J., Behle, R.W., Jackson, M.A. 2004. Formulation of *Bacillus* spp. for biological control of plant diseases. *Phytopathology* 94:1267–1271.

Shahzad, R., Khan, A.L., Bilal, S., Waqas, M., Kang, S.M., Lee, I.J. 2017. Inoculation of abscisic acid-producing endophytic bacteria enhances salinity stress tolerance in *Oryza sativa. Environmental and Experimental Botany* 136:68–77.

Sharma, S.B., Sayyed, R.Z., Trivedi, M.H., Gobi, T.A. 2013. Phosphate solubilizing microbes: Sustainable approach for managing phosphorus deficiency in agricultural soils. *SpringerPlus* 2:587.

Sharma, L., Oliveira, I., Raimundo, F., Torres, L., Marques, G. 2018. Soil chemical properties barely perturb the abundance of entomopathogenic *Fusarium oxysporum*: A case study using a generalized linear mixed model for microbial pathogen occurrence count data. *Pathogens* 7(4):89.

Sharma, L., Bohra, N., Singh, R.K., Marques, G. 2019. Potential of entomopathogenic bacteria and fungi. In: Khan, M., Ahmad, W. (eds) *Microbes for sustainable insect pest management. Sustainability in plant and crop protection*, Vol. 1. Switzerland: Springer Nature, pp. 115–149.

Silva, N.I., Brooks, S., Lumyong, S., Hyde, K.D. 2019. Use of endophytes as biocontrol agents. *Fungal Biology Reviews* 33(2):133–148.

Souza, R., Ambrosini, A., Passaglia, L.M.P. 2015. Plant growth-promoting bacteria as inoculants in agricultural soils. *Genetics and Molecular Biology* 38:401–419.

Stein, T. 2005. *Bacillus subtilis* antibiotics: Structures, syntheses and specific functions. *Molecular Microbiology* 56(4):845–857.

Su, Y., Liu, C., Fang, H., Zhang, D. 2020. *Bacillus subtilis*: A universal cell factory for industry, agriculture, biomaterials and medicine. *Microbial Cell Factories* 19:173.

Szilagyi-Zecchin, V.J., Ikeda, A.C., Hungria, M., Adamoski, D., Kava-Cordeiro, V.K., Glienke, C., Galli-Terasawa, L.V. 2014. Identification and characterization of endophytic bacteria from corn (*Zea mays* L.) roots with biotechnological potential in agriculture. *AMB Express* 4:1–9.

Tendulkar, S.R., Saikumari, Y.K., Patel, V., Raghotama, S., Munshi, T.K., Balaram, P., Chattoo, B.B. 2007. Isolation, purification and characterization of an antifungal molecule produced by *Bacillus licheniformis* BC98, and its effect on phytopathogen *Magnaporthe grisea*. *Journal of Applied Microbiology* 103(6):2331–2339.

Trevor, J.T., van Elsas, J.D., Lee, H., van Overbeek, L.S. 1992. Use of alginate and other carriers for encapsulation of microbial cells for use in soil. *Microbial Releases* 1:61–69.

Tripathi, M.K., Giri, S.K. 2014. Probiotic functional foods: Survival of probiotics during processing and storage. *Journal of Functional Foods* 9:225–241.

van Dijl, J.M., Hecker, M. 2013. *Bacillus subtilis*: From soil bacterium to super-secreting cell factory. *Microbial Cell Factories* 12:3.

Vásconez, R.D.A., Tenorio, E.M.L., Collaguazo, M.L.A., Yépez, L.A.C., Chiluisa-Utreras, V.P., Suquillo, I.A.V. 2020. Evaluation of *Bacillus megaterium* strain AB4 as a potential biocontrol agent of *Alternaria japonica*, a mycopathogen of *Brassica oleracea* var. *italica*. *Biotechnology Reports* 26:1–6.

Vemmer, M., Patel, A.V. 2013. Review of encapsulation methods suitable for biological control agents. *Biological Control* 67:380–389.

Villa-Rodríguez, E., Parra-Cota, F., Castro-Longoria, E., López-Cervantes, J., de los Santos-Villalobos, S. 2019. *Bacillus subtilis* TE3: A promising biological control agent against *Bipolaris sorokiniana*, the causal agent of spot blotch in wheat (*Triticum turgidum* L. subsp. *durum*). *Biological Control* 132:135–143.

Wei, Y.H., Wang, L.F., Chang, J.S., Kung, S.S. 2003. Identification of induced acidification in iron-enriched cultures of *Bacillus subtilis* during biosurfactant fermentation. *Journal of Bioscience and Bioengineering* 96:174–178.

Wu, L., Wu, H.J., Qiao, J., Gao, X., Borriss, R. 2015. Novel routes for improving biocontrol activity of *Bacillus* based bio-inoculants. *Frontiers in Microbiology* 6:1395.

Xie, G., Su, B., Cui, Z., Wei, S., Wu, X.B. 1998. Isolation and identification of N_2-fixing strains of *Bacillus* in rice rhizosphere of the Yangtze river valley. *Wei Sheng Wu Xue Bao* 38(6):480–483.

Xu, M., Sheng, J., Chen, L., Men, Y., Gan, L., Guo, S., Shen, L. 2014. Bacterial community compositions of tomato (*Lycopersicum esculentum* Mill.) seeds and plant growth promoting activity of ACC deaminase producing *Bacillus subtilis* (HYT-12-1) on tomato seedlings. *World Journal of Microbiology and Biotechnology* 30:835–845.

Yabur, R., Bashan, Y., Hernández-Carmona, G. 2007. Alginate from the macroalgae *Sargassum sinicola* as a novel source for microbial immobilization material in wastewater treatment and plant growth promotion. *Journal of Applied Phycology* 19:43–53.

Yang, L., Han, X., Zhang, F., Goodwin, P.H., Yang, Y., Li, J., Xia, M., Sun, R., Jia, B., Zhang, J. 2018. Screening *Bacillus* species as biological control agents of *Gaeumannomyces graminis* var. *tritici* on wheat. *Biological Control* 118:1–9.

Yeh, M., Wei, Y., Chang, J. 2006. Bioreactor design for enhanced carrier-assisted surfactin production with *Bacillus subtilis*. *Process Biochemistry* 4:1799–1805.

Yi, G., Liu, Q., Lin, J., Wang, W., Huang, H., Lia, S. 2016. Repeated batch fermentation for surfactin production with immobilized *Bacillus subtilis* BS-37: Two-stage pH control and foam fractionation. *Journal of Chemicals Technology and Biotechnology* 92:530–535.

Young, C.C., Rekha, P.D., Lai, W.A., Arun, A.B. 2006. Encapsulation of plant growth-promoting bacteria in alginate beads enriched with humic acid. *Biotechnology and Bioengineering* 95(1):76–83.

Yu, X., Ai, C., Xin, L., Zhou, G. 2011. The siderophore-producing bacterium, *Bacillus subtilis* CAS15, has a biocontrol effect on *Fusarium* wilt and promotes the growth of pepper. *European Journal of Soil Biology* 47(2):138–145.

Zalila-Kolsi, I., Ben Mahmoud, A., Ali, H., Sellami, S., Nasfi, Z., Tounsi, S., Jamoussi, K. 2016. Antagonist effects of *Bacillus* spp. strains against *Fusarium graminearum* for protection of durum wheat (*Triticum turgidum* L. subsp. *durum*). *Microbiological Research* 192:148–158.

Zeriouh, H., Romero, D., García-Gutiérrez, L., Cazorla, F.M., de Vicente, A., Pérez-García, A. 2011. The iturin-like lipopeptides are essential components in the biological control arsenal of *Bacillus subtilis* against bacterial diseases of cucurbits. *Molecular Plant-Microbe Interaction* 24:1540–1552.

Zhu, J., Tan, T., Shen, A., Yang, X., Yu, Y., Gao, C., Li, Z., Cheng, Y., Chen, J., Guo, L., Sun, X., Yan, Z., Li, J., Zeng, L. 2020. Biocontrol potential of *Bacillus subtilis* IBFCBF-4 against *Fusarium* wilt of watermelon. *Journal of Plant Pathology* 102:433–441.

8 Recognizing Insight Interaction, Adaptation and Potentialities of Phyllosphere Microbiome under Various Distress towards Sustainable Plant Growth

Rakesh Yonzone and Pulak Bhaumik
Uttar Banga Krishi Viswavidyalaya, Majhian, India

M. Soniya Devi
Rani Lakshmi Bai Central Agricultural
University, Jhansi, India

CONTENTS

DOI: 10.1201/9781003147077-8

8.1 INTRODUCTION

The term phyllosphere is generally used to describe the above-ground surface of the plant which is in contact with the outer environment above the soil. It has been derived from a Greek word "Phyllos" meaning "leaf" and "Sphaira" meaning "glow" in analogy to the term rhizosphere. The phyllosphere consist of the unique and most dynamic environment where hyperdiverse microbial communities harbour (Vacher et al. 2006). It is also considered as heterogenous habitat as most of the parts of the plant are direct in contact with the surrounding environment and are very sensitive towards slight changes in the environmental parameters. The phyllosphere mostly includes the aerial parts of plants such as leaves, stem, flowers and fruits which act

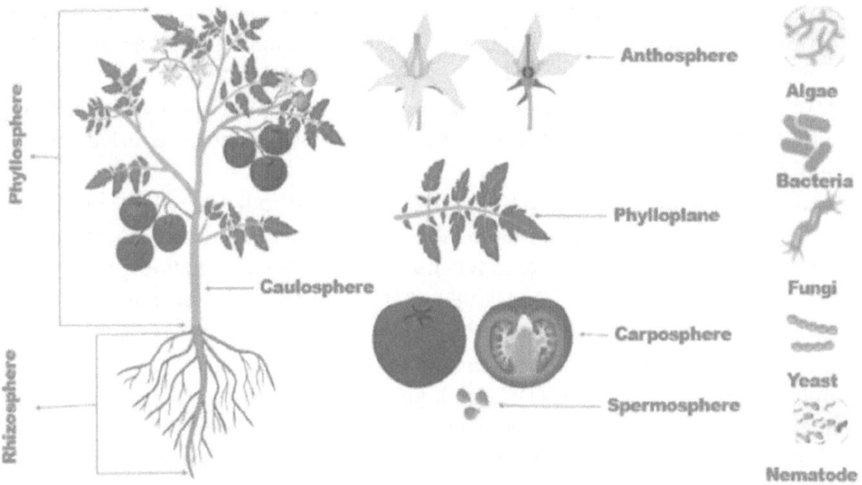

FIGURE 8.1 Different regions of the phyllosphere and epiphytes commonly found on/in plants.

as the micro-habitat for growth, development and survival of a variety of microorganisms. The phyllospheres have been a habitat of numerous microorganisms that establish in this region and make symbiotic relationship with the plant (Figure 8.1). The phyllosphere can be subdivided into five sections (Rastogi et al. 2013):

- *Caulosphere:* It refers to the stem section of the above-ground plant parts.
- *Phylloplane:* The term used to refer to the leaves portion of the phyllosphere.
- *Anthosphere:* It refers to the flower section of the phyllosphere.
- *Carposphere:* It denotes mostly the fruit section of the phyllosphere.
- *Spermosphere:* It denotes the seed portion of the phyllosphere.

The microorganisms which colonize in these aerial regions are known as epiphytes. They are mostly microorganisms of various origins, the most important among the epiphytes inhibiting this region are fungi, algae, protozoa and to some extent the nematodes. However, the most widely and abundantly colonizing epiphytes are bacteria. One of the important reasons behind the establishment of these epiphytes in the phyllosphere is the microclimate and hospitable environment found in this region (Figure 8.1). This unique microenvironment allows all the epiphytes to not only grow and reproduce but also help them to interact with the host plant, thereby improving the plant's health which can be measured by plant vigour, free from infection and ultimately the yield.

Even though less intensive studies have been carried out about microorganism-inhabiting phyllosphere and their interaction with the host, their functions in improving plant health as compared to rhizosphere inhabiting microorganisms have received a considerable amount of attention over the past few years. Also, the interest of epiphytes dominating the aerial surface of plant has been now well acknowledging to the extent of its importance beyond pathogens. The interest to study epiphyte existence and its role in improving plant health has been receiving a lot of attention among the researchers as various functions such as modulation to access nutrients from tissues of leaves (Bulgarelli et al. 2013), protection from more or less incoming sunlight (Atamna-Ismaeel et al. 2012a) or providing gateway to entre to the plant endosphere (Schreiber et al. 2004) has been reported through dynamic interaction.

Their traditional coexistence in the phyllosphere led to the new level of interaction with host plants which is far beyond the communalistic relationships. One of the important reasons that they share a dynamic interaction with the host plant is their ability to colonize in extreme, stressful and changing micro environment provided by the host. In spite of having many limitations in the phyllosphere such as availability of nutrients, carbon and nitrogen and the effect of extreme fluctuating physiochemical factors such as light, UV rays, highly volatile organic compounds (VOCs) pollutants, rainfall, temperature and noxious gases, these epiphytes have evolved and developed highly fascinating mechanisms to overcome these challenges and survive (Sivakumar et al. 2020).

8.2 PHYLLOSPHERE: THE WORKING INHABITANT OF THE MICROBIOME

Phyllosphere provides an excellent hospitable environment to many epiphytes. However, the phases that the phyllosphere inhabiting microorganism goes through for successful colonization and functioning in the host are not an easy task. They undergo various challenges such as resistance/tolerance against the antimicrobial and/or immunity compounds produced by plant tissues or by other competing microorganisms (Trouvelot et al. 2014) that lead to the survival for the fittest microorganisms.

The important factors of phyllosphere that lead to the unsuccessful establishment of various microbiomes are as follows:

- The phyllosphere governs the extreme and unstable habitat which makes it unsuitable for successful establishment of many microorganisms.
- It bears deficiency in many nutrients that are required by the microorganism for their growth and development.
- The dynamic environment at the leaf interface such as various levels of VOCs, changing the level of plant hormones and pollutants makes an important limiting factor (Jackson et al. 2006).
- The complex bi-dimensional and tri-dimensional heterogeneous structure of the leaf surface of different plant species and nutrient availability (Delmotte et al. 2009) also affects the interaction between microorganism and the host (Miller et al. 2001).
- The growth and development of epiphytes are also affected by abiotic factors such as fluctuation of seasonal cycle, day/night cycle, variation in rainfall and temperature (Rastogi et al. 2013) also affect the buds, leaves, flowers and fruits to senescence.
- The most important factor limiting the establishment of the epiphytes is the unavailability of the disseminating nutrient that are utilized by the

microorganisms such as sugar, amino acids, polyols, isoprenoids, alcohols, CO_2-nitrogen-containing compound, sesquiterpenoid, aldehydes and long-chain hydrocarbons (Whipps et al. 2008).

- Most of the host leaf has thick cuticular wax, lipids and metabolic flux which greatly limit the availability of free surface water which is an important factor for initial attachment for establishment of various fungi and bacteria.
- The physiological characteristics of the plant such as opening and closing of stomata, biochemical exchange, exudation from the wounds, guttation and presence of waxy cuticle (Burch et al. 2014) which varies from species to species of host plants adds a major limiting factor towards the growth and development of the microorganism.

8.3 MICROBIOME HARBOURING THE PHYLLOSPHERE

Different genera of microorganisms have been documented to harbour the phyllosphere, among which the epiphytes such as filamentous fungi, yeast and algae are found most commonly. The microorganisms which are less frequently found are the protozoa and nematodes. However, the most dominating microorganisms are the different genera of bacteria with population ranging between 10^{-2} and 10^{-12} per gram of leaves (Lindow and Brandl 2003).

8.3.1 Fungi Inhabiting the Phyllosphere

Only a handful of information is available about the phyllosphere inhabiting fungi. For any given fungal epiphyte in order to establish on the host surface successfully and to function properly, one full cycle has to be completed starting from spore dissemination to establishment and finally reproduction. After landing on the host plant, the spores require a thin film of water on the host surface to germinate which is to be provided by the micro environment. Apart from this, the influence of temperature, sunlight, host genotype and the right time to germinate, i.e., the favourable period, plays its own importance in fungal germination. Their influence and function on the host plant can be taken negatively or positively. They negatively act as a pathogen harming the host and producing disease with visible symptoms or positively by increasing the stress tolerance of the host (Guerreiro et al. 2018; Yadav et al. 2018). Their positive impact may be subjected to competition for space and food, hyperparasitism or the production of various types of poisonous toxins that are harmful to other competing microorganisms (Kohl et al. 2019). These types of mechanisms have led to their importance which has grown among researchers and have played a major role in changing the dynamic and diversity of different plant populations. Being early colonizers, their presence has a great influence on the suppression of the plant pathogen population (Heimpel and Mills 2017).

The majority of the fungi found on the phyllosphere have been hypothesized to have originated on the surface of the plant by accidentally falling in the process of dispersal through the means of air and water. These spores stick to the host's external surface and eventually germinate by producing germ tubes and aspersorium after a certain period of dormancy and establish there. *Colletotrichum*, *Alternaria* and *Chaetomium* are major dormant genera, which have been isolated and identified from the leaves and stem of *Catharanthus roseus* by Dhayanithy et al. (2019). Only after the part of a host falls off and onto the ground, these fungi due to their low saprophytic ability cannot survive and decompose under the influence of other microorganism having high saprophytic ability. However, such theories need to be backed up with proper evidence which is indeed a shortage of the hour. Many researchers have examined the micro-flora in phyllosphere using different techniques such as leaf washing, spore fall, leaf impression on nutrient media, balloon technique, examination of leaf surface *in situ* including leaf bleaching, cuticle print with collodion or nail varnish, scanning electron icroscopy and use of light microscope, which has contributed a significant finding regarding epiphytic fungi (Arunkumar et al. 2019). The important predominant genera found in the phyllosphere are *Cladosporium*, *Acremonium*, *Aspergillus*, *Alternaria*, *Penicillium* and *Mucor*, which colonize the host epiphytically as well as endophytically. Their population has been found to range between 10^2 and 10^{10} CFU/g of leaves based on the methods followed for culturing the fungal epiphytes (Inacio et al. 2002).

8.3.2 Bacteria Inhabiting the Phyllosphere

Among the diverse microbiome inhabiting the phyllosphere, bacteria hold a vital position and are mostly dominating (Table 8.1). Their presence and function have been well documented by various researchers. The most important classes of bacteria that are dominant in phyllosphere are α-*Proteobacteria*, γ-*Proteobacteria*, *Bacteroidetes*, *Actinobacteria*, β-*Proteobacteria* and *Sphingobacteria* (Whipps et al. 2008). Important characteristic features of the colonizing epiphytic bacteria are as follows:

- They are mostly aerobic and organo-heterotrophs in nature.
- The epiphytic bacterial species possess a characteristic pigmentation.
- Most of the bacteria are highly antagonist in nature.

TABLE 8.1
Examples of the Commonly Dominating Microorganisms in a Different Host

Host Species	Dominating Species	References
Tropical tree species	*Alphaproteobacteria, Acidobacteria, Gammaproteobacteria*	Kim et al. (2012)
Lettuce	*Proteobacteria, Firmicutes, Actinobacteria*	Williams et al. (2013)
Soybean, clover, arabidopsis	*Sphingomonas* and *Methylobacterium*	Delmotte et al. (2009)
Spinach	*Proteobacteria* and *Firmicutes*	Lopez-Velasco et al. (2011)
Lettuce	*Pseudomonas, Bacillus, Massilia, Arthrobacter* and *Pantoea*	Rastogi et al. (2012)

- Members of *Proteobacteria* are characterized with important functional activities such as methyltrophy, nitrification and nitrogen fixation (Kohl et al. 2019), for example, various species of *Pseudomonas*, *Methylobacterium*, *Beijerinekia*, *Azotobacter*, *Klebsiella* and *Cyanobacter* (nostoc, scytonema and stigonema).
- Most of the bacteria of these classes possess an oxygenic photosynthesis where sulphur is used as a reducing agent in the process of photosynthesis resulting in terminal reductant as hydrogen sulphide instead of water in oxygenic photosynthesis and by-products are elemental sulphur instead of molecular oxygen, for example, green and sulphur bacteria.
- The most dominating family of Bacterioidetesis the Cytophagacea or Chitinophagacea which are aerobic and pigmented. Apart from these bacteria, other predominating genera inhabiting the phyllosphere include *Methylibium*, *Hyphomibrobium*, *Methylocella*, *Massilia*, *Flavobacterium*, *Rathayibacter*, *Cryptococcus*, *Rhodotorula* and *Sporobolomyces* (Glushakova and Chernov 2004). However, the low colonizers in the phyllosphere are *Acidobacteria*, *Cyanobacteria* and *Actinobacteria*.

The findings of many researchers suggest that the presence of phyllosphere bacteria not only plays an important role in influencing a plant's ecosystem but also helps in managing the geography of the ecosystem by increasing performance under various distressing environmental conditions through their interaction with hosts. Even after undergoing stressful situations for survival, these bacteria have made a remarkable impression by their adaptation and their usefulness in plant defence mechanism enhancement (Pascazio et al. 2015). However in spite of these, there is still a lack of information regarding what factors actually drive the variation in the bacterial biodiversity on different host plants and how adaptation has been possible for a single cell is still unknown which needs more study.

8.3.3 Yeast Inhabiting the Phyllosphere

The research findings on the phyllosphere inhabiting yeast has focussed more on the phylloplane (leaf) inhabiting population. They occur as dormant spores rather than actively rowing mycelia and are present on all the leaves except the older once (de Jager et al. 2001). The phylloplane allows numerous genera of saprophytic yeast to grow and develop, which in return inhibits the invading pathogenic organism to establish and helps in improving the plant's health. The most important characteristics are the ability to grow most actively which make it difficult to other competitors making shortage in nutrient and space to flourish and act. Their antagonistic and symbiotic associate character helps them to grow rapidly and interact with host plants more easily. More recently, their occurrence has also been recorded in the interior of flowers and fruits of higher plants. The dynamic characteristic feature of the phyllosphere makes an excellent habitat for the growth and development of yeast and in return makes us easy for studying its ecology. It also provides effective protection to the host from various invading pathogens and external environmental damage. The most important culturable yeast genera found in the phyllosphere include *Cryptococcus*, *Geotrichum*, *Candida*, *Rhodotorula*, *Sporobolomyces* and *Aureobasidium* which are mostly restricted to the leaves (Glushakova and Chernov 2004). The preference in colonization and establishment of the epiphytes varies from genus to genus and species to species. The plant species also has a huge impact on the establishment of the epiphytes in the phyllosphere.

On the basis of the preference of different regions, the epiphytes harbouring the phyllosphere can be classified as follows.

8.3.3.1 Microorganisms Harbouring the Caulosphere (Stems)

The caulosphere is characterized by the presence of hard wooden aerial parts and harbours a diverse group of microbial communities. Most of the microorganisms do not find their ideal habitat in the caulosphere as the stems are characterized by the presence of chitin and wax layering on its surface which makes it hydrophobic in nature (Beattie 2002). That microorganism residing in the caulosphere needs to have developed an advanced adaptation technique to withstand and overcome the challenges of the environmental parameters which include mostly temperature and moisture content (Stone et al. 2018). The predominant microorganism residing in the caulosphere is the fungi followed by bacteria and nematodes. The dominant fungal species are *Saccharomyces*, *Hanseniaspora*, *Candida* and *Lachancea*. Among bacterial species inhabiting the caulosphere, *Proteobacteria*, *Flavobaterium* and *Pseudomonas* are the dominating ones.

8.3.3.2 Microorganisms Harbouring the Phylloplane (Leaves)

The phylloplane is the most important inhabitant of the microorganism. The physiology of the leaves which includes mostly the leaf arrangement, epidermal cells and the micro environment determines the presence and distribution of the population of the epiphytes on the surface of leaves. The presence of organisms is influence by different physical characteristic of the leaf surface that restricts or allows the microbe to inhabit. For example, the presence or absence of the outer cuticle, leaf physiology, helps the microorganism to colonize and the nutrient availability along with leaching water allows this organism to utilize it and develop colonies for growth and reproduction on the phyllosphere (Whipps et al. 2008). The epiphyte also makes biofilm-like growth in the form of large bacterial aggregates on the epidermal cell, veins and trichomes and the presence of leaf exudates creates the nutrient-rich region which helps this microbe to utilize it and establish in the leaves (Fokkema et al. 1983). Among the various microorganisms, bacteria are most abundantly found in this region. On average 10^{-6}–10^{-7} bacterial cell/cm^2 are present on the leaf surface (Inacio et al. 2002). They are present in larger aggregate unlike a solitary cell or small group of cells like that of fungi. Their aggregates are restricted to the depression formed on the junction of the epidermal cell, veins and at the base of trichomes. The size of bacterial aggregates can be positively correlated with the availability of water and can vary as little as 0.1 mm^2 in a small scale (Kinkel et al. 2000). However, the presence of fungal communities is hypervariable based on the region as temperate phyllosphere exhibits greater diversity and presence of more microbial communities than that of plants phylloplane present in tropical areas. Apart from these, the epiphyte present is also influenced by the types of plant species found in the particular region and occurs as aggregates or biofilms (Jacques et al. 2005).

8.3.3.3 Microorganisms Harbouring the Anthrosphere (Flowers)

The anthrosphere is the region mostly characterized around flowers and is an important habitat for the diverse microorganism. They are mostly inhabited by the vast diversity of microbial communities which are mostly specific to flowers. The most important characteristic feature of the anthrosphere is its durability. The permanence of this habitat has a shorter life span as compared to the other sphere of phyllosphere. The most important components present in the anthrosphere are pollen, nectar, petals, sepals, styles, stamens, ovary and stigma which act as the microsites for short span of colonization for the microorganisms. Several studies have reported that dominant microorganism inhabiting the anthrosphere is fungi followed by bacteria (Arunkumar et al. 2019). Among fungal communities, the most abundant are the basidiomycetes yeast which are present on the floral surface and ascomycetous yeast species on the pollen surface. The fungal population are mostly found in the nectar parts rather than the floral parts. However, the bacterial population count is most abundant in the pollen region ranging from 10^6 to 10^9 (Arunkumar et al. 2019). Its composition and diversity vary with the availability of different nutrients, pollen structure, viability and moisture from species to species. They exist in the atmosphere either singly or in clusters along with the formation of particularly thin biofilm in many habitats. The fungi isolated from some medicinal flowers possessing therapeutic metabolites are *Pestalotiopsis*, *Phomopsis* and *Coelomycete* sp. found in temperate regions (Arunkumar et al. 2019).

8.3.3.4 Microorganisms Harbouring the Carposphere (Fruit)

The carposphere is the most unique and dynamic habitat of microorganisms. They consist of a diverse group of microbial communities which are mostly present in the fruit skin. Most of the dominant microorganisms found in the carposphere are fungi and bacteria (Pascazio et al. 2015). Their population mostly depends upon the presence of nutrient. Since fruits are rich in sugar, they serve as a source of food for microbial communities harbouring the carposphere. Based on the nutrient availability, the population of the microbial community varies with the composition of the fruits. For example, in fruits such as lemon and orange, large populations of yeasts are found, while in grapes and apple, bacterial cells are found in higher numbers (Pascazio et al. 2015). Apart from these, the weather parameter has a great influence on microbial population build up as they are mostly affected by fluctuating temperatures, relative humidity and also UV radiation which mostly determines the restriction of growth or building up of populations of microbial communities (Pascazio et al. 2015). Also, the presence of a tough water proof layer covering the fruit restricts the entry and colonization of many microorganisms in the fruits.

8.4 WORKING STRATEGIES OF THE MICROBIOME UNDER A DISTRESSING ENVIRONMENT

The organisms surviving in the phyllosphere have to undergo various extreme, adverse biotic and abiotic conditions in order to develop the strategies to survive and function with the plant (Bringel and Couée 2015). During the process of survival, they need to overcome the challenges to develop various mechanisms and to perform effectively for the benefit of the host. Most of the fungi have established themselves on the different spheres of the phyllosphere making them high-valued as they have a significant impact on plant health (Bringel and Couée 2015). Their strategies or the mechanisms for growth, development and function to improve plant health and overcome the challenges of other microorganisms are mostly based on the following factors:

- *Antibiosis:* This is the mechanism in which one of the two organisms living in close vicinity becomes affected by production of metabolic substances which are harmful such as enzymes, toxins and antibiotics to the competing organism.
- *Hyperparasitism/Direct parasitism:* This is the mechanism in which one organism develops within the other limiting its growth and development which often leads to death such as necrotrophs and biotrophs (Kohl et al. 2019).
- *Competition:* In this mechanism, the organism competes for nutrients and space for the growth and development where the fittest survives whereas the others are thrown out of the race.
- *Induced resistance:* This is the indirect means of working strategy performed by the epiphyte where the inhabiting organism helps the plant to induce different responses locally or systematically withstand various biotic and abiotic stresses (Pieterse et al. 2014; Conrath et al. 2015).

8.4.1 Positive Effect of Epiphytes Working Strategy

The epiphytes inhabiting the phyllosphere often have a direct positive influence on the host. They help in altering the properties of plant surface and enhance the fixation of various compounds such as nitrogen, thereby promoting growth and development of the plants. They also help in controlling various plant pathogens by inhibiting them through various mechanisms such as by parasitizing them or through inducing resistance in the host (Karlsson et al. 2017; Mauch-Mani et al. 2017; Ghorbanpour et al. 2018). They also act as a good source for degradation of organic pollutants and help in managing them.

Epiphytes have a great influence on the ecological relationship to improve growth, adaptation, infection and resistance of the host plant. They have a great influence on the functioning of leaves, longevity of seed mass, growth of apical parts, its flowering and even the development of fruits. Apart from these, this epiphyte also plays a vital role in increasing yield, removes various contaminants and thereby helps in producing the novel substances which are required to activate plant defence mechanisms against various biotic and abiotic stresses. They also help in producing different growth hormones or growth-promoting factors such as indole acetic acid (IAA), cytokinins, etc., thereby enhancing crop growth, nutrient uptake and ultimately a good yield (Kaur et al. 2019; Kumar et al. 2019; Yadav et al. 2019).

8.4.2 Negative Effect of Epiphytes Working Strategy

Even though lots of positive effects are reported by various researchers, the presence of epiphytes on the host plant and its negative effect has also been reported in the past. One of the important harmful effects can be seen in the competition for nutrients, water and space with many plants because of the presence of a large number of microbial population. The negative effect can also be seen in the interaction between the microorganisms present in the phyllosphere which leads to losses of many important beneficial organisms, the useful properties of which cannot be known due to these losses. The competing microorganism present in the phyllosphere may sometimes act as a pathogen causing damage to the host plant, thereby resulting in yield losses due to disease (Lindow and Brandl 2003).

8.5 INTERACTING MECHANISMS BETWEEN PLANTS AND THE MICROBIOME

There has been a handful of studies on how the microorganism gets in contact to the plant in the phyllosphere. Some researchers claim that attachment of microorganisms takes place by means of air while others claim it is by water dispersal. However, the knowledge gap on this aspect needs to be thoroughly examined to cover the loophole of information shortage (Lindow and Brandl 2003). But once the microorganisms come into contact with the host plants, their further interaction have been well studied and are known. In the case of fungi, once the spore lands on the leaf surface, they provide the area and environment for its germination. Only after this stage, the interaction with the host plant begins and shows the most conspicuous response against the changing abiotic stress.

8.5.1 Classification of Fungi Based on Interaction Mechanisms

Based on the interaction mechanisms of fungi with plants, they can be categorized into the following:

Ectophytic: They are also known as superficial growing fungi which tend to grow externally without showing any aggressiveness to the absorption of nutrients.

Ectendophytic: They are mostly parasitic in nature to the plants which penetrate into the cells of the host and grow there, drawing the nutrients and producing external visible symptoms.

Endophytic: These microorganisms mostly grow internally within the host without producing external symptoms.

Apart from these, the microbial fluctuating system and interaction with the host in the phyllosphere can be categorized into two components:

Residential community: The term residential community is used to describe the microorganism which grows and multiplies on the external surface of the host such as leaves, without showing any visual or noticeable effects on the host plant.

Casual community: The term casual community is used to describe the communities of microorganisms which are accidentally present on the host surface and are mostly inactive.

The residential communities are more important than the casual community with relation to plant health. As their effect on the host plant is thought to be either hindrance to the pathogen development that causes disease in the host through direct interaction or through indirect interaction with the pathogen. The entry of the bacterial species on the host plant is mainly through natural openings such as stomata, hydrathodes, nectarthodes, lenticels, wounds and cracks in the plants. Being the dominating colonies among the microorganism, their persistence has been mostly on the leaves. On an average 10^6–10^7 cells/cm^2 or even higher have been reported by various researchers (Lindow and Brandl 2003). Due to this, most of the research have focused on bacteria and the information available are more compared to fungi, yeast or algae. However, their establishment in order to function in plants is not always an easy task. Their communities are strongly influenced by certain antagonistic bacteria during the course of plant development.

8.5.2 Factors Affecting Interaction of Epiphytes with Hosts

The factors affecting the interaction of the epiphyte with the host are as follows.

8.5.2.1 Presence/Absence of Light

Light plays a vital role in growth, development and interaction of the epiphytes with the host. Unlike the rhizosphere, the presence of light on the phyllosphere is high and microorganisms like bacteria and fungi utilize it to produce different chemical substances that activate various metabolic pathways which help them to grow and survive. Apart from these, various aspects of physiological changes are brought about by light in plants which influence the secretion from plants, production of secondary metabolites which have a large impact on the diversity of available microorganisms (Schaller et al. 2015). It also has an influence on the biological control behaviour of the epiphytes towards the pathogens.

8.5.2.2 Temperature

Besides various abiotic stresses, temperature also has a great influence on the colonization and establishment of the microorganism in the phyllosphere. The fluctuating temperature which changes in night and day regimes in the phyllosphere brings about large impact on the available microbial population (Meena et al. 2015). The temperature not only affects microbial colonization but also has great impact on plant growth which ultimately hinders the interaction between microorganisms with plants. The microbial population colonizing in the phyllosphere is always subjected to seasonal and diurnal fluctuation of moisture and heat which also has a large impact on the microbial population of the phyllosphere.

8.5.2.3 Species of Plants

The presence and absence of types of plant species also have a significant influence on the identity of the microorganisms present. The microbial population present on plants varies from species to species. Even if different species are grown adjacent to one another, the microorganism harbouring the species will differ. Their presence mostly depends on various factors such as fruit content, flower present and stem which affect the growth and activities of the microorganisms (Kinkel et al. 2000).

Apart from the age of the plants, the growing phase of the plant or developmental stage determines the interaction of microbial organism with the plant. Also, the defence system activates through response of various elicitors of biotic or abiotic origin that have a great influence on determining the colonization of the microorganisms (Boller and Felix 2009).

8.5.2.4 Interaction between Microorganisms

The interaction between microorganisms also has a great influence in the colonization of the microorganism in the phyllosphere. Even though there is not much understanding about the interaction, the extent of an outcome of these interactions has been well explained by many researchers (Spadaro and Droby 2016; Karlsson et al. 2017; Mauch-Mani et al. 2017; Ghorbanpour et al. 2018). These interactions between microorganisms are parasitism, competition and mutualism. Each group of microorganisms plays a vital role in the existence of the microbiome; absence of these can cause significant changes in their composition and function (Karlsson et al. 2017). The direct interaction of microorganisms by the primary colonizer and other opportunistic organisms utilizes the host and causes metabolic activities and malfunction, thereby suppressing plant defence mechanisms, colonizing in the host plant and thereby causing the disease.

8.5.2.5 Leaf Exudates

The presence of leaf exudates has a huge impact on the colonization of the epiphytes as they contain many microbial growth-promoting factors such as glucose, fructose, sucrose, amino acids, etc., which facilitate the establishment of the epiphytes (Hirano and Upper 1983). The presence of moisture in the leaves or release of moisture during the process of transpiration provides water which is beneficially utilized by the microorganism.

8.5.2.6 Appendages of Leaves

The leaf appendages have a great influence on the establishment of the epiphytes. The toughness in the surface of leaves and along the veins affects the growth and development of the microorganism present in the phyllosphere. The presence and absence of shallow depressions on leaf surfaces also influence the establishment of microorganisms. As more depression areas are present, the microorganisms benefit from easy attachment and establishment (Hirano and Upper 1983). Therefore, if more depression areas are present on the leaf surface, the higher will be the colonization of the microorganism.

8.5.2.7 Leaf Position

The arrangement of the leaves on the host also has a great influence on the establishment of microorganism in the phyllosphere, as the surface of the leaves which are in direct contact to sunlight is more exposed and contains relatively low epiphytes population as compared to other leaves which are in shade. Also, the exposed leaves are mostly dry and do not contain moisture which makes it difficult for microorganisms to establish on the exposed parts of the leaves at an early stage of attachment and grow in that microenvironment (Shiraishi et al. 2015; Esser et al. 2015).

8.5.2.8 Types of Hosts and Their Age

Most of the epiphytes choose their host for establishment depending upon the availability of nutrients which make them specific to some particular host. However, with the change in species and ages of the plant secretion of exudates, the number of stomata and rate of transpiration differs greatly which limits the growth and establishment of the epiphytes.

8.5.2.9 Stomatal Opening

The presence of the number of stomata and its opening also affects the establishment of epiphytes in the phyllosphere. The stomata mostly contain water in the form of droplets which support the growth of microorganisms in the phyllosphere. Therefore, in and around stomata, the populations of microorganisms are comparatively high.

8.5.2.10 Pollutants

The presence and absence of pollutants on the leaf surface also inhibit the establishment and colonization of the epiphytes. The deposition of pollutants in the form of dust on the phyllosphere makes the microorganism difficult in settlement and inhibits the interaction with the host to function properly.

8.5.2.11 Pesticide Usage

The use of pesticides such as insecticides, herbicides, antibiotics or other chemicals to the plants for managing various pests or weeds causes a negative impact on the population build up of beneficial microorganisms. This causes a reduction in the epiphytic population and also hinders their functioning with the host plant.

8.5.2.12 Defence Compounds Produced by Plants

The plant has various defensive compounds which are present within the host either before the pathogen attack or are activated after attack. These compounds are known as phyto-allexin or pathogenesis-related proteins. This naturally occurring compound varies from species to species of plant and is produced in response to the elicitors. Elicitors can be classified into two categories such as biotic, which includes various chemical compounds excreted or produced by the colonizing phyllospheric microorganisms; and abiotic elicitors, which include mostly UV light damage of tissues, temperature fluctuation, etc., and have a large impact on the production of defensive compounds, thereby activating defensive genes by the host which hinders the establishment of the epiphytes (Hirano and Upper 1983).

8.6 EXPLORING THE PHYLLOSPHERE MICROBIOME THROUGH MOLECULAR APPROACHES

Scientists across the world have evaluated various techniques for exploring the diversity and affluence of the microbiome and their community dynamics prevailing on the phyllosphere. Basically, two approaches are elucidated by the scientists, firstly *in situ*, the microorganisms are explored on the surface of the leaf directly or indirectly by any kind of microscope using any advanced technology (microscopic evaluation by bright field, confocal microscope, fluorescence, TEM and SEM), and secondly *ex situ*, the microorganisms are washed or removed from the surface for further exploring by other means, viz., different culture-dependent (leaf washings plating and impression techniques) or -independent processes (PCR-DGGE, PLFA, CLPP, ARDRA, RAPD, Next Gen sequencing techniques like Pyrosequencing, Illumina Sequencing). All these techniques have been successfully utilized to study the dynamics of both rhizosphere and phyllosphere (Thapa et al. 2017).

Previously so-called first-generation molecular techniques like Sanger sequencing, terminal restriction fragment length polymorphism (TFLP) and denaturing gradient gel electrophoresis (DGGE) have been applied to study the difference in microbiome structure in the context of plant genotype, plant phenotype and geographical location. There are lots of disadvantage like low output and more expensive and allow only a superficial comparison of microbial communities (Rastogi and Sani 2011).

Nowadays, the culture-independent mass sequencing methods have been improved with advanced high throughput molecular methods for detecting the superficial microbial community of the environment. Now next-generation DNA sequencing is being extensively used with reduced cost and time by permitting hundreds of samples in a single run. The 454 pyrosequencing technique was widely implemented to study the microbial community through rRNA or ITS amplicon sequencing, whole-genome sequencing, shotgun metagenomics and transcriptional profiling (Delmotte et al. 2009). The Illumina platform is used for sequencing of microbial communities and produces large amounts of sequence data that are several orders of magnitude higher. The change in landscape of microbial ecology has offered a new window of "omic exploration". The proteogenomics and convergence of metagenomic with metaproteomic analysis focuses on new insights into the structure, function and variability of microbiota in the phyllosphere and other environments by facilitating comparative ecological analysis (Rastogi et al. 2013). Through these techniques, scientists have disclosed that *Actinobacteria*, *Firmicutes*, *Pseudomonas*, *Bacteroidetes*, *Sphingomonas*, *Bacillus*, *Massilia*, *Methylobacterium*, *Arthrobacter*, *Pantoea* and *Proteobacteria* genera consistently strengthen in the phyllosphere (Bulgarelli et al. 2013) and the variation of the microbial community structure is based on the genotypic nature of the plant species and also its geographical location (Sivakumar et al. 2020).

Microbial diversity on several host plants such as arabidopsis, apple tree, oak, beech, poplar, grapevine, Prunus, rice, soybean, spinach, tomato wheat, etc., have been documented through those methods. The fundamental facts of growth behaviour, colonization ability, genus-level community structure formation (or) association, low and high index of diversity and the host genotype effects on the self-defence as well as the cell wall integrity can be explored from the metadata produced from metagenomic studies (Sivakumar et al. 2020).

8.6.1 Sanger Method

Frederick Sanger developed a method where he opted for involving a third form of the ribose sugar, rather than using chemical cleavage reactions for sequencing (Sanger et al. 1977). Ribose is having a hydroxyl group on both the 20 and 30 carbons, where there is only 1-hydroxyl group on the 30 carbons of deoxyribose. But hydroxyl group is missing from both the 20 and 30 carbons in the third form of ribose, dideoxyribose. The chain irreversibly stops or terminates whenever a dideoxynucleotide is incorporated into a polynucleotide. Sanger developed the basic idea behind the termination method in 1974 for generating all the possible single-stranded DNA molecules complementary to a template, starting at a common 50 base and extends up to 1 kb in the 30 directions. These DNAs are labelled to identity of the 30-end base in each molecule. With the help of Electrophoresis, the molecules are separated according to size and each class of molecule produces a band differing in length by one nucleotide from the adjacent band (Bisht and Panda 2014).

Feil et al. (2005) used the Sanger capillary sequencing for sequencing the genome of *Pseudomonas syringae* pv. *syringae* B728, a common epiphyte that under certain conditions can cause bacterial brown spot disease of bean.

8.6.2 Illumina Sequencing

Single-stranded library fragments are amplified through a process called "bridge amplification". In Illumina, more than 50 million clusters/flow cells, each with 1000 copies of the same template, 1 billion bases per run and 1% of the cost of a capillary-based method make it cost-effective, highly accurate, straightforward sample preparation and well-established open-source software community and less time-consuming method (Bisht and Panda 2014). Remus-Emsermann et al. (2013) used the high-throughput Illumina sequencing to discover the genes associated with phyllospheric fitness of *Pantoea agglomerans* 299R, which was used as a model bacterium to the physiological adaptation of phyllosphere to the environment. Different adaptation mechanisms like sugar utilization, DNA repair systems and production of osmoprotectants have been revealed through this sequencing. The genome of *Pseudomonas syringae* pv. *syringae* B728 has been resequenced using paired-end Illumina sequencing with 25× coverage. It revealed that genes encode for DNA repair, UV resistance, reactive oxygen species, siderophores and indole-3-acetic acid and they may contribute to epiphytic growth during colonization. The B728a genome also contained many biosynthetic pathways which code for osmoprotectants (e.g., trehalose, betaine, ectoine) as an adaptive strategy to withstand osmotic stress on the leaf surface (Farrer et al. 2009). Xie et al. (2015) explored the phyllosphere bacterial community of *Wolffia australiana* in the paddy soil ecosystem; using Illumina sequencing revealed *Proteobacteria* and *Bacteroidetes* as the predominant phyla.

8.6.3 Pyrosequencing

The method pyrosequencing is based on the "sequencing by synthesis" principle to determine the order of nucleotides in DNA. The pyrosequencing relies on enzymes ATP sulphurylase and luciferase. Advantages of pyrosequencing are very fast process and read length up to 500 bp. It is different from Sanger sequencing as it relies on the detection of pyrophosphate release on nucleotide incorporation, rather than chain termination with dideoxynucleotides (Patricia et al. 2015). A limitation of the pyrosequencing method is that the lengths of individual reads of DNA sequence are approximately 300–500 nucleotides, but in Sanger sequencing method, the lengths of individual reads are 800–1000 nucleotides, making the process more difficult, particularly for sequences which contain large amount of repetitive DNA (Bisht and Panda 2014).

Greatest variability in community composed of generalist fungal species: *Lalaria*, *Woollsia* and *Taphrina* were observed at very small spatial scales, that is, at the leaf level of the European beech (*Fagus sylvatica*) was found using 454 pyrosequencing of the ITS1 region (Cordier et al. 2012). A strong correlation between genetic distance of beech trees and differences in fungal community composition has been found indicating that host genetics is a determinant of fungal community assembly on beech leaves. Balint et al. (2013) used ITS pyrosequencing analysis to explore balsam poplar (*Populus balsamifera* L.) phyllosphere and found the plant genotype has an important role in fungal community composition.

8.6.4 Denaturing Gradient Gel Electrophoresis

Bacterial community composition in the phyllosphere of *Deschampsia antarctica* illustrated by the sequencing of DGGE bands revealed significant differences in the community composition at different locations of Antarctica (Cid et al. 2017). Irrespective of different locations, Pseudomonadales (*Pseudomonas* and *Psychrobacter*) and Rhizobiales (*Agrobacterium* and *Aurantiomonas*) orders were found to be dominant over others. Thapa et al. (2018) analysed the composition of rice (variety Pusa Basmati 1509) phyllosphere under different fertilizer application and the method of rice cultivation (conventional-flooded, direct seeded rice/DSR and system of rice intensification/SRI) microbiome by PCR-DGGE. They found a distinct modulation in the communities belonging to phyla such as Bacteroidetes, Firmicutes, and Planctomyces, besides Proteobacteria (Table 8.2).

8.6.5 Community-Level Physiological Profiling

This is a different form of studying phyllosheric bacteria based on catabolic diversity of microorganisms. Microtitre plate format which consists of a wide range of organic substrates varying in structural complexity are evaluated by recording

TABLE 8.2
Dominating Phyllospheric Microorganisms on a Different Host

Host Species	Dominating Species	References
Cultivated rice	*Proteobacteria, Firmicutes, Planctomycetes, Cyanobacteria, Actinobacteria*	Thapa et al. (2018)
Maize	*Sphingomonas, Acinetobacter*	Kadivar and Stapleton (2003)
Wheat	*γ-Proteobacteria*	Gu et al. (2010)
Plant species (*Amygdalus communis, Citrus paradisi, Nicotiana glauca*)	*Bacilli, Actinobacteria*	Izhaki et al. (2013)

the ability to metabolize a range of individual carbon sources in community-level physiological profiling (CLPP). CLPP along with DGGE revealed the phyllosphere microbial community of powdery mildew-infected cucumber and Japanese spindle with their functional diversity, species richness and evenness of infection (Suda et al. 2009).

8.6.6 Genome-Wide Association Studies

Genome-wide association studies or whole genome association studies are carried out to investigate the contribution of host plants in shaping the leaf microbial community of *Arabidopsis thaliana* using single nucleotide polymorphism data. According to Horton et al. (2014), plant loci responsible for defence and cell wall integrity have a wide effect in shaping the bacterial community of the phyllosphere (Table 8.3). The methods of rice cultivation and fertilizer application along with host genotype significantly modulated both the structural (taxonomical) and functional attributes of the rice phyllosphere microbiome (Thapa et al. 2017, 2018). *Proteobacteria*, *Bacteroidetes* and *Actinobacteria* are the dominating species on *Arabidopsis* (Horton et al. 2014).

8.6.7 Overview of the Omic Strategies to Explore the Phyllosphere Microbiome

In recent years, scientists have sequenced thousands of bacterial and fungal isolate genome from different plant environments. It helps in identifying new genes belonging to unexplored microorganisms and revealing the facts behind several adaptations of plants including roots and shoots, nodulation and nitrogen fixation, bio control activity and quorum sensing (Yin et al. 2015). The meta-prefix in metagenomics indicates that the data which were represented reveal the whole microbial community, and were not from any single isolate. Genes present in the endosphere, rhizosphere of different plants and discovery of novel metabolic enzymes can be sequenced by the metagenome sequence (Levy et al. 2018a).

The genes that are differently expressed under certain conditions can be detected by transcriptomic analysis of plant-associated bacteria using RNA sequencing technology or gene expression micro-array approaches (Levy et al. 2018b). Recently, two highly correlated approaches have been developed by Nobori et al. (2018) to significantly enrich the transcriptome of *Pseudomonas syringae* in *Arabidopsis* leaf infection. Transcripts of the entire community from environmental samples are directly sequenced by metatranscriptomics. Chapparro et al. (2014) and Chapelle et al. (2016) have applied metatranscriptomics to explore and identify bacterial and fungal genomes, respectively, from *Arabidopsis* plant.

The basis of the proteomic and metaproteomic approaches is the liquid chromatography-tandem mass spectrometry technology which helps in exploring the diversity of the microbial protein within an environment of semi-quantitative manner. The technique consists of sample collection, protein extraction, isolation and fractionation, mass spectroscopy and comparison with proteome database (Levy et al. 2018b).

TABLE 8.3
Microbe-Mediated Abiotic Stress Tolerance in Plants

Abiotic Stress Plant	Microbial Inoculants	Plant	Tolerance Strategy
Salt	*Bacillus subtilis* GB03	*Arabidopsis thaliana*	Tissue-specific regulation of sodium transporter HKT1
	Pseudomonas simiae	*Glycine max*	4-Nitroguaiacol and quinoline promote seed germination in soybean
	Pseudomonas syringae DC3000, *Bacillus* sp. strain L81, *Arthrobacter oxidans*	*A. thaliana*	Salicylic acid-dependent pathway
	Root-associated plant growth-promoting rhizobacteria (PGPR)	*Oryza sativa*	Salt stress-related RAB18 plant gene expression
	Cyanobacteria and cyanobacterial extracts	*O. sativa*, *Triticum aestivum*, *Zea mays*, *Gossypium hirsutum*	Phytohormones (elicitor)
	Pseudomonas koreensis strain AK-1	*G. max* L. Merrill	Reduction in Na^+ level and increase in K^+ level
	Glomus etunicatum	*G. max*	Increased root but decreased shoot proline concentrations
	Burkholderia, *Arthrobacter* and *Bacillus*	*Vitis vinifera*, *Capsicum annuum*	Increased accumulation of proline
	Glomus fasciculatum	*Phragmites australis*	Carbohydrates accumulation
	Glomus intraradices	*G. max*	Carbohydrates accumulation
Osmotic stress	*Bacillus megaterium*	*Z. mays*	High hydraulic conductance, increased root expression of two ZmPIP isoforms
	Glomus intraradices BEG 123	*Phaseolus vulgaris*	High osmotic root hydraulic conductance due to increased active solute transport through roots

(Continued)

TABLE 8.3 *(Continued)*
Microbe-Mediated Abiotic Stress Tolerance in Plants

Abiotic Stress Plant	Microbial Inoculants	Plant	Tolerance Strategy
Drought	*Rhizobium tropici* and *Paenibacillus polymyxa* (co-inoculation)	*P. vulgaris*	Stress-tolerant gene upregulation
	Burkholderia phytofirmans, *Enterobacter* sp. FD17	*Z. mays*	Increase of photosynthesis, root and shoot biomass under drought conditions
	Bacillus thuringiensis AZP2	*T. aestivum*	Volatile organic compounds production
	Pseudomonas chlororaphis O6	*A. thaliana*	Production of 2R,3R butanediol: A volatile compound
	Pseudomonas putida strain GAP-P45	*Helianthus annuus*	production of epoxypolysaccharide
	Bacillus licheformis strain K11	*Capsicum annum*	Stress-related genes and proteins
	Bacillus cereus AR156, *B. subtilis* SM21 and *Serratia* sp. XY21	*Cucumis sativa*	Production of monodehydro-ascorbate, proline and antioxidant enzyme, gene expression
Salinity	*Azospirillum brasilense* and *Pantoea dispersa* (co-inoculation)	*C. annuum*	High stomatal conductance and photosynthesis
	Glomus intraradices BAFC 3108	*Lotus glaber*	Decrease of root and shoot Na^+ accumulation and enhanced root K^+ concentrations
	Glomus clarum and *G. etunicatum*	*Vigna radiata*, *C. annuum*, *T. aestivum*	Decrease of Na^+ in root and shoot and increase in concentration of K^+ in root
	Bacillus subtilis	*Arabidopsis*	Decrease in root transcriptional expression of a high-affinity K^+ transporter (AtHKT1) decreasing root Na^+ import
	Glomus intraradices BEG121	*Lactuca sativa*	Reduce in concentration of ABA
	Pseudomonas putida Rs-198	*G. hirsutum*	Prevention of salinity-induced ABA accumulation in seedlings
	Azospirillum brasilense strain Cd	*P. vulgaris*	Stimulation of persistent exudation of flavonoids
	B. subtilis	*L. sativa*	Root-to-shoot cytokinin signalling and stimulation of shoot biomass

(Continued)

TABLE 8.3 ***(Continued)***
Microbe-Mediated Abiotic Stress Tolerance in Plants

Abiotic Stress Plant	Microbial Inoculants	Plant	Tolerance Strategy
Heat	*Bacillus amyloliquefaciens, Azospirillum brasilence*	*T. aestivum*	Reduction in regeneration of reactive oxygen species, pre-activation of heat shock transcription factors, changes in metabolome
Heat and drought	*Curvularia proturberata* isolate Cp4666D	*Dichanthelium lanuginosum, Solanum lycopersicum*	Colonization of roots
Arsenic toxicity	*Staphylococcus arlettae*	*Brassica juncea*	Increase in soil dehydrogenase, phosphatase and available phosphorus
Pb/Zn toxicity	*Phyllobacterium myrsinacearum*	*Sedum plumbizincicola*	Resistance to 350 mg/L Cd, 1000 mg/L Zn, 1200 mg/L Pb
Zn toxicity	*Pseudomonas aeruginosa*	*T. aestivum*	Improved biomass, N and P uptake and total soluble protein
	Enterobacter intermedius MH8b	*Sinapis alba*	ACC deaminase, IAA, hydrocyanic acid, P solubilization
	Pseudomonas brassicacearum, Rhizobium leguminosarum	*B. juncea*	Metal-chelating molecules
Hg toxicity	*Photobacterium* sp.	*P. australis*	IAA, mercury reductase activity

Metabolomics, a powerful tool, detect and quantify small molecules and molecular dynamics, significantly boost plant health, growth and resilience to stress at the plant bacterial interface (Levy et al. 2018b). Metaproteogenomics helps in the characterization of *Rhizobium*, *Methylobacterium* and *Microbacterium*, colonizing on rice plant (Knief et al. 2012).

8.7 ADAPTATION MECHANISMS UNDER VARIOUS STRESSES

8.7.1 Genetic and Metabolic Adaptations of Phyllosphere Microbiota

A plant has to fight against several biotic and abiotic stresses in the environment. Different defence signals are produced due to the stress in the plant and therefore interaction between plant and microbe can be decided by prioritization of physiological pathways in plants (Schenk et al. 2012). Scientists have revealed the facts of the microbial community composition and their functional behaviour in complex environments like the rhizosphere and phyllosphere by the application of the data-driven science of multi-omics.

Recently, meta-omics approaches like metagenomics, metatranscriptomics and metaproteomics have advanced as promising tools to explore microbial communities and functions within a given environment at a deeper level (de Castro et al. 2013). Genomic analysis of both the host and associated microbial communities especially phyllosphere-associated microbial communities helps to identify the system of interaction between plant and microorganisms. Functional potential of microbial communities in terms of the abundance of the genes involved in particular metabolic processes linked with stresses or stress alleviation mechanisms can be detected by metagenomics.

Metatranscriptomics can unveil the dynamics of rhizosphere microbiome structure. Community-wide gene expression, protein abundance and putative proteins can be revealed by metaproteomics (Turner et al. 2013). Comparison of transcriptome profiles is used to identify different sets of transcripts responsible for differences between two biologically different expressions in various conditions (Bräutigam and Gowik 2010). Plant-microbe interactions can be reflected by the use of mRNA sequencing analysis and microarray techniques which produce transcriptome level information (Budak et al. 2015; Wang et al. 2016). Under abiotic stresses like drought, salinity and cold, miRNAs play an important role in rice, *Medicago*, *Phaseolus*, *Arabidopsis* and other plants (Trindade et al. 2010).

Proteomic, being a powerful tool for the exploration of physiological metabolism and protein-protein interactions in microorganisms and plants, leads to generate a knowledge of the regulation of biological system by identifying several proteins as signal of changes in physiological status due to biotic and abiotic stress or factors responsible for stress alleviation (Silva-Sanchez et al. 2015). The proteins produced in stressed, non-stressed and microbe-associated plants can be explored by the comparative analysis of proteomics. Stress responses in *Arabidopsis*, wheat (*Triticum aestivum*), durum wheat (*Triticum durum*), barley (*Hordeum vulgare*), maize (*Zea maize*), rice (*Oryza sativa*), soybean (*Soybean max*), common bean (*Phaseolus vulgaris*), pea (*Pisum sativum*), oilseed rape (*Brassica napus*), potato (*Solanum tuberosum*) and tomato (*Lycopersicon esculentum*) have been widely studied with the application of proteomics (Liu et al. 2015; Kosova et al. 2015; Xu et al. 2015; Wang et al. 2016). Scientists are giving more focus to the proteomic study of methylotrophic bacteria and constitute a major portion of phylosphere community, typically leaf surface, where one-carbon substrate, methanol is easily available via transpiration activity (Table 8.3). Plant growth regulation molecules are released by the pink-pigmented facultative methylotrophic (PPFM) and the potentiality of these PPFMs under various stress is explored and revealed by proteomic studies bacteria (Meena et al. 2012; Tani et al. 2012; Yim et al. 2013; Araujo et al. 2015). The detailed proteomic study of these phyllosphere communities helps to reveal novel ideas regarding the involvement of proteins in the survival mechanisms of microorganisms under relatively inadequate environments, generally encountered on leaf surfaces of plants, where in addition to intense solar radiation, there exists frequent scarcity of several nutrients. The identification of proteins involved in these processes is sufficient to create a boom in stress alleviation strategies at the molecular aspects where direct implementation of the active molecules was thought upon rather than employing the whole organism. The high-throughput mass spectrometric profiling of metabolites of plant-associated microorganisms under the influence of stressors can reveal the level of interference by the stressor in the overall cellular homeostasis (Meena et al. 2017).

Metaproteogenomic analysis of leaf surface communities identified porins and transporter-related proteins (TonB transporters), which permits the transport of substrates (e.g., sugars, vitamins, siderophores) by *Sphingomonas* allowing *Sphingomonas* to be competitive with other colonizers. Methanol is a by-product of pectin demethylation during plant cell wall metabolism abundantly present on plant leaf surfaces (Galbally and Kirstine 2002). Proteome profiling of the proteins belonging to *Methylobacterium* spp. reveals that the proteins are involved in the assimilation of methanol, found in high abundance in the phyllosphere (Delmotte et al. 2009).

The identification of proteins involved in methylotrophy and their function in *Methylobacterium* spp. seems to explain to a large extent of epiphytic fitness of this bacterium in the phyllosphere of several plants. In another study (Knief et al. 2012), a comparison between foliar microbiota and rhizosphere microbiota helps in identification of unique metabolic processes that are specific to microbiota from the two compartments.

At the genomic level, phyllosphere communities residing upon rice leaf consist of *Rhizobium*, *Methylobacterium* and *Microbacterium*. Proteome analysis reveals the occurrence of many methylotrophic enzymes (e.g., methanol dehydrogenase, formaldehyde-activating enzyme) that were assigned to *Methylobacterium*. Genes associated with nitrogen fixation were found in both the phyllosphere and rhizosphere samples (Rastogi et al. 2013).

Gourion et al. (2006) applied proteomic profiling of *Methylobacterium extorquens* colonizing leaves or roots or growing on synthetic medium. Proteins involved in methylotrophic metabolism (e.g., MxaF and Fae) and stress responses (e.g., PhaA) were found to be upregulated during epiphytic growth. This study also reveals the facts on the two-component regulatory mechanisms involving response regulator PhyR, which plays an important role in controlling many proteins (e.g., SodA, KatE) that contributes to phyllosphere colonization.

8.8 POTENTIALITIES OF THE PHYLLOSPHERE MICROBIOME TOWARDS SUSTAINABLE PLANT GROWTH

In comparison with the rhizosphere microorganism, the phyllosphere microorganism has always been recognized as least important even though they have a huge influence on the growth and development of plants and resistance towards various invading pathogens. They are less species rich in microorganism assemblages as compared to the rhizosphere (Knief et al. 2012). The most commonly reported bacterial species are those belonging to class gamma proteobacteria that are actively involved in functioning various mechanisms such as methyl trophy, nitrification, anoxygenic photosynthesis and nitrogen fixation in plants. Other bacterial species belonging to family such as cytophagaceae or chitinophagaceae, and actinobacteria are mostly antagonistic to fungi, well decomposer and nitrogen fixer (Palaniyandi et al. 2013). The role of actinobacteria has also been well recognized to promote the growth of plant when foliar application is done. Fungi are also an important component of the phyllosphere inhibiting microorganisms and help in functioning various defence mechanisms in plants. Their population mostly fluctuates depending upon growing season and host genotype. The important fungi that are dominant in the leaf surface are yeast belonging to class ascomycetes and basidiomycetes.

8.8.1 Functions of the Phyllosphere Microbiome in Sustainable Plant Health

The phyllosphere microbiome helps the host plant to overcome the challenges in the following ways.

8.8.1.1 Plant Nutrient Acquisition

Many bacteria have been reported in the phyllosphere that have the capacity to fix nitrogen (Holland 2011). However, the mechanistic study is still lacking and needs to be thoroughly studied about the plant phyllosphere nitrogen dynamic. Sattelmacher et al. (1998) reported that the nitrogen fixation mostly occurs in the internal tissues of the leaf rather than on the leaf surface and allows nitrogenous products to reach the plant cells in temperate regions. However, tropical region nitrogen fixation has been reported at the surface of leaves due to the presence of high moisture availability that allows the activation of nitrogen-fixing bacteria (Fürnkranz et al. 2008). Hietz et al. (2002) reported that due to the presence of high microbial nitrogen fixer in the canopies of higher plants in tropical areas, the plants have been found to fix nitrogen in the phyllosphere. Such nitrogen-fixing microorganisms can help in promoting plant growth and nitrogen content subjected that the microorganisms are present internally. In spite of the importance and huge potentialities of microorganisms in providing nutrients, the direct mechanistic demonstration is indeed a need of the hour of the phyllosphere nitrogen fixer.

8.8.1.2 Host-Stress Tolerance

Only a handful of information is available on how phyllosphere microorganisms promote robust plant stress and responds and promotes growth. However, its impact has been well known towards overcoming extreme drought and temperature effects on the plant. The phyllospheric bacteria produce exopolysaccharide that helps in protecting the plants from UV radiation and desiccation. Some of the bacteria such as *Pseudomonas putida* produce bioflames that helps to resist water oxygen and nutrient diffusion. They also help plants to overcome the high effect of UV radiation, water losses, plant development and genetic damage. The microorganisms present in the phyllosphere have also been reported to increase resistance against drought and tolerant towards through hormonal stimulation in plants. Rico et al. (2014) reported that phyllosphere microbiome can increase nitrogen fixation that is triggered by drought. Apart from these, they are found to play an important role in heat tolerance such as in case of wheat where fungal endophyte induces better seed germination and yield in heat-stress conditions (Hubbard et al. 2014). Some of the microbiomes are also found to produce proteins that are triggered by heat shock with various temperature fluctuations that helps to overcome the host plant reaction from such abiotic stresses. Several studies have been documented regarding formation of ice crystals that are generated by bacterial cell wall protein in temperate regions which help to raise the temperature and protect the plant from ice damage that allows plants to access nutrients more easily (Lindow et al. 1982a, 1982b; Attard et al. 2012). Another important mechanism employed by phyllosphere colonizing microorganisms is tolerance towards chemical pollutants by promoting roots to grow deeply and reducing uptake of various chemicals, thereby limiting their harmful effect in the aerial portion. Some of the species of *Pseudomonas* help in preventing such toxic effect in many plants. The impact of global changes in climate, temperature and air increases carbon dioxide concentration and drought may have a huge impact on host-microbe interaction in the phyllosphere, where a lot of research and understanding has to be devoted.

8.8.1.3 Plant Hormonal Changes

The impact of phyllosphere microbiome in inducing plant hormone is comparatively less as compared to rhizosphere inhibiting microorganism. The phyllosphere inhabiting microorganisms also have the capacity to synthesize growth-promoting hormones, such as IAA, thereby enhancing plant growth. Apart from these, some microorganisms have also been reported to produce cytokinin which helps in stimulating nitrogen transportation to the plant tissues and thereby increasing growth of plant biomass. Some phyllospheric microorganisms have also been reported to produce abscisic acid (Cao et al. 2011) that causes stomatal closure to overcome the effect of drought in some plants.

8.8.1.4 Plant-Pathogen Interactions

The phyllosphere inhabiting microorganisms are very active and have the potential to overcome the effect of invading pathogens. They are an active colonizer and compete with pathogens for food and nutrients that inhibit directly and induce defence mechanisms in plants. Another mechanism in which the bacteria restricts the growth of fungal infection is through the production of enzymes such as beta-1,3-glucanase, chitinase and peroxidase (Fernando et al. 2007).

8.9 CONCLUSION AND FUTURE PROSPECTS

The phyllosphere microbiome even though it has a huge potentiality, there has been much less research done to understand the mechanisms of host-microbe and microbe-microbe interactions. The use of highly effective techniques such as omics, transcriptomic, metagenomics, high-throughput sequencing and molecular understanding needs to be upregulated to know the mechanisms of the phyllosphere microbiome in future research. Such approaches could be helpful in better understanding interaction between host microorganisms and pathogen in a broader sense. They can also be used to investigate temporal and spatial population dynamics of the phyllosphere inhabiting microorganism as well as the competition with the other microbial community and their dominance. Since the interaction is not very well understood, exploring it in the right way can help us to understand whether the phyllosphere members act as an ecosystem engineer in modifying various physiological changes in plants or by producing various exopolysaccharide. Therefore, the identification of key species of microorganisms will definitely help in plant protection and production in a sustainable way. Apart from these, applications of various modelling techniques can help complex microbiome to trace and predict the role of phyllosphere microbiome more accurately. These approaches can help in identifying microorganisms that are actively involved in functioning as a network hub in improving plant health which can be a strong ecological keystone species. The functioning of microbiome is mostly dependent on various abiotic factors and development of proper modules can help us to predict even a single variable in response to data generated non-parametric microbe and environment. Therefore, proper research and understanding is indeed a need of the hour to explore further mechanisms of the phyllosphere microorganism.

REFERENCES

Araujo, W.L., Santos, D.S., Dini-Andreote, F., Salgueiro-Londono, J.K., Camargo-Neves, A.A., Andreote, F.D. 2015. Genes related to antioxidant metabolism are involved in *Methylobacterium mesophilicum* soybean interaction. *Antonie Van Leeuwenhoek* 108:951–963.

Arunkumar, N., Rakesh, S., Rajaram, K., Kumar, N.R., Durairajan, S.S.K. 2019. Anthosphere microbiome and their associated interactions at the aromatic interface. In: Varma, A., Tripathi, S., Prasad, R. (Eds.) *Plant Microbe Interface*. Springer, Cham, 309–324. doi: 10.1007/978-3-030-19831-2_14.

Atamna-Ismaeel, N., Finkel, O. M., Glaser, F., Sharon, I., Schneider, R., Post, A. F., et al. (2012a). Microbial rhodopsins on leaf surfaces of terrestrial plants. *Environmental Microbiology* 14, 140–146. doi: 10.1111/j.1462-2920.2011.02554

Attard, E., Yang, H., Delort, A.M., et al. 2012. Effects of atmospheric conditions on ice nucleation activity of *Pseudomonas*. *Atmospheric Chemistry and Physics* 12: 10667–10677.

Balint, M., Tiffin, P., Hallstrom, B., O'Hara, R.B., Olson, M.S., Fankhauser, J.D., Piepenbring, M., Schmitt, I. 2013. Host genotype shapes the foliar fungal microbiome of balsam poplar (*Populus balsamifera*). *PLoS ONE* 8: e53987.

Beattie, G.A. 2002. Leaf surface waxes and the process of leaf colonization by microorganisms. In: Lindow, S.E., Hecht-Poinar, E.I., Elliott, V.J. (Eds.) *Phyllosphere Microbiology*. APS Press, St. Paul, pp.3–26.

Bisht, S.S., Panda, A.K. 2014. DNA sequencing: methods and applications. In: Ravi, I., Baunthiyal, M., Saxena, J. (Eds.) *Advances in Biotechnology*. Springer, New Delhi.

Boller, T., Felix, G. 2009. A renaissance of elicitors: perception of microbe-associated molecular patterns and danger signals by pattern-recognition receptors. *Annual Review of Plant Biology* 60: 379–407. doi: 10.1146/annurev.arplant.57. 032905.105346.

Bräutigam, A., Gowik, U. 2010. What can next generation sequencing do for you? Next generation sequencing as a valuable tool in plant research. *Plant Biology* 12: 831–841.

Bringel, F., Couée, I. 2015. Pivotal roles of phyllosphere microorganisms at the interface between plant functioning and atmospheric trace gas dynamics. *Frontiers in Microbiology* 6: 486.

Budak, H., Kantar, M., Bulut, R., Akpinar, B.A. 2015. Stress responsive miRNAs and iso miRs in cereals. *Plant Science* 235: 1–13.

Bulgarelli, D., Schlaeppi, K., Spaepen, S., Loren, V., van Themaat, E., Schulze-Lefert, P. 2013. Structure and functions of the bacterial microbiota of plants. *Annual Review of Plant Biology* 64: 807–838.

Burch, A.Y., Zeisler, V., Yokota, K., Schreiber, L., Lindow S.E. 2014.The hygroscopic biosurfactant syringafactin produced by *Pseudomonas syringae* enhances fitness on leaf surfaces during fluctuating humidity. *Environmental Microbiology* 16:2086–2098.

Cao, F.Y., Yoshioka, K., Desveau, D. 2011. The roles of ABA in plant-pathogen interactions. *Journal of Plant Research* 124: 489–499.

Chapelle, E., Mendes, R., Bakker, P.A.H., Raajimakers, J.M. 2016. Fungal invasion of rhizosphere microbiome. *The ISME Journal* 1: 265–268.

Chapparro, J.M., Badri, D.V., Vivanco, J.M. 2014. Rhizosphere micro-biome assemblage is affected by plant development. *The ISME Journal* 8: 790–803.

Cid, F.P., Inostroza, N.G., Graether, S.P., Bravo, L.A., Jorquera, M.A. 2017. Bacterial community structures and ice recrystallization inhibition activity of bacteria isolated from the phyllosphere of the Antarctic vascular plant *Deschampsia antarctica. Polar Biology* 40: 1319–1331.

Conrath, U., Beckers, G.J.M., Langenbach, C.J.G., Jaskiewicz, M.R. 2015. Priming for enhanced defense. *Annual Review of Phytopathology* 53: 97–119.

Cordier, T., Robin, C., Capdevielle, X., Fabreguettes, O., Desprez-Loustau, M.L., Vacher, C. 2012. The composition of phyllosphere fungal assemblages of European beech (*Fagussylvatica*) varies significantly along an elevation gradient. *New Phytopathology* 196: 510–519.

de Castro, A.P., Sartori, A., Silva, M.R., Quirino, B.F., Kruger, R.H. 2013. Combining "omics" strategies to analyze the biotechnological potential of complex microbial environments. *Current Protein and Peptide Science* 14: 447–458.

De Jager, E.S., Wehner, F.C., Korsten, L. 2001. Microbial ecology of the mango phylloplane. *Microbiological Ecology* 42: 201–207.

Delmotte, N., Knief, C., Chaffron, S., Innerebner, G., Roschitzki, B., Schlapbach, R., Mering, C.V., Vorholt, J.A. 2009. Community proteogenomics reveals insights into the physiology of phyllosphere bacteria. *Proceedings of the National Academy of Sciences* 106: 16428–16433.

Dhayanithy, G., Subban, K., Chelliah, J. 2019. Diversity and biological activities of endophytic fungi associated with *Catharanthus roseus. BMC Microbiology* 19: 22.

Esser, D.S., Leveau, J.H.J., Meyer, K.M., Wiegand, K. 2015. Spatial scales of interactions among bacteria and between bacteria and the leaf surface. *FEMS Microbiological Ecology*. doi: 10.1093/femsec/fiu034.

Farrer, R.A., Kemen, E., Jones, J.D., Studholme, D.J. 2009. De novo assembly of the *Pseudomonas syringae* pv. *syringae* B728a genome using Illumina/Solexa short sequence reads. *FEMS Microbiological Letter* 291: 103–111.

Feil, H., Feil, W.S., Chain, P. 2005. Comparison of the complete genome sequences of *Pseudomonas syringae* pv. *syringae* B728a and pv. tomato DC3000. *Proceedings of the National Academy of Sciences* 102: 11064–11069.

Fernando, W.G.D., Nakkeeran, S., Zhang, Y., Savchuk, S. 2007. Biological control of *Sclerotinia sclerotiorum* (Lib.) de Bary by *Pseudomonas* and *Bacillus* species on canola petals. *Crop Protection* 26: 100–107.

Fokkema, N.J., Riphagen, I., Poot, R.J., De Jong, C. 1983. Aphid honeydew, a potential stimulant of *Cochliobolus sativus* and *Septoria nodorum* and the competitive role of saprophytic mycoflora. *Transactions of the British Mycological Society* 81: 355–363.

Fürnkranz, M., Wanek, W., Richter, A., Abell, G., Rasche, F., Sessitsch, A. 2008. Nitrogen fixation by phyllosphere bacteria associated with higher plants and their colonizing epiphytes of a tropical lowland rainforest of Costa Rica. *The ISME Journal* 2: 561–570.

Galbally, I.E., Kirstine, W. 2002. The production of methanol by flowering plants and the global cycle of methanol. *Journal of Atmospheric Chemistry* 43: 195–229.

Ghorbanpour, M., Omidvari, M., Abbaszadeh-Dahaji, P., Omidvar, R., Kariman, K. 2018. Mechanisms underlying the protective effects of beneficial fungi against plant diseases. *Biological Control* 117: 147–157.

Glushakova, A.M., Chernov, I.Y. 2004. Seasonal dynamics in a yeast population on leaves of the common wood sorrel *Oxalis acetosella* L. *Microbiology* 73: 184–188.

Gourion, B., Rossignol, M., Vorholt, J.A. 2006. A proteomic study of *Methylobacterium extorquens* reveals a response regulator essential for epiphytic growth. *Proceedings of the National Academy of Sciences* 103: 13186–13191.

Gu, L., Bai, Z., Jin, B., Hu, Q., Wang, H., Zhuang, G., Zhang, H. 2010. Assessing the impact of fungicide enostroburin application on bacterial community in wheat phyllosphere. *Journal of Environmental Science* 22: 134–141.

Guerreiro, M.A., Brachmann, A., Begerow, D., Peršoh, D. 2018. Transient leaf endophytes are the most active fungi in 1-year-old beech leaf litter. *Fungal Diversity* 89: 237–251.

Heimpel, G. E., Mills, N. 2017. *Biological Control – Ecology and Applications.* Cambridge University Press, Cambridge.

Hietz, P., Wanek, W., Wania, R., Nadkarni, N.M. 2002. Nitrogen-15 natural abundance in a montane cloud forest canopy as an indicator of nitrogen cycling and epiphyte nutrition. *Oecologia* 131: 350–355.

Hirano, S.S., Upper, C.D. 1983. Ecology and epidemiology of foliar bacterial plant pathogens. *Annual Review of Phytopathology* 21: 243–269.

Holland, M. 2011. Nitrogen: give and take from phyllosphere microbes. In: Ploacco, J.C., Todd, C.D. (Eds.) *Ecological Aspects of Nitrogen Metabolism in Plants.* 10: 217–230. John Wiley & Sons, Inc. New York.

Horton, M.W., Bodenhausen, N., Beilsmith, K., Meng, D., Muegge, B.D., Subramanian, S., Vetter, M.M., Vilhjálmsson, B.J., Nordborg, M., Gordon, J.I., Bergelson, J. 2014. Genome-wide association study of *Arabidopsis thaliana*'s leaf microbial community. *Natural Communication* 5: 5320.

Hubbard, M., Germida, J.J., Vujanovic, V. 2014. Fungal endophytes enhance wheat heat and drought tolerance in terms of grain yield and second-generation seed viability. *Journal of Applied Microbiology* 116: 109–122.

Inacio, J., Pereira, P., de Carvalho, M., Fonseca, A., Amaral-Collaco, M.T., Spencer-Martins, I. 2002. Estimation and diversity of phylloplane mycobiota on selected plants in a Mediterranean-type ecosystem in Portugal. *Microbial Ecology* 44: 344–353.

Izhaki, I., Fridman, S., Gerchman, Y., Halpern, M. 2013. Variability of bacterial community composition on leaves between and within plant species. *Current Microbiology* 66: 227–235.

Jackson, E.F., Echlin, H.L., Jackson, C.R. 2006. Changes in the phyllosphere community of the resurrection fern, *Polypodium polypodioides*, associated with rainfall and wetting. *FEMS Microbiological Ecology* 58: 236–246.

Jacques, M.A., Josi, K., Darrasse, A., Samson, R. 2005. *Xanthomonas axonopodis* pv. *phaseoli* var. *fuscans* is aggregated in stable biofilm population sizes in the phyllosphere of field-grown bean. *Applied Environmental Microbiology* 71: 2008–20015.

Kadivar, H., Stapleton, A.E. 2003. Ultraviolet radiation alters maize phyllosphere bacterial diversity. *Microbiological Ecology* 45: 353–361.

Karlsson, M., Atanasova, L., Jensen, D.F., Zeilinger, S. 2017. Necrotrophic mycoparasites and their genomes. *Microbiology Spectrum* 5: FUNK0016-2016.

Kaur, G., Zueweller, B., Motarvalli, P.P., Nelson, K. A. 2019. Screening corn hybrids for soil waterlogging tolerance at an early growth stage. *AGRIS* 9 (2). doi: 10.3390/agriculture9020033

Kim, M., Singh, D., Lai-Hoe, A., Go, R., Rahim, R.A., Ainuddin, A.N., Chun, J., Adams, J.M. 2012. Distinctive phyllosphere bacterial communities in tropical trees. *Microbiological Ecology* 63: 674–681.

Kinkel, L.L., Wilson, M., Lindow, S.E. 2000. Plant species and plant incubation conditions influence variability in epiphytic bacterial population size. *Microbiological Ecology* 39: 1–11.

Knief, C., Delmotte, N., Chaffron, S., Stark, M., Innerebner, G., Wassmann, R., Mering, C., Vorholt, J.A.2012. Metaproteogenomic analysis of microbial communities in the phyllosphere and rhizosphere of rice. *The ISME Journal* 6: 1378–1390.

Kohl, J., Kolnaar, R., Ravensberg,W.J. 2019. Mode of action of microbial biological control agents against plant diseases: relevance beyond efficacy. *Frontiers in Plant Science*. Article: 845.

Kosova, K., Vitamvas, P., Urban, M.O., Klima, M., Roy, A., Prasil, I.T. 2015. Biological networks underlying abiotic stress tolerance in temperate crops–a proteomic perspective. *International Journal of Molecular Science* 16: 20913–20942.

Kumar, M., Kour, D., Yadav, A.N., Saxena, R., Rai, P.K., Jyoti, A., Tomar, R.S. 2019. Biodiversity of methylotrophic microbial communities and their potential role in mitigation of abiotic stresses in plants. *Biologia* 74: 287–308.

Levy, A., Jonathan, M.C., Jeffery, L.D., Woyke, T. 2018a. Elucidating bacterial gene functions in the plant microbiome. *Cell Host and Microbe* 24: 475–485.

Levy, I., Salas-Gonzalez, M., Mittelviefhaus, S., Clingenpeel, S., Herrera-Paredes, J., Miao, K., Wang,G., Devescovi, K., Stillman, F.M. 2018b. Genomic features of bacterial adaptation to plants. *Nature Genetics* 50: 138–150.

Lindow, S.E., Brandl, M.T. 2003. Microbiology of the phyllosphere. *Applied and Environmental Microbiology* 69: 1875–1883.

Lindow, S.E., Arny, D.C., Upper, C.D. 1982a. Bacterial ice nucleation: a factor in frost injury to plants. *Plant Physiology* 70: 1084–1089.

Lindow, S.E., Hirano, S.S., Barchet, W.R. et al. 1982b. Relationship between ice nucleation frequency of bacteria and frost injury. *Plant Physiology* 70: 1090–1093.

Liu, Z., Li, Y., Cao, H., Ren, D. 2015. Comparative phospho-proteomics analysis of salt-responsive phosphoproteins regulated by the MKK9-MPK6 cascade in *Arabidopsis*. *Plant Science* 241: 138–150.

Lopez-Velasco, G., Welbaum, G.E., Boyer, R.R., Mane, S.P., Ponder, M.A. 2011. Changes in spinach phylloepiphytic bacteria communities following minimal processing and refrigerated storage described using pyrosequencingof16SrRNAamplicons. *Journal of Applied Microbiology* 110: 1203–1214.

Mauch-Mani, B., Baccelli, I., Luna, E., Flors, V. 2017. Defense priming: an adaptive part of induced resistance. *Annual Review of Plant Biology* 68: 485–512.

Meena, K.K., Kumar, M., Kalyuzhnaya, M.G., Yandigeri, M.S., Singh, D.P., Saxena, A.K. 2012. Epiphytic pink-pigmented methylotrophic bacteria enhance germination and seedling growth of wheat (*Triticum aestivum*) by producing phytohormone. *Antonie Van Leeuwenhoek* 101: 777–786.

Meena, R.K., Singh, R.K., Singh, N.P.et al. 2015. Isolation of low temperature surviving plant growth–promoting rhizobacteria (PGPR) from pea (*Pisum sativum* L.) and documentation of their plant growth promoting traits. *Biocatalysis and Agricultural Biotechnology* 4: 806–811.

Meena, K.K., Sorty, A.M., Bitla, U.M., Choudhary, K., Gupta, P., Pareek, A., Singh, D.P., Prabha, R., Sahu, P.K., Gupta, V.K., Singh, H.B., Krishanani, K.K., Minhas, P.S. 2017. Abiotic stress responses and microbe-mediated mitigation in plants: the omics strategies. *Frontiers in Plant Science* 8: 172.

Miller, M.B., Bassler, B.L. 2001. Quorum sensing in bacteria. *Annual Review Microbiology* 55: 165–99.

Nobori, T., Velásquez, A.C., Wu, J., Kvitko, B.H., Kremer, J.M., Wang, Y., He, S.Y., Tsuda, K. 2018. Transcriptome landscape of a bacterial pathogen under plant immunity. *Proceedings of the National Academy of Sciences of the United States of America* 115: 3055–3064.

Palaniyandi, S.A., Yang, S.H., Zhang, L., Suh, J.W. 2013. Effects of actinobacteria on plant disease suppression and growth promotion. *Applied Microbiology and Biotechnology* 97: 9621–9636.

Pascazio, S., Crecchio, C., Ricciuti, P., Palese, A.M., Xiloyannis, C., Sofo, A. 2015. Phyllosphere and carposphere bacterial communities in olive plants subjected to different cultural practices. *International Journal of Plant Biology* 6(1): 6011. doi: 10.4081/pb.2015.6011.

Patricia, J.S., Khare, R., Nancy, L.W. 2015. Rapidly growing mycobacteria. *Molecular Medical Microbiology*, pp.1679–1690. Elsevier Ltd. Academic Press, Amsterdam.

Pieterse, C.M.J., Zamioudis, C., Berendsen, R.L., Weller, D.M., Van Wees, S.C.M., Bakker, P.A.H.M. 2014. Induced systemic resistance by beneficial microbes. *Annual Review of Phytopathology* 52: 347–375.

Rastogi, G., Sani, R.K. 2011. Molecular techniques to assess microbial community structure, function, and dynamics in the environment. In: Ahmad, I., Ahmad, F., Pichtel, J. (Eds.) *Microbes and Microbial Technology*. Springer, New York, pp. 29–57.

Rastogi, G., Sbodio, A., Tech, J.J., Suslow, T.V., Coaker, G.L., Leveau, J.H.J. 2012. Leaf microbiota in an agroecosystem: spatiotemporal variation in bacterial community composition on field-grown lettuce. *The ISME Journal* 6: 1812–1822.

Rastogi, G., Coaker, G.L., Leveau, J.H.J. 2013. New insights into the structure and function of phyllosphere microbiota through high-throughput molecular approaches. *FEMS Microbiology Letters* 348: 1–10.

Remus-Emsermann, M.N.P., Kim, E.B., Marco, M.L., Tecon, R., Leveau, J.H.J. 2013. Draft genome sequence of the phyllosphere model bacterium *Pantoea agglomerans* 299R. *Genome Announcements* 1: e00036-13.

Rico, L., Ogaya, R., Terradas, J., Peñuelas, J. 2014. Community structures of N_2-fixing bacteria associated with the phyllosphere of a Holm oak forest and their response to drought. *Plant Biology* 16: 586–593.

Sanger, F., Nicklen, S., Coulson, A.R. 1977. DNA sequencing with chain-terminating inhibitors. *Proceedings of Natural Academic Science* 74: 5463–5467.

Sattelmacher, B., Mühling, K.H., Pennewiss, K. 1998. The apoplast – its significance for the nutrition of higher plants. *Journal of Plant Nutrition and Plant Science* 161: 485–498.

Schaller, G.E., Bishopp, A., Kieber, J.J. 2015. The yin-yang of hormones: cytokinin and auxin interactions in plant development. *Plant Cell* 27: 44–63.

Schenk, P.M., Carvalhais, L.C., Kazan, K. 2012. Unraveling plant–microbe interactions: can multi-species transcriptomics help? *Trends in Biotechnology* 30: 177–184.

Schreiber, L., Krimm, U., Knoll, D. 2004. Interactions between epiphyllic microorganisms and leaf cuticles. In: Varma, A., Abbott, D., Hampp, R. (Eds.) *Plant Surface Microbiology*, Springer-Verlag, Berlin-Heidelberg, pp. 145–156.

Shiraishi, K., Oku, M., Kawaguchi, K., Uchida, D., Yurimoto, H., Sakai, Y. 2015. Yeast nitrogen utilization in the phyllosphere during plant lifespan under regulation of autophagy. *Science Report* 5. doi: 10.1038/srep09719.

Silva-Sanchez, C., Li, H., Chen, S. 2015. Recent advances and challenges in plant phosphoproteomics. *Proteomics* 15: 1127–1141.

Sivakumar, N., Sathishkumar, R., Gopal, S., Shyamkumar, R., Arjunekumar, K. 2020. Phyllospheric microbiomes: diversity, ecological significance, and biotechnological applications. In: Yadav, A.N. et al. (Eds.) *Plant Microbiomes for Sustainable Agriculture*. Springer, Nature Switzerland, vol. 25, pp. 113–172.

Spadaro, D., Droby, S. 2016. Development of biocontrol products for postharvest diseases of fruit: the importance of elucidating the mechanisms of action of yeast antagonists. *Trends Food Science and Technology* 47: 39–49.

Stone, B.W.G., Weingarten, E.A., Jackson, C.R. 2018. The role of the phyllosphere microbiome in plant health and function. *Annual Plant Reviews* 1: 1–24.

Suda, W., Nagasaki, A., Shishido, M. 2009. Powdery mildew-infection changes bacterial community composition in the phyllosphere. *Microbes and Environments* 24: 217–223.

Tani, A., Takai, Y., Suzukawa, I., Akita, M., Murase, H., Kimbara, K. 2012. Practical application of methanol-mediated mutualistic symbiosis between *Methylobacterium* species and a roof greening moss, *Racomitrium japonicum*. *PLoS ONE* 7: e33800.

Thapa, S., Prasanna, R., Ranjan, K., Velmourougane, K., Ramakrishnan, B. 2017. Nutrients and host attributes modulate the abundance and functional traits of phyllosphere microbiome in rice. *Microbiological Research* 204: 55–64.

Thapa, S., Ranjan, K., Ramakrishnan, B., Velmourougane, K., Prasanna, R.2018. Influence of fertilizers and rice cultivation methods on the abundance and diversity of phyllosphere microbiome. *Journal of Basic Microbiology* 58: 172–186.

Trindade, I., Capitao, C., Dalmay, T., Fevereiro, M.P., Santos, D.M. 2010. miR398 and miR408 are up-regulated in response to water deficit in *Medicago truncatula*. *Planta* 231: 705–716.

Trouvelot, S., Héloir, M.C., Poinssot, B., Gauthier, A., Paris, F., Guillier, C., et al. 2014. Carbohydrates in plant immunity and plant protection: roles and potential application as foliar sprays. *Frontier in Plant Science* 5: 592.

Turner, T.R., Ramakrishnan, K., Walshaw, J., Heavens, D., Alston, M., Swarbreck, D. 2013. Comparative meta-transcriptomics reveals kingdom level changes in the rhizosphere microbiome of plants. *The ISME Journal* 7: 2248–2258.

Vacher, C., Hampe, A., Porte, A.J., Sauer, U., Compant, S., Morris, C.E. 2006. The phyllosphere: microbial jungle at the plant-climate interface. *Annual Review of Ecological Evolutionary System* 47: 1–24.

Wang, Y., Hu, B., Du, S., Gao, S., Chen, X., Chen, D. 2016. Proteomic analyses reveal the mechanism of *Dunaliella salina* Ds-26-16 gene enhancing salt tolerance in *Escherichia coli*. *PLoS ONE* 11: e0153640.

Whipps, J.M., Hand, P., Pink, D., Bending, G.D. 2008. Phyllosphere microbiology with special reference to diversity and plant genotype. *Journal of Applied Microbiology* 105: 1744–1755.

Williams, T.R., Moyne, A.L., Harris, L.J., Marco, M.L.2013. Season, irrigation, leaf age, and *Escherichia coli* inoculation influence the bacterial diversity in the lettuce phyllosphere. *PLoS ONE* 7: 1–10.

Xie, W.Y., Su, J.Q., Zhu, Y.G. 2015. Phyllosphere bacterial community of floating macrophytes in paddy soil environments as revealed by Illumina high-throughput sequencing. *Applied and Environmental Microbiology* 81: 522–532.

Xu, J., Lan, H., Fang, H., Huang, X., Zhang, H., Huang, J. 2015. Quantitative proteomic analysis of the rice (*Oryza sativa* L.) salt response. *PLoS ONE* 10: e0120978.

Yadav, A.N., Kumar, V., Prasad, R., Saxena, A.K., Dhaliwal, H.S. 2018. Microbiome in crops: diversity, distribution and potential role in crops improvements. In: Prasad, R., Gill, S.S., Tuteja, N. (Eds.) *Crop Improvement through Microbial Biotechnology*. Elsevier, USA, pp. 305–332.

Yadav, A.N, Kour, D., Sharma, S., Sachan, S.G., Singh, B., Singh, C.V., Sayyed, R.Z., Kaushik, R., Saxena, A.K. 2019. Plant growth promoting rhizobacteria for sustainable stress management. *Psychrotrophic Microbes: Biodiversity, Mechanisms of Adaptation, and Biotechnological Implications in Alleviation of Cold Stress in Plants*. Springer, Singapore, pp. 219–253.

Yim, W., Seshadri, S., Kim, K., Lee, G., Sa, T. 2013. Ethylene emission and PR protein synthesis in ACC deaminase producing *Methylobacterium* spp. inoculated tomato plants (*Lycopersicon esculentum* Mill.) challenged with *Ralstonia solanacearum* under greenhouse conditions. *Plant Physiological Biochemical* 67: 95–104.

Yin, M., Ma, Z., Cai, Z., Lin, G., Zhou, J. 2015. Genome sequence analysis reveals evidence of quorum-sensing genes present in *Aeromonas hydrophila* strain KOR1, isolated from a mangrove plant (*Kandelia obovata*). *Genome Announcement* 3(6): e01461-15.

9 Importance of Biofertilizers in Turmeric (*Curcuma longa*)

Prospects and Challenges

Bandi Arpitha Shankar
Sardar Vallabhai Patel University, Meerut, India

Prashant Kaushik
Kikugawa Research Station, Shizuoka, Japan
Instituto de Conservación y Mejora de la Agrodiversidad, Valencia, Spain

CONTENTS

9.1 INTRODUCTION

Turmeric is a miracle spice which was used from ancient India till today because of its valuable properties. Turmeric is a rhizomatous plant belonging to Zingiberaceae where ginger belongs. Turmeric is known to have several therapeutic properties and medicinal properties to treat certain ailments. Along with its properties, turmeric is used as a mild digestive, stimulant, and carminative making it one of the best healers in the world. Turmeric is commonly called *haldi* in several parts of India. This plant is well grown between 20 and 30°C and requires annual rainfall as irrigation. The rhizomes of turmeric are collected annually after the growing season and these

DOI: 10.1201/9781003147077-9

rhizomes are reseeded in the following season. These plants grow up to a maximum height of 1 m having oblong leaves. Leaves dry up at the time of harvest. Rhizomes are brown-yellowish in colour on the outside and dull orange inside with small tubers branching off sideways. Dried rhizome forms a yellow powder on grinding, which is used as a spice and flavouring agent containing many health and therapeutic properties (Manimegalai et al. 2011).

Over 2500 years ago, turmeric was first used as a spice, later its importance was known. Turmeric is proven to be beneficial against cancers and Alzheimer's diseases due to its phytochemical contents (Sharma and Kaushik). Certain ointments are produced by using turmeric which were used as antiseptics against skin allergies and wounds. Also dried roasted turmeric powder is used in preparation of toothpaste and very beneficial against dysentery (Nelson et al. 2017). Along with these important uses, turmeric is known for its greatness in the beauty industry as essential oils; powder of turmeric is also used a lot (Rajesh et al. 2013).

Curcumin is the major antioxidant found in turmeric (Joseph 2012). Certain volatile oils like cinol, zingiberene, sesquiterpene borneol, etc., are also helpful in treating several diseases and having major benefits. Zingiberene along with other compounds like turmerone and artermerone were responsible for the aroma in turmeric. It was well known that turmeric contains a good amount of omega-3 fatty acids and alpha-linoleic acids in good amounts (Arutselvi et al. 2012).

Although there are several uses with turmeric, it is easily available in nature and was also considered as the cheapest spice, due to its low-cost production. Several biochemicals and phytochemicals in turmeric are very useful and important in turmeric due to their health benefits for humans (Nisar et al. 2011). Antioxidant property in turmeric is very high and is well known to reduce several cancers, especially breast cancer in females. These antioxidants were known to neutralize the free radicals which usually damage the cells and cause cancers. Moreover, very important phytochemical like curcumin is a well-known component in turmeric as it imparts the yellow colour and wonderful flavour. Along with colour and flavour, this curcumin is well known for its several advantages as a powerful antioxidant, muscle relaxant, against arthritis, and muscle soreness. It was well proven to reduce several ailments in humans (Mastura et al. 2017).

Certain beneficial microorganisms are used in turmeric to enhance the yield and growth of the plant, ultimately leading to an increase in all the good components. Arbuscular mycorrhiza fungi (AMF) is a very beneficial fungi and has proven to increase the uptake of nutrients and water from the soil (Baum et al. 2015; Pankaj et al. 2017). These fungi act as mediators between the plant and the soil and are very helpful for the benefit of the plants. This sort of symbiotic association was found in plants having corms, bulbs rhizomes, etc., facilitating the growth of the plants under stressful conditions by increasing the photosynthetic rates and gas exchange (Bonneau et al. 2013; Singh et al. 2019).

These fungi are simply sprinkled along with the seed material or by adding some granules below the seed material. These fungi grow very easily in soil that is well maintained without many chemicals. It is beneficial to limit the use of phosphorus and pesticides and completely stop the tillage as it breaks the chain of fungal hyphae. The process of soil aggregation and stimulation of microbial activity is possible with the help of fungi where it helps the plants to tolerate different environments (Giovannetti et al. 2013).

Certain bacterial strains like *Pseudomonas fluorescens* and *Bacillus megaterium* are very helpful in reducing plant diseases caused by harmful fungi. These biocontrol agents are added to the soil before planting and have proven to be very beneficial as they promote growth of the plants by controlling the severity of several fungal infections. In recent years, it was proven that the use of AMF and *Pseudomonas* was creating wonders by reducing several harmful effects caused by both abiotic and biotic stresses (Gul and Bakht 2015; Sharma et al. 2022).

The present study investigates the use of AMF and *Pseudomonas* together in turmeric against certain abiotic stresses. Salinity is the major stress these days in turmeric leading to loss of yield and stunted growth of the plant, sometimes eventually leading to the death of the plant. Certain plants were given salt treatment under different saline conditions; these plants were associated symbiotically with the AMF, and *Pseudomonas* was incorporated in the soil and the effect of the fungi and bacteria together on the growth and yield of the turmeric plant was investigated (Hart et al. 2015). The method of sustainable production by using both the fungi and bacteria would surely impact the yield and growth of the plants in counter to the use of chemical fertilizers which would have several disadvantages in long-term usage (Dar et al. 2019). Coupled cultivation will not only increase the growth of the plants but also increase the growth and availability of soil microorganisms. Surely, coupled cultivation would lead to new connections in vascular tissues of plants by encouraging the supply of nutrients from soil to the plants (Kohler et al. 2007).

Several other crops and certain combinations of fungi and bacteria were also reported to be very effective in turmeric as they are very helpful in nutrient uptake and in the growth of plants. Certain bacteria were effective and those found in the turmeric rhizome and rhizosphere were very helpful in not increasing nutrient uptake but also in IAA production with phosphate solubilization (Table 9.1). Also, some fungal species like *Eurotium* and *Penicillium* alone without any combinations were very effective in asparginase production and antibacterial activity which were available from turmeric rhizome and leaves.

9.2 AGRONOMY OF TURMERIC

India is currently the world's biggest producer, user and exporter of turmeric, followed by China and a number of other Indian subcontinent nations. Turmeric cultivars come in a wide range of regional variations, the bulk of which are identified by the location in which they are grown or produced. There are significant variations in the size and colour of the rhizomes, as well as the amount of curcumin present in different varieties. Madras and Alleppey are the two most popular turmeric cultivars on the global market, with the former holding a substantial market share advantage over the latter. Turmeric plants' morphological characteristics, such as the amount of leaves, shoots, and rhizomes produced, are affected by a variety of variables, including nutrition, management methods, and genotype (Astinfeshan et al. 2019).

TABLE 9.1
Important Microorganisms and Their Properties

Bacterial Strains	Plant Growth-Promoting Properties
Klebsiella sp.	IAA production, phosphate solubilization, and ACC deaminase enzyme production
Penibacillus sp.	IAA production
AM fungi + PGPR	Enhancement in antioxidant properties, flavonoids, and total phenol content
Eurotium sp.	Asparaginase enzyme production
Bacillus cereus, *Bacillus thuringiensis*, *Bacillus* sp., *Bacillus pumilis*, *Pseudomonas putida*, and *Clavibacter michiganensis*	Antifungal activity, IAA production, and phosphate solubilization
Azotobacter chroococcum	Enhancement in curcumin content
Bacillus subtilis, *Bacillus* sp., *Burkholderia thailandensis*, *Agrobacterium vtumefaciens*, *Klebsiella* sp., *Bacillus cereus*, *Pseudomonas putida*, and *Pseudomonas fluorescens*	Antibacterial activity, antifungal activity, IAA production, and phosphate solubilization
Penicillium sp.	Nanoparticle-mediated antibacterial activity

Source: Kumar et al. (2017).

Curcuma plant development is influenced by a number of variables, including the weather, the kind of turmeric utilized and the type of planting media used. It may, however, be cultivated in a variety of tropical climates with temperatures ranging from 20 to 30°C and annual rainfall of at least 1500 mm, providing the environment is warm enough. Turmeric flourishes and produces a significant quantity of turmeric in a variety of soil types, from red soils and light black loam to clay loams as well as rich loamy soils with good irrigation and drainage facilities (Bandi and Kaushik 2022). This turmeric crop may be grown in a variety of soil types, including sandy loam. Turmeric harvesting season in India starts in mid-April and lasts until the first week of July, depending on the region. In certain areas of the globe, turmeric is sometimes cultivated as an intercrop with other crops such as mango, jackfruit, and litchi (Angel et al. 2014).

Turmeric's rhizosphere is home to a diverse range of microorganisms, all of which contribute to the savour and seasoning of the spice. It has been shown that interactions between highly gram-negative bacterial populations in rhizosphere are preferred for interactions between low gram-negative microbial populations in the bulk soil when the root produces rhizodeposition or carbon compounds. Each of these subgroups was divided into three subgroups: *Proteobacteria*, *Agrobacterium*, *Azotobacter*, and *Burkholderia*. In this study, six important bacterial species were discovered, accounting for almost two-thirds of all bacterial isolates: *Pseudomonas*, *Klebsiella*, *Agrobacterium*, *Azotobacter*, and *Burkholderia* (*Bacillus* sp.) (Akarchariya et al. 2017).

In the field of plant growth-promoting bacteria (PGPB), bacterial species such as *Pseudomonas*, *Bacillus*, *Azotobacter*, *Enterobacter*, *Klebsiella*, *Alcaligenes*,

Arthrobacter, *Burkholderia*, *Azospirillum*, and *Serratia* that are discovered are known to possess plant growth-stimulating activity and are referred to as PGPB. Bacteria such as algae, *Pseudomonas*, *Azospirillum*, *Azotobacter*, *Klebsiella*, and *Enterobacter* are among those that have been identified (Hart et al. 2015). Many academics are interested in understanding the fate of these connections, as well as their use in plant growth and the medicinal potential of herbs such as turmeric. In order to ensure the efficacy of PGPR therapy, it is necessary to create a significant population size of microbially active bacterial cells near the plant roots. It is necessary to do root dipping and soil inoculations using PGPR bacterial solutions. As a result of the variability in whether the soil was sterilized before to inoculation, the cell population in certain cases may quickly decrease following inoculation. Also discussed are those aspects of the inoculation's efficacy that are reliant on the rhizobacteria's competence for the particular plant (Bahadori and Demiray 2017).

To achieve the intended effects on the turmeric plant, different bacterial strains were injected into the rhizome of the turmeric plant to regulate growth and metabolite synthesis, respectively, in order to achieve the desired effects on the plant. As reported by Kumar et al. (2014), inoculating the turmeric rhizome with *Azotobacter chroococcum* enhanced the number of leaves, as well as the height and biomass of shoots, and the amount of rhizome biomass and curcumin content. Kumar et al. (2016) discovered that inoculating turmeric with *Pseudomonas fluorescens* led in an increase in morphological yields along with curcumin content in the plant itself. When mycorrhizal fungus and plant growth-promoting bacteria were grown together in the turmeric rhizomes, it was found that the turmeric plant's antioxidant activity, flavanoids, phenolic content, and curcumin content were significantly increased (Dosoky and Setzer 2018).

Due to soil salinity the plants lose their moisture content, they become pale dry losing their chlorophyll content and ultimately plants die. There are several changes in the leaves and the stems of the plants due to unfavourable conditions. Several effects were seen in case of the leaves and stems of the plants. The rhizomes were not fully developed during the initial stage of the plant. In general, plants exhibit certain physical symptoms during their initial stages, i.e., vegetative stages of the plant. This is the main stage of the plant because it draws all the essential nutrients and synthesizes several important components. There was a huge effect in the growth of the plant due to salinity at this stage even though after coupled inoculation.

Several plant parameters were increased due to the addition of fungi and bacteria together (Figure 9.1). The size of the plant and colour is drastically changed from its vegetative stage. The chlorophyll content was increased, and necrosis of the plant was reduced to some extent. It was also observed that the physiochemical and biochemical components were increased to a certain level. The change in the plant was clearly observed and can be differentiated from the vegetative stage and later stages only due to coupled inoculation.

Research found that comparing the applications of combinations like nitrogen and potassium (N&K) or nitrogen, phosphorus, and potassium, the combination treatment resulted in increased shoot weight (4–6 times) as well as maximum yield (8–9 times) when compared to the separate treatments (Gulati et al. 2010). The results of an additional research have corroborated this trend, revealing that

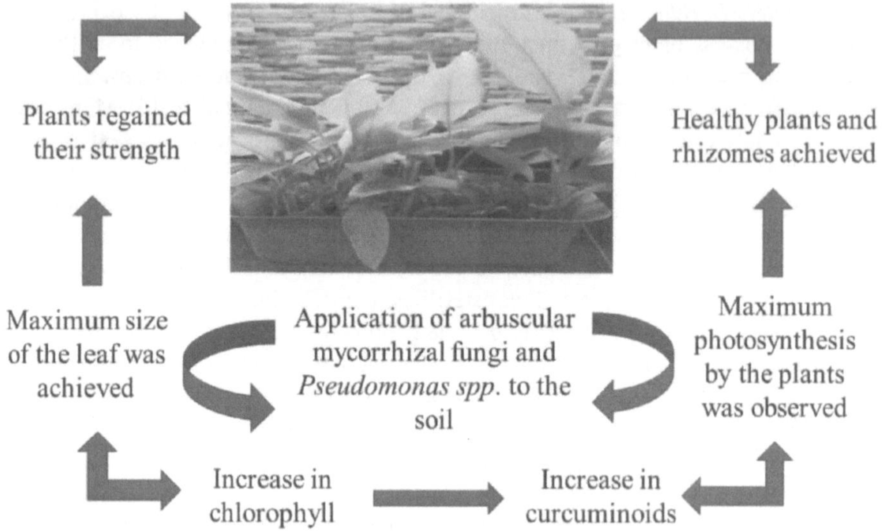

FIGURE 9.1 Healthy growth of the plant due to coupled inoculation.

the administration of nitrogen, either alone or in combinations of P or K, increases crop growth. Prolonged as well as inappropriate use of chemical fertilizers has been proven to have a detrimental effect on the productivity of soil and turmeric production, according to several studies (curcumin). A study conducted by PGPB found that this compound is very beneficial for both the soil and growth of the turmeric plant during turmeric production and growth (Edwards et al. 2017). Most likely, the increase in P availability that has occurred because of mycorrhizal interaction has been responsible for the increase in curcumin levels. We are unable to explain how AMF impacts turmeric curcumin synthesis in humans currently owing to a scarcity of data on the subject. The fact that turmeric types may have a substantial influence on curcumin concentration even though the rise in curcumin concentration is a result of AMF inoculation, which affects the quality and price of turmeric on the worldwide market, is a notable discovery in this study. Over a long period of time, relationships between mycorrhizal organisms and plants have developed under a variety of complicated but generally in stable environmental circumstances. In agricultural systems, we are now attempting to manage this symbiotic relationship via a variety of methods. Japanese turmeric grows without the use of chemical biocides or soluble inorganic fertilizers, resulting in the nation generating an overwhelming majority (Dutta and Neog 2017). According to many specialists in the industry, AMF is anticipated to play a more important role in alternative energy systems than it does in traditional energy systems in the future. It was found that using mycorrhizal inoculation for turmeric production is a viable method of increasing yields where it also preserves quality (Malhi et al. 2021). The organism margarita, which was discovered in commercial inoculums and was utilized as a model organism, was discovered. The genus *Glomus*, on the other hand, is more often found in agricultural soils that have a high concentration of nutritional elements (Astinfeshan et al. 2019).

The genus *Gigaspora*, on the other hand, is more common in soils that are poor in nutrients or that are nutrient binding. Various AMF isolates, it is hypothesized, have varied effects on the development of the host plant and produce a diverse range of qualitative secondary metabolites with variable compositions. As a result, more research is required to determine the most effective AMF for turmeric growth and curcumin synthesis (Farzaei et al. 2018).

9.3 PRODUCTION OF BIOFERTILIZERS

For the large-scale production of biofertilizers, first the specific microorganism or fungi should be separated from the soil. This is the primary step for the production of biofertilizers. Several microorganisms, i.e., bacteria or fungi, were isolated from the soil or from the roots of the plants that are known to have a symbiotic association with the microorganisms like legumes, turmeric, etc. Proper screening of microorganisms is the most important step to understand its effect on the growth of the plant. Different plant species are tested using these biofertilizers under proper laboratory conditions. These selected microorganisms that are very helpful to the plants are added to the soil in pots in green house to check the ability of the biofertilizers (Ward and Brender 2019; Pankaj 2020).

Later, these potted plants with added biofertilizers were tested in different field, soil, and climatic conditions. The effect of biofertilizers on different plant species under different soil conditions has been thoroughly investigated to conclude the area, soil type, and plant species on which these biofertilizers are effective (Blankson et al. 2016).

The microorganisms that are proven to be very effective and promote growth in plant species were selected and separated from the other microorganisms and this method is called refinement of inoculum. The particular species of bacteria and fungi that are proven to be very beneficial for the plants and can adapt to different soil conditions are selected. These microorganisms are separated from others and are sent for production unit for large-scale production. Several mechanisms and machinery are involved in the production and refinement of the biofertilizers (Domínguez et al. 2016; Pankaj 2020).

These selected biofertilizers are called potential type and can be used as mother cultures. Broth is prepared using these mother cultures and is allowed to go to the fermentation chamber for further growth. Maximum production of these microorganisms takes place in the fermentation chamber as the cultures are maintained in controlled conditions. After fermentation, these individual cultures or mixed cultures are sent for proper packaging. Proper care should be taken while packing to reduce the entry of air along with foreign microorganisms and leakage of the microorganism particles. The packaging of the biofertilizers is done in double-layer packaging properly. These biofertilizers are quality checked well in advance before packaging. The biofertilizer packet was properly labelled mentioning the type of microorganisms used and their quantity. Curing of these packed biofertilizers is done in room temperature for proper growth conditions and is stored in low temperatures later to restrict any further growth. These packages are taken and can be directly used in field or pot application (Bhardwaj et al. 2014).

Out of all the processes, the condition for fermentation is a major step and has a high impact on the growth of the microorganisms. The growth curve was well observed in this stage with a continuous supply of nutrients (Hossain et al. 2015).

9.4 FORMULATION OF BIOFERTILIZERS

The form in which the biofertilizers are to be applied to the soil depends upon the quality of the inoculant. Formulation depends upon the standard of the inoculant. There are four different formulations of bio-inoculants depending upon the nature and quality of the microorganism (Naher et al. 2018).

9.4.1 Liquid Formulations

These formulations depend upon the broth cultures, not only used for microorganisms delivery but also for nutrient supply to the soil and the plant (Pankaj 2020). These are very useful for seed treatment, root dipping, and as soil application. In liquid formulations, a quantity of 10^9 cells is available per 1 mL.

9.4.2 Carrier-Based Formulations

Different materials can be used as a carrier for soil and seed inoculation. A proper carrier formulation requires a suitable carrier. These formulations have good absorption capacity and are available in adequate amount at very lower prices, but the presence of microorganisms is comparatively low than that of other formulations.

The carrier materials that are mixed along with bio-inoculants may involve peat, lignite, organic soil, FYM, vermiculites and press mud (Pankaj 2020).

9.4.3 Granular Formulations

Granular formulations are coated with the target microorganisms. These granules are made up of silica grains, peat, etc., and are wetted with an adhesive for proper binding of microorganisms. The microorganisms to be bound to the granules are usually in powder form. These wonderful microbial granules are placed in a furrow near the seed to be available for both the soil and the roots.

9.4.4 Encapsulated Formulations

The microorganisms are present within the encapsulations made with natural or chemical formulations. These microorganisms are protected from harsh environments and improper drains in the encapsulations. The effect microorganisms are very high by this method because of their slow release into the environment in less and adequate quantities. This encapsulation is made up of alginate material, but the preparation of these encapsulations from polymers is still experimental under bacterial inoculation technique (Yanti et al. 2018). In another way, solid carrier formulations involve microgranules, granules, wettable powders, wettable granules, pellets, and dusts, whereas liquid formulations include ultra-low volume suspension, oil dispersion, oil concentrates, and suspension concentrates. The granular formulations are quite advantageous compared to others due to the nature of granules, as they are easy to store and apply. Along with granular formulations, encapsulated formulation is also advantageous due to slow release of the microorganisms or bio-inoculants into the soil. This helps in the availability of the microorganisms throughout the crop's life cycle providing adequate nutrients by symbiotic relationships. The dust formulations and pellets are not too effective because of their very dry nature. Due to this dry nature, they tend to disperse away from the field and dry up too quickly (Batool and Iqbal 2019).

9.5 ASSESSMENT OF BIOFERTILIZER EFFICACY

Biofertilizers are affected by several parameters like carrier material, formulation type, soil conditions, environment, and plant species type. The proper efficiency of the biofertilizers is well maintained by the influence of soil fertility, cropping patterns, host-plant interaction, and biological and environmental factors (Nelson et al. 2018).

The use of chemical fertilizers is well known to reduce the microbial count in the soil; alternative use of microbial organisms increases bacterial growth by decomposition and mineralization. The use of biofertilizers on soil is very useful as they increase the microbial content by proving suitable conditions to the microorganisms. Maximum vegetative or forage growth is achieved by the proper relationship of the plant with suitable microorganisms. By the prolonged use of biofertilizers, the nutrient contents in the soils are increased, thereby increasing the overall fertility of the soil (Gulati et al. 2010).

Also, growing crops other than legumes along with legumes can increase the growth of suitable bacterial species like rhizobia, which increase the nutrient content in the soils. Not all the plant species are friendly towards the biofertilizers but a few plant species like legumes, rice, turmeric, and few solanaceous species are very well adapted to the use of biofertilizers. The host-plant relation with these biofertilizers is proven to be highly beneficial to the plants, soil, and microorganisms as well for their multiplication. Therefore, growing plants that are not friendly with biofertilizers along with friendly plants is very beneficial for other plants as well due to the presence of the biofertilizers in the soil (Sung et al. 2017).

Suitable environmental conditions are also important for the growth of microorganisms in the soil. Proper sunlight and moisture are very important for proper growth and multiplication of microorganisms in the soil. Moreover, the uptake of nutrients by plants along with their symbiotic relationship with the microorganisms varies depending upon the environmental conditions (Halim et al. 2018).

9.6 SHELF LIFE OF BIOFERTILIZERS

The biofertilizers are very effective in maintaining plant growth and increasing the yield by their symbiotic relationship with plants. But when it comes to storage, these biofertilizers have a very small shelf life of about 6 months in all the formulations except for liquid formulations which have a shelf life of about 2 years. The solid-based formulations have little thermotolerant capacity, whereas liquid formulations can a tolerate temperature of nearly 55°C (Saeid et al. 2018). These biofertilizers are to be stored in cool, dry places with low or room temperature conditions. Also, these biofertilizers are not to be mixed with other chemicals as the chemicals are very harmful to the microorganisms as well as to the plants (Rajput et al. 2012).

9.7 IMPORTANCE OF BIOFERTILIZERS WITH RESPECT TO TURMERIC

Turmeric which was considered as a wonderful spice in Indian tradition is not only used for cooking purpose but also as a condiment, cosmetic, and a holy product. It was well known that this turmeric contains several anti-cancerous, anti-inflammatory, anti-ageing properties, making it more useful in the medical, drug, and beauty world. Curcumin is the major compound that acts as a backbone for the immense greatness of turmeric (Akbik et al. 2014). Curcumin is a phytochemical compound containing all the important properties and imparts colour to the rhizomes. In recent studies, it was noted that Indian women were less prone to breast cancer and men were less prone to prostate cancers due to the regular usage of this miracle spice and more importantly the curcumin content in it. Along with curcumin, several other biochemical compounds were also present in turmeric that act as important and essential compounds which include flavonoids, proteins, tannins, carbohydrates, saponins, etc. These compounds have a major impact on the quality of the turmeric rhizome and the curcumin content values the medicinal properties present in the rhizomes (Olatunde et al. 2014).

Recently, there was a drastic reduction in the growth, yield, and curcumin content of the turmeric plants due to salinity stress. These plants are dried and the growth of rhizomes was delayed. In our study, the effect of AMF and pseudomonas on the growth of turmeric against abiotic stresses, especially salt stress, was studied. It was proven that both the biological agents are well known for their role in nutrient uptake from soil to plants (Saini et al. 2021; Srivastav et al. 2022). This use of biological agents gave immense scope regarding the growth of turmeric plants in response to salt stress (Aroca et al. 2013). This AMF aka endomycorrhiza is a symbiotic fungus which penetrates the roots of the plants forming vascular connections with the plants (Sharma et al. 2021). These fungi produce contains structures after invading the roots of turmeric plant called arbuscules, i.e., branches for nutrient exchange. These fungi help plants uptake the nutrition in two ways, physically the fungal mycelia is thin and very minute so that it can go deeper into soil and draw nutrients, by chemically secreting certain organic acids that were chelated from ion exchange method (Pankaj et al. 2017).

The mixture of AMF and pseudomonas was added directly to the soil in field after sowing or pot above the seed material. Salt-like sodium chloride was added to the pots a few days after the coupled inoculation. It was observed very well that the vegetative growth of the plants was greater than the controls and an increase in certain biochemicals was also observed to be high as compared to that of the control (Bettoni et al. 2014).

In the present study, it was observed that this coupled inoculation was highly beneficial by improving plant growth and yield under salinity conditions. It was also observed that the chlorophyll and curcumin content were less during initial stage, i.e., vegetative period, and later it was increased during treatment phase and gradually it decreased, but with the use of AMF and *Pseudomonas* together, i.e., coupled inoculation, there was not much reduction in both the chloroplast and curcumin content. This change was observed only in the plants that were inoculated with these biological agents but not in non-AMF and *Pseudomonas*-inoculated varieties (Hass et al. 2011).

AMF is very helpful in maintaining root health by associating with the roots over 90%. This coupled inoculation with AMF and *Pseudomonas* was well known to increase the salinity tolerance by various mechanisms like increased nutrient uptake, producing suitable growth hormones, enhancing root health, and improving soil conditions (Saini et al. 2020). This coupled inoculation was also proven to improve photosynthetic activity and responsible for proper water use efficiency (Tanwar et al. 2013a). It was also observed that the biochemical constituents in turmeric plants would gradually be reduced in case of salt conditions, but with the help of coupled inoculation, the levels of these were maintained without much reduction (Tanwar et al. 2013b). Also, it was well known that the effect of chemical fertilizers would affect plant growth and would reduce the nutrients in the soil which is ultimately a disadvantage to humans. But the use of these AMF and *Pseudomonas* together is not only beneficial for the plant itself but very helpful to maintain soil health. For this reason, these biological agents are known as friendly fertilizers towards the environment because they are proven to reduce pollution (Wertheim et al. 2014). Hence, this technique of coupled inoculation with AMF and *Pseudomonas* is proven to be highly beneficial for the growth of turmeric even in severe abiotic stress conditions like salinity (He et al. 2010).

9.8 CONCLUSION AND FUTURE PROSPECTS

Several microorganisms are known to increase soil fertility and enhance nutrition status of the soil. Sometimes the mixture of fungi and bacteria would be very appropriate rather than the single species alone. These mixed combinations are called coupled cultivations of both the fungi and bacteria. These studies on coupled cultivation have gained greater importance in recent years due to their efficiency in solving many problems. The effects of certain abiotic stresses are significantly reduced due to the presence of these microorganisms. In this review, the effect of salinity on the growth of turmeric plant under the influence of AMF and *Pseudomonas* was studied carefully and it was understood that the growth, yield, and biochemical parameters of the turmeric plants were maintained even after salinity stress. As known turmeric is a wonderful spice and used every day in Indian culture for cooking and several other uses; it is also known to possess many medicinal and therapeutic properties. The curcumin content is very effective in treating several ailments and it is well known in cancer treatment.

It was well observed that due to the usage of microorganisms, there was not much loss in the growth and yield of the crop. It was observed that there was reduction in the yield during the initial stages of plant growth, i.e., vegetative stage, and in later stage, the growth of the plant remained stable but never decreased due to coupled cultivation. Several combinations of these microorganisms are used in coupled cultivations; they can be only fungi or bacteria or both fungi and bacteria can be involved. These microorganisms are simply added to the soil prior to sowing, or are added along with seed material by simply sprinkling the microorganisms in the surface. The effect of these microorganisms was very well observed only after a few days of inoculation. This coupled cultivation was so effective that there is no maximum change in chlorophyll and curcumin content.

This coupled cultivation is not only beneficial for nutrient uptake but also very helpful in increasing the microorganism content in the soil. This would create good soil conditions for the growth of plants. Also, the soil environment is well maintained due to the presence of microorganisms. Importantly, the use of these microorganisms proving to be very advantageous in all aspects helps us to reduce the use of harmful chemical fertilizers which would result in disturbing soil texture and structure. Chemical fertilizers are very effective initially but would damage the soil health later after prolonged use. At present, efficient and sustainable biofertilization production for plants, where inorganic fertilizer applications can be significantly reduced to avoid further pollution problems, represents interest in the main research.

Future prospects for the use of biofertilizers may include:

1. Production of multifunctional biofertilizers.
2. Maximum effect of the biofertilizers immediately after application.
3. Perfect quality of the inoculants.
4. Further investigations on the persistence of biofertilizers in different soil conditions.
5. Mass production of the biofertilizers.
6. Technical production of biofertilizers on an industrial level.
7. Factors related to economic conditions.
8. Proper regulation board to check the quality of these biofertilizers.
9. Production of biofertilizers suitable for all crop plants.
10. Educating farmers and research groups regarding the importance of biofertilizers.

There is a large need to study the combinations of fungi and bacteria as coupled cultivation to reduce the damage caused to both plants and soil. This method is not only effective in turmeric but in several other crops belonging to *Solanaceae*, *Cucurbitaceae*, etc., and in some tree species.

REFERENCES

Akarchariya, N., Sirilun, S., Julsrigival, J., Chansakaowa, S. 2017. Chemical profiling and antimicrobial activity of essential oil from *Curcuma aeruginosa*, Curcuma glans K. Larsen & J. Mood and *Curcuma xanthorrhiza* collected in Thailand. *Asian Pacific Journal on Tropical Biomedicine* 7(10): 881–885.

Akbik, D., Ghadiri, M., Chrzanowski, W., Rohanizadeh, R. 2014. Curcumin as a wound healing agent. *Life Sciences* 116(1): 1–7.

Angel, G.R., Menon, N., Vimala, B., Nambisan, B. 2014. Essential oil composition of eight starchy *Curcuma* species. *Indian Crops Production Journal* 60: 233–238.

Aroca, R., Ruiz-Lozano, J.M., Zamarreno, A.M., Paz, J.A., Garcia-Mina, J.M., Pozo, M.J., Lopez-Raez, J.A. 2013. Arbuscular mycorrhizal symbiosis influences strigolactone production under salinity and alleviates salt stress in lettuce plants. *Journal of Plant Physiology* 170: 47–55.

Arutselvi, R., Balasaravanan, T., Ponmurugan, P., Muthu Suranji, N., Suresh, P. 2012. Phytochemical screening and comparative study of antimicrobial activity of leaves and rhizomes of turmeric varieties. *Asian Journal of Plant Science and Research* 2(2): 212–219.

Astinfeshan, M., Rasmi, Y., Kheradmand, F., Karimipour, M., Rahbarghazi, R., Aramwit, P., et al. 2019. Curcumin inhibits angiogenesis in endothelial cells using downregulation of the PI3K/Akt signalling pathway. *Food Biosciences Journal* 29: 86–93.

Bahadori, F., Demiray, M. 2017. A realistic view on "the essential medicinal chemistry of curcumin". *ACS Journal of Medical Chemistry Letters* 8: 893–896.

Bandi, A., Kaushik, P. 2022. Mechanisms and approaches for salt tolerance in turmeric: A breeding perspective. *OBM Genetics* 6(2): 1.

Batool, S., Iqbal, A. 2019. Phosphate solubilizing rhizobacteria as alternative of chemical fertilizer for growth and yield of *Triticum aestivum* (Var. Galaxy 2013). *Saudi Journal of Biological Sciences* 26: 1400–1410.

Baum, C., El-Tohamy, W., Grudac, N. 2015. Increasing the productivity and product quality of vegetable crops using arbuscular mycorrhizal fungi: A review. *Scientific Horticulture Journal* 187: 131–141.

Bettoni, M.M., Mogor, Á.F., Pauletti, V., Goicoechea, N. 2014. Growth and metabolism of onion seedlings as affected by application of humic substances, mycorrhizal inoculation and elevated CO_2. *Scientific Horticulture Journal* 180: 227–235.

Bhardwaj, D., Ansari, M.W., Sahoo, R.K., Tuteja, N. 2014. Biofertilizers function as key player in sustainable agriculture by improving soil fertility, plant tolerance and crop productivity. *Microbial Cell Factories* 13: 66.

Blankson, G.K., Osei-Fosu, P., Adeendze, E., Ashie, D. 2016. Contamination levels of organophosphorus and synthetic pyrethroid pesticides in vegetables marketed in Accra, Ghana. *Food Control Journal* 68: 174–180.

Bonneau, L., Huguet, S., Wipf, D., Pauly, N., Truong, H.N. 2013. Combined phosphate and nitrogen limitation generates a nutrient stress transcriptome favourable for arbuscular mycorrhizal symbiosis in *Medicago truncatula*. *New Phytology Journal* 199: 188–202.

Dar, K.K., Ali, S., Ejaz, M., Nasreen, S. et al. 2019. In vivo induction of hepatocellular carcinoma by diethylnitrosoamine and pharmacological intervention in Balb C mice using *Bergenia ciliata* extracts. *Brazilian Journal of Biology – Revista Brasileira de Biologia* 79(4): 629–638.

Domínguez, I., Romero-González, R., Liébanas, F.J.A., Vidal, J.L.M., Frenich, A.G. 2016. Automated and semi-automated extraction methods for GC–MS determination of pesticides in environmental samples. *Trends in Environmental Annual Chemistry* 12: 1–12.

Dosoky, N.S., Setzer, W.N. 2018. Chemical composition and biological activities of essential oils of *Curcuma* species. *Nutrients Journal* 10(9): 1–42.

Dutta, S.C., Neog, B. 2017. Inoculation of arbuscular mycorrhizal fungi and plant growth promoting rhizobacteria in modulating phosphorus dynamics in turmeric rhizosphere. *National Academic Scientific Letters* 40: 445–449.

Edwards, R.L., Luis, P.B., Varuzza, P.V. et al. 2017. The anti-inflammatory activity of curcumin is mediated by its oxidative metabolites. *Journal of Biological Chemistry* 292: 21243–21252.

Farzaei, M.H., Zobeiri, M., Parvizi, F., El-Senduny, F.F. et al. 2018. Curcumin in liver diseases: A systematic review of the cellular mechanisms of oxidative stress and clinical perspective. *Nutrients Journal* 10(7): 855.

Giovannetti, M., Avio, L., Sbrana, C. 2013. Improvement of nutraceutical value of food by plant symbionts. In: Ramawat, K.G., Me'rillon, J.M. (Eds.), *Natural Products*. Springer: Berlin, Heidelberg, pp. 2641–2662.

Gul, P., Bakht, J. 2015. Antimicrobial activity of turmeric extract and its potential use in food industry. *Journal of Food Science and Technology* 52: 2272–2279.

Gulati, A., Sharma, N., Vyas, P., Sood, S., Rahi, P., Pathania, V., Prasad, R. 2010. Organic acid production and plant growth promotion as a function of phosphate solubilization by *Acinetobacter rhizosphaerae* strain BIHB 723 isolated from the cold deserts of the Trans-Himalayas. *Archives of Microbiology* 192: 975–983.

Halim, N.S.A., Abdullah, R., Karsani, S.A., Osman, N., Panhwar, Q.A., Ishak, C.F. 2018. Influence of soil amendments on the growth and yield of rice in acidic soil. *Agronomy Journal* 8: 165.

Hart, M., Ehret, D.L., Krumbein, A., Leung, C., Murch, S., Turi, C., Franken, P. 2015. Inoculation with arbuscular mycorrhizal fungi improves the nutritional value of tomatoes. *Journal of Mycorrhiza* 25: 359–376.

Hass, J.H., Bar-Yosef, B., Krikun, J., Barak, R., Markovitz, T., Kramer, S. 2011. Vesicular-arbuscular mycorrhizal fungus infection and phosphorus fertilization to overcome pepper stunting after methyl bromide fumigation. *Agronomy Journal* 79: 905–910.

He, Z.Q., Tang, H.R., Li, H.X., He, C.X., Zhang, Z.B., Wang, H.S. 2010. Arbuscular mycorrhizal alleviated ion toxicity, oxidative damage and enhanced osmotic adjustment in tomato subjected to NaCl stress. *American Eurasian Journal of Agricultural and Environmental Sciences* 7: 676–683.

Hossain, M.J., Ran, C., Liu, K., Ryu, C.M. et al. 2015. Deciphering the conserved genetic loci implicated in plant disease control through comparative genomics of *Bacillus amyloliquefaciens* subsp. *plantarum*. *Frontiers in Plant Science* 6: 631.

Joseph, R. 2012. Water soluble complexes of curcumin with cyclodextrins: Characterization by FT-Raman spectroscopy. *Vibration Spectroscopy* 62: 77–84.

Kohler, J., Caravaca, F., Carrasco, L., Roldan, A. 2007. Interactions between a plant growth promoting rhizobacterium, an AM fungus and a phosphate-solubilising fungus in the rhizosphere of *Lactuca sativa*. *Applied Soil Ecology* 35: 480–487.

Kumar, A., Singh, R., Giri, D.D., Singh, P.K., Pandey, K.D. 2014. Effect of *Azotobacter chroococcum* CL13 inoculation on growth and curcumin content of turmeric (*Curcuma longa* L.). *International Journal of Current Microbiology and Applied Sciences* 3(9): 275–283.

Kumar, A., Vandana, R.S., Singh, M., Singh, P.P., Singh, S.K., Singh, P.K., Pandey, K.D. 2016. Isolation of plant growth promoting rhizobacteria and their impact on growth and curcumin content in *Curcuma longa* L. *Biocatalysis and Agriculture Biotechnology* 8: 1–7.

Kumar, A., Singh, A.K., Kaushik, M.K., Mishra, S.K., Raj, S., Singh, P.K., Pandey, K.D. 2017. Interaction of turmeric (*Curcuma longa* L.) with beneficial microbes: A review. *Biotech Journal* 7: 357.

Malhi, G.S., Kaur, M., Kaushik, P., Alyemeni, M.N., Alsahli, A.A., Ahmad, P. 2021. Arbuscular mycorrhiza in combating abiotic stresses in vegetables: An eco-friendly approach. *Saudi Journal of Biological Sciences* 28(2): 1465–1476.

Manimegalai, V., Selvaraj, T., Ambikapathy, V. 2011. Studies on isolation and identification of VAM fungi in *Solanum viarum* Dunal of medicinal plants. *Advances in Applied Science Research* 2 (4): 621–628.

Mastura, H., Hasnah, Y., Dang, T.N. 2017. Total phenolic content and antioxidant capacity of beans: Organic vs. inorganic. *International Food Research Journal* 24(2): 510–517.

Naher, U.A., Panhwar, Q.A., Othman, R., Shamshuddin, J., Ismail, M.R., Zhou, E. 2018. Proteomic study on growth promotion of PGPR inoculated aerobic rice (*Oryza sativa* L.) cultivar MR219-9. *Pakistan Journal of Botany* 50: 1843–1852.

Nelson, M.K., Dahlin, J.L., Bisson, J., Graham, J., Pauli, G.F., Walters, M.A. 2017. The essential medicinal chemistry of curcumin. *Journal of Medicinal Chemistry* 60(5): 1620–1637.

Nelson, D.W., Sommers, L.E., Sparks, D.L., Page, A., Helmke, P., Loeppert, R. 2018. Total carbon, organic carbon, and organic matter. In: *Soil Science Society of America*. Soil Science Society of America and American Society of Agronomy: Madison, WI, pp. 961–1010.

Nisar, M., Ali, S., Qaisar, M. 2011. Preliminary phytochemical screening of flowers, leaves, bark, stem and roots of *Rhododendron arboretum*. *Middle East Journal of Scientific Research* 10: 472–476.

Olatunde, A., Joel, E.B., Tijjani, H., Obidola, S.M., Luka, C.D. 2014. Anti-diabetic activity of aqueous extract of *Curcuma longa* (Linn.) rhizome in normal and alloxan induced diabetic rats. *Researcher* 6: 58–65.

Pankaj, U. 2020. Bio-fertilizers for management of soil, crop and human health. In: Singh, C., Tiwari, S., Singh, J.S., Yadav, A.N. (Eds.), *Microbes in Agriculture and Environmental Development*. CRC Press/Taylor & Francis: New York, NY, pp. 71–85.

Pankaj, U., Verma, S.K., Semwal, S., Verma, R.K. 2017. Assessment of natural mycorrhizal colonization and soil fertility status of lemongrass [(*Cymbopogon flexuosus*, Nees ex Steud) W. Watson] crop in subtropical India. *Journal of Applied Research on Medicinal and Aromatic Plants* 5: 41–46.

Rajput, M.S., Kumar, G.N., Rajkumar, S. 2012. Repression of oxalic acid-mediated mineral phosphate solubilization in rhizospheric isolates of *Klebsiella pneumoniae* by succinate. *Archives of Microbiology* 195: 81–88.

Rajesh, H., Rao, S.N., Megha Rani, N., Prathima, K., Shetty, Rajesh, E.P., Chandrashekhar, R. 2013. Phytochemical analysis of methanolic extract of *Curcuma longa* Linn. *International Journal of Universal Pharmacy and Bio Sciences* 2(2): 39–45.

Saeid, A., Prochownik, E., Dobrowolska-Iwanek, J. 2018. Phosphorus solubilization by *Bacillus* species. *Molecules* 23: 2897.

Saini, I., Kaushik, P., Al-Huqail, A.A., Khan, F., Siddiqui, M.H. 2021. Effect of the diverse combinations of useful microbes and chemical fertilizers on important traits of potato. *Saudi Journal of Biological Sciences* 28(5): 2641–2648.

Saini, I., Rani, K., Gill, N., Sandhu, K., Bisht, N., Kumar, T., Kaushik, P. 2020. Significance of arbuscular mycorrhizal fungi for Acacia: A review. *Pakistan Journal of Biological Sciences: PJBS* 23(10): 1231–1236.

Sharma, M., Kaushik, P. 2021. Vegetable phytochemicals: An update on extraction and analysis techniques. *Biocatalysis and Agricultural Biotechnology* 36: 102149.

Sharma, M., Delta, A.K., Kaushik, P. 2022. Effects of *Funneliformis mosseae* and potassium silicate on morphological and biochemical traits of onion cultivated under water stress. *Horticulturae* 8(7): 663.

Sharma, M., Delta, A.K., Kaushik, P. 2021. *Glomus mosseae* and *Pseudomonas fluorescens* application sustains yield and promote tolerance to water stress in *Helianthus annuus* L. *Stresses* 1(4): 305–316.

Singh, G., Pankaj, U., Chand, C., Verma, R.K. 2019. Arbuscular mycorrhizal fungi-assisted phytoextraction of toxic metals by *Zea mays* L. from tannery sludge. *Soil and Sediment Contamination: An International Journal* 28(8): 729–746.

Srivastava, A., Sharma, V.K., Kaushik, P., El-Sheikh, M.A., Qadir, S., Mansoor, S. 2022. Effect of silicon application with mycorrhizal inoculation on *Brassica juncea* cultivated under water stress. *Plos One* 17(4):e0261569.

Sung, C.T., Ishak, C.F., Abdullah, R., Othman, R., Qurban Panhwar, A.O. 2017. Soil properties (physical, chemical, biological, mechanical). In: Ashraf, M.A., Othman, R., Ishak, C.F. (Eds.), *Soils of Malaysia*. CRC Press/Taylor & Francis: New York, NY.

Tanwar, A., Aggarwal, A., Kadian, N., Gupta, A. 2013a. Arbuscular mycorrhizal inoculation and super phosphate application influence plant growth and yield of *Capsicum annuum*. *Journal of Soil Science and Plant Nutrition* 13: 55–66.

Tanwar, A., Aggarwal, A., Kaushish, S., Chauhan, S. 2013b. Interactive effects of AM fungi with *Trichoderma viride* and *Pseudomonas fluorescens* on growth and yield of broccoli. *Plant Protection Science* 49: 137–145.

Ward, M.H., Brender, J.D. 2019. Drinking water nitrate and human health. In *Encyclopaedia of Environmental Health*. Elsevier BV: Amsterdam, The Netherlands, pp. 173–186.

Wertheim, F., Douds, D., Handley, D., Hutton, M. 2014. Evaluating the potential of arbuscular mycorrhizal fungi to boost yields in field-grown leeks. *Journal of National Association of Country's Agricultural Agents* 7: 1.

Yanti, Y., Warnita, W., Reflin, R., Nasution, C.R. 2018. Characterizations of endophytic *Bacillus* strains from tomato roots as growth promoter and biocontrol of *Ralstonia solanacearum*. *Biodiversitas Journal of Biological Diversity* 19: 906–911.

10 Role of Soil Organisms in Maintaining Soil Health

Thounaojam Thomas Meetei
Lovely Professional University, Phagwara, India

Yumnam Bijilaxmi Devi
Rani Lakshmi Bai Central Agricultural University
Jhansi, India

Thorny Chanu Thounaojam
Cotton University, Guwahati, India

CONTENTS

10.1 INTRODUCTION

Soil is the fundamentally biologically active component of the environment, playing a significant role in support of the Earth's ecosystem. After physical disintegration and chemical decomposition of rocks and minerals, it is formed in the uppermost layer of the earth crust, therefore, also termed as 'skin of the earth' (Kutílek and Nielsen 2015). Soil compositions include mineral matter, organic matter, air, and water. Mineral particle is the major components of soil (45%), consisting of sand, silt, and clay that occupy around more than half the volume of soil. The organic matter (5%) of soil comprises living and non-living components, such as earthworms and billions of microorganisms as living components and humus as a non-living component, which is formed from the decomposition of dead plant and animal. These living and non-living components of soil together will improve soil health sustainability. Maintaining a proper percentage for different components of soil is essential

DOI: 10.1201/9781003147077-10

for the soil to be healthy and can be achieved by proper microbiological activities (Sahu et al. 2017).

Soil has different layers or horizons (O, A, E, B, and C horizons), each having different characteristics. The 'O' horizon is mostly humus, which may be thick or thin or even may not present in some soils. 'A' horizon, the topsoil, is formed by organic material and minerals released from weathering, creating a suitable zone for organisms to live. 'E' horizon includes those minerals lost from it and deposited somewhere in the deeper depth (in B horizon). 'B' horizon, or subsoil, is rich in minerals that leached from above horizons. 'C' horizon (regolith), or soil base, is the horizon where predominantly no soil formation occurs. Depending on the presence of organic matter and soil porosity, the microbial diversity varies with soil horizons (Bhattarai et al. 2015). So, with an increase in depth, the organic matter and porosity decrease, resulting in a low microbial population count (Vermeire et al. 2018). Since different microorganisms present in soil require food and nutrients for their growth and development, the maximum diversity lies on the horizon with high organic matter content which is 'O' and 'A' horizon.

Within the next few decades, food production has to be increased to feed the world's growing populations, where the production of food is largely depending on the soil health. All the physical, chemical, and biological properties of the soils are related with the soil health (Fierer et al. 2021). So, any instability in those properties will lead to reduced capacity of the soil to function, leading to a lower crop production. In the early years, soil health is normally related with crop productions and yields for better profits, but in recent years, due to its contribution in biosphere, soil health sustainability is an important topic to consider. Since the 1960s with the commencement of the Green Revolution, there has been an extensive increase in food production following highly input-intensive agriculture system with the introduction of new crop varieties and extensive use of synthetic agrochemicals. These practices of agricultural productions lead to deteriorated soil health in long run. Also with increasing food demands of the increasing population globally, very less amount of biomass returns to the soil during harvesting, making the soil organic matter content low (Urra et al. 2019). Successively with the application of excess quantities of inorganic fertilizers, the carbon (C) and other nutrient ratio have also been disturbed. With such activities, the microbial growth and their activities are also disturbed. And for a proper nutrient cycling in soil, an optimum ratio of C to other nutrients is essential for a proper mineralization process. Earlier in agriculture, soil health was solely related with the physical and chemical properties neglecting the biological properties. But soil is considered living with the microorganisms present and their effect on soil. Without life, the soil will be just an aggregation of different primary particles (sand, silt and clay) released from the weathering of rocks and minerals. During microbial decomposition of organic matter, humic and non-humic substances are produced. These humic substances are the stable fraction of organic matter which has an important role of soil aggregation acting as a cementing agent. So, a stable aggregates' formation depends on the microbial activities in soil which have significant roles on good soil structure. Different types of enzymes released in soil are directly or indirectly from where the microorganisms reside. So, high microbial diversity in soils is essential to promote proper functioning and nutrient cycling by decomposing different fractions of soil organic matter (Sahu et al. 2017).

Soil performs many important roles by serving as a medium for plant growth, a habitat for soil organisms and systems for recycling nutrients, and water supply and purification as well as a system for removing greenhouse gases from the atmosphere (Tahat et al. 2020). Therefore, a healthy soil environment is quite essential in the biosphere. Soil health is the capacity of a soil to function, which is highly influenced by all the different soil properties, namely, physical, chemical, and biological properties; however, the biological components of soil, that is the soil, organism are found to be a vital aspect for soil health, having a strong correlation with soil microbiota. Different important soil health indicators of both physical and chemical properties of soil influence directly or indirectly the microbial diversity in soil. Physical properties of soil like water-holding capacity (WHC), texture, soil porosity, aggregate stability, bulk density (BD), etc., and chemical properties, pH, cation exchange capacity, charge site, etc., affect the microbial diversities in soil. Moreover, it has been suggested that measures of microbial abundance may serve as early indicators of soil health degradation before the physicochemical properties of soil are detected (Trivedi et al. 2016).

Soil organisms represent the highest biodiversity on earth containing large varieties of both micro- and macro organisms (Figure 10.1). The soil microorganisms such as bacteria, fungi, and nematodes increase soil nutrient cycling, thereby enhancing plant nutrient availability and uptake, and increase plant hormones' production and plant resistance to various environmental stresses. Moreover, soil microorganisms give a connection between soil and plant root, helping in decomposition of organic

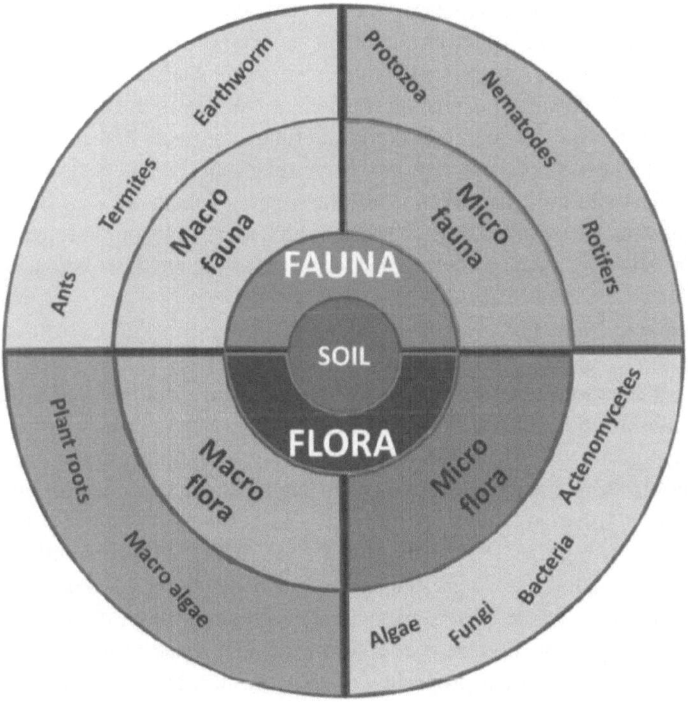

FIGURE 10.1 Common macro- and micro-flora and fauna present in soil.

matter in nutrient recycling, and serving a good indicator of any changes occurring in the soil environment. Soil macroorganisms such as earthworm and arthropods also serve positive role in soil fertility and health (Tahat et al. 2020). Therefore, proper maintenance of the micro- and macro-organism population in soil can minimize the residual effect of harmful chemicals which are used excessively for soil fertility and plant growth such as fertilizer, pesticides, and artificial plant growth regulators. The residual effects of xenobiotic chemicals used in agriculture have brought serious harm to human health and the environment. So, it is now vital to either reduce the use of it or degrade the residue present in the soil by different microbial populations (Mishra et al. 2019).

The relationship between soil and soil microorganism is significant for sustainable soil health and agriculture. The problems of modern agriculture, food insecurity for the growing population, and degradation of the environment can be alleviated by proper understanding of the different roles and contribution of soil organisms in soil health.

10.2 TYPES OF SOIL ORGANISMS

Soil organisms can be divided into the following points.

10.2.1 Size of the Organisms

Soil microorganisms: As the name indicates, the size of the organism is in μm which includes bacteria, fungi, mycorrhizae, protozoa, and nematodes. Bacteria have been categorized as fungi, and micorrhiza as microflora while protozoa and nematodes as microfauna. Microbial biomass returned to the soil is an important labile fraction of soil organic matter. In spite of their small quantity return, their availability in soil is very important for its significant effect on soil quality. Bacteria and fungi contribute higher amounts of biomass returns to the soil but these may vary from soil to soil (Rashid et al. 2016). Bacteria are the smallest and most numerous of all the microorganisms in soil. Because of its larger population, bacterial biomass could be as high as 1500 kg/ha. In fungi because of their larger body size, the biomass return is estimated as high as 3500 kg/ha.

Soil macroorganisms: The organisms having their size from mm to cm are kept in this group, for example, arthropods and earthworms. They can be categorized into mesofauna and macrofauna where microarthropods are grouped under mesofauna and macroarthropods and earthworms are grouped under macrofauna (Fortuna 2012). These groups of organisms play a vital role in maintaining the soil ecosystem and controlling microbial activities through decomposition, nutrient cycling, increased mineralization, and many more. The total biomass of earthworm in a soil furrow slice is as high as 1100 kg/ha (Saha et al. 2017). Different cellulolytic and hemi-cellulolytic bacteria are present in the gut of earthworm, making the earthworm cast rich in nutrient content particularly nitrogen, phosphorus, calcium, etc., and easily decomposable organic matter. Different arthropod influences soil properties by affecting the physical properties of soil (through burrowing), changing soil texture, increasing infiltration rate, reducing soil erosion, decomposing or reducing the size of highly resistant fraction of organic matter, etc. (Xin et al. 2018).

10.2.2 Role of the Organisms

Decomposers: They are the organisms that break down dead and decaying organic matter. They feed on dead remains of plants and animals, decomposing the remains and returning back the nutrients present in them to the soil. They play an important role in maintaining soil ecology by decomposing different organic fractions and nutrient transformation in soil (Griffiths et al. 2021), for example, saprophytic bacteria and fungi, earthworm.

Mutualist: Association between two different species with mutual benefits. They form a symbiotic relationship with root of plants, enhancing soil fertility, and stabilize soil aggregates, for example, *Rhizobium* bacteria with leguminous plants and mycorrhizal fungi with roots of higher plants.

Pathogenic: They are harmful organisms that infect plant and cause diseases. They reduced soil fertility that leads to poor soil health and ultimately poor crop yield. A different group of bacteria, viruses, fungi, and Protista can infect and impart diseases to plants, for example, pathogenic bacteria and fungi.

Parasitic: It is an association where one partner lives upon another and derived food from the body of host. They prey upon organisms. This interaction is widespread in the soil community. Bacteria, fungi, virus, etc., parasite on earthworm and other macroorganisms, for example, nematode and insect-feeding fungi.

Bioremediators: They are the biosorbent of heavy metals which are harmful to the environment. They absorb the toxic metals from soil and accumulate them in their bodies. Microorganisms required C and energy for growth and multiplication. The required C and energy are derived from the mineralization of the toxic substances. In this process, the parent substances become detoxicated by the reactions initiated through enzymes secreted by microorganisms (Bulgariu and Bulgariu 2020). Several bacteria and fungi such as *Achromobacter*, *Aerobacter*, *Bacillus*, *Clostridium*, *Aspergillus*, etc., can reduce the harmful effect of certain toxic substance and are rendered harmless to the environment.

Biocontrol agents: They suppress and control pathogens in soil ecosystem by living organisms. They protect crops from damage by a different mode of actions such as induced resistance against infection, competition for space and nutrients, secretion of antibiotics, antimicrobial secondary metabolites, etc.). Different genera of bacteria and fungi can suppress the pathogens in soil (*Bacillus*, *Lycobacter*, *Trichoderma*, *Rhodotorula*, *Sporobolomyces*, and *Cryptococcus*).

10.2.3 Prominent Role of Soil Organisms on Soil Health

The different soil organisms play significant roles at different levels in soil ecosystems including organic matter turnover, nutrient transfer, soil structure improvement, and disease prevention and pollutant degradation (Fortuna 2012). However, not all the soil organisms are beneficial to soil; some are friends while some are foes. Depending on the significant roles on soil health, different soil organisms are discussed below.

10.2.3.1 Bacteria

Soil bacteria are an important component of the soil ecosystem. The bacteria which are beneficial to the soil ecosystem are termed as plant growth-promoting rhizobacteria (PGPR). They inhabit plants and promote their growth. PGPR can be categorized into two categories on the basis of their residing sites in plants. They are symbiotic bacteria and free-living bacteria. Symbiotic bacteria are those bacteria that are localized inside the plant cell and are mutually beneficial, whereas free-living bacteria live outside the plant cell. PGPR functions in three ways, biofertilization which includes nitrogen (N_2) fixation and phosphate solubilization, phytostimulation, and biocontrol, plant prevention from disease.

Nitrogen (N) is one of the key essential nutrients required by plant for its growth and development, where biological N_2 fixation contributes more than half of total nitrogen requirement in crop plants. Nitrogen concentration in the earth atmosphere is made up of approximately 78% in the form of di-nitrogen boned with a triple bond. Different genera of microorganisms have the ability to break the triple bond and assimilate the N from the atmosphere which is termed as biological N fixation. Biological N fixation can be mediated by both symbiotic and non-symbiotic microorganisms including both bacteria (autotrophic and heterotrophic) and archaebacteria (Gaby and Buckley 2011). The symbiotic N_2-fixing bacteria, *Rhizobia*, form intimate symbiotic relationship with legumes and also with some of the non-legumes species such as rice, maize, sweet corn, cotton, etc., and fixed atmospheric N_2 (Hayat et al. 2010). The legumes plant released flavonoid molecules as a signal for inducing nodulation genes (nod) expression in *Rhizobia* that leads to the stimulation of mitotic cell division in root and formed nodules, which is the site of N_2 fixation.

Actinobacteria Frankia can fix atmospheric N_2 under free-living conditions and also by forming symbiotic associations with dicotyledonous plants. They form nitrogen-fixing root nodules of different genera (25) of dicotyledonous plants where N_2 fixation occurs and eight plant families belonging to three different orders (Fagales, Rosales, Cucurbitales) collectively called actinorhizal plants; hence, they are also called actinorhizal nitrogen fixer (Van Nguyen and Pawlowski 2017).

Another symbiotic nitrogen-fixing bacterium is *Azospirillum*, which is free-living N_2-fixing rhizosphere bacteria and also forms symbiotic association with staple food crops like rice, maize, sorghum, wheat, and millets and fixes atmospheric nitrogen. *Azospirillum* is considered as the crucial biofertilizer in the cultivation of rice that can fix significant amount of atmospheric nitrogen as well as solubilize the phosphorus within the soil, thereby improving soil fertility and crop production (Suhameena et al. 2020). *Azospirillum* influences plant growth and increased yield by fixing N_2 and producing growth-promoting substances. The contribution of N to soil by *Azospirillum* is questionable when compared with other diazotrophs. But its availability in soil improves soil health and crop production by secreting plant growth-promoting substances and increasing uptake of nutrients from the soil (Steenhoudt and Vanderleyden 2000).

Unlike symbiotic bacteria, non-symbiotic N_2-fixing microorganisms, cyanobacteria and genera like *Azotobacter*, *Beijerinckia*, and *Clostridium*, are identified as free-living within soil or associated with plant organs where fixation occurs without forming plant-microorganism symbiosis. Cyanobacteria have the ability to colonize different plants, from division Bryophyta to Angiosperms, either extracellularly (liverworts, hornworts, *Azolla*, and Cycadaceae) or intracellularly (*Gunneraceae*), where the plant structure colonized by the symbiotic cyanobacteria develops independently of cyanobacterial infection (Santi et al. 2013). Cyanobacteria can be grouped into two types, heterocystous cyanobacteria and non-heterocystous cyanobacteria. Heterocystous cyanobacteria, such as Nostoc, Anabaena, possess specialized enlarged cell, which is called heterocyst, where fixation of nitrogen is done, whereas in non-heterocystous cyanobacteria like *Trichodesmium erythraenum*, some other specialized cells called diazocytes are present, which are responsible for N_2 fixation (Kulasooriya and Magana-Arachchi 2016). Moreover, the subterranean layers of thick mats of *Oscillatoria*, *Lyngbya*, *Microcoleus*, and *Plectonema* also enable fixation of atmospheric nitrogen.

The essential macronutrient of plant phosphorous occurs in organic and inorganic phosphates in soil, which are in insoluble forms. Bacterial strains belonging to the genera of *Pseudomonas*, *Bacillus*, *Rhizobium*, etc., solubilize insoluble inorganic phosphate compounds, enhancing the phosphorus status of plant. In addition to N_2 fixation and phosphate solubilization, PGPR promote phytohormone (auxins, cytokinins, abscicic acids, etc.) production which is essential for plant growth and development and also increase resistivity of plant indirectly against various stresses. They also act as a biocontrol agent, which is involved in controlling diseases of crop plants through two strategies, direct inhibition and indirect inhibition of the pathogens (Köhl et al. 2019). Direct inhibition involved hyperparasitism, where cells of the pathogen are directly killed by invading, and antibiosis, where antimicrobial secondary metabolites are produced that inhibited the pathogens. In indirect inhibition, pathogens are made to compete for nutrients, oxygen, and space. Another mode of indirect interaction and inhibition of the pathogen involved stimulation of plant resistance against the pathogen (Tahat et al. 2020).

10.2.3.2 Fungi

Fungi are eukaryotic organisms which are widely distributed on earth and are of great importance in soil health. As reviewed by Frąc et al. (2018), soil fungi are categorized into three groups as follows.

Biological controller fungi: They are those fungi which are responsible for the regulation of diseases, pests, and growth of other organisms. Mycorrhizal fungi are grouped under this category as they protect plant from pathogens in addition to the enhancement of nutrient uptake (Frąc et al. 2018). As a result, an alternative approach to maintain the pests in agriculture is through biological control agents which is an eco-friendly approach. Among all the biological control agents used, fungi offer certain advantages (Latz et al. 2018). Few of them are omnipresent, host specificity, easy to culture, cause destruction to the host, etc.

Example: *Ampelomyces quisqualis*

Decompositor fungi: Most fungi are characterized by the presence of hyphae and aggregation of hyphae is called mycelium. Being filamentous, saprophytic fungi in soil perform a very important role in decomposing the most resistant organic residues (Al-Ani et al. 2021). They can penetrate inside the rigid cell wall with the pressure exerted by hyphae and produce hydrolytic enzymes inside the cell making way for the bacteria to decompose easily. They are responsible for organic matter decomposition and compound transformation.

Ecosystem regulator fungi: Different genera of fungi produce organic substrate that act as a cementing agent aggregating the primary soil particles (sand, silt, and clay) involved in soil structure formation. A good soil structure maintains a proper ratio of solid, liquid, and gas, providing a suitable environment for high microbial diversity. With its ability as a biocontrol agent and decomposition of highly resistance fractions of soil organic matter, it is involved in nutrient cycling and energy flow from one tropic level to another. They can modify habitats of other organisms by regulating the dynamics of physiological processes in the soil environment.

Fungi are critically important for a healthy soil ecosystem as they perform different significant roles in soil and plant including decomposition and stabilization of organic matter, nitrogen fixation, hormone production, and protection from biotic and abiotic stresses in plants.

The most important fungi that play a promising role in soil and plants are the arbuscular mycorrhizal fungi (AMF) where they form a symbiotic association with roots of higher plants, favour soil structure development and stability, establish mycorhizosphere which improve nutrient availability, absorption, mobility, and plant growth, and stimulate nutrient cycling, plant nutrition, and production of growth-promoting substances that enhance plant tolerance to both biotic and abiotic stresses (Tahat et al. 2020). Moreover, AMF with PGPR improved soil ecosystem and plant health, under various environmental stress conditions (Bagyaraj and Ashwin 2017; Pankaj et al. 2019, 2020, 2021).

10.2.3.3 Protozoa

Soil protozoa belong to microfauna and are largest in size when compared with different other microorganisms present. They are found in greater abundance in soil with high organic matter content, playing important roles in maintaining soil health. There are two phases of the life cycle; first, an actively growing phase and second, a resting phase under adverse environmental conditions. According to Hoorman (2011), protozoa are categorized into three major groups. They are as follows:

Flagellates: The smallest and most numerous protozoa of soil that mainly feed on bacteria and move with the help of whip-like flagella. They are numerous in number and highly active in predation of bacteria with high nutrient transformation in soil (Khanipour Roshan et al. 2021).

Ciliates: The largest and the least numerous protozoa, which feed on the other two types of protozoa, as well as bacteria (Chinnarajan et al. 2021). They move by means of hair-like cilia. They are confined to moist or very moist habitat.

Amoeboid: They inhabit in rhizosphere and graze on bacteria populations. They can explore tiny pore spaces that support soil nutrient recycling (Kuppardt-Kirmse and Chatzinotas 2020). They actually are not a N_2-fixing organism, but the required nitrogen is derived from the bacteria they feed upon. So, the microbial biomass nitrogen (MBN) returns to the soil after death and decay will be very high which improves the nitrogen content in soil. Of its ability to explore tine spores, the micropores present in soil increase, which is important for nutrient movement and gaseous exchange, an essential condition for proper soil health.

Protozoa play important roles in soil ecosystems by controlling bacterial abundance in soil environments by grazing actively on them (Finlay et al. 2000). They help in maintaining bacterial equilibrium in soil and controlling many plant diseases. The carbon and nitrogen content of protozoa cells are low when compared with bacteria, but when they feed on bacteria containing high nitrogen levels, their body contains too much N for the amount of C protozoa needed. So, the excess nitrogen from the cells is released in the form of ammonium (NH_4^+), thereby helping in mineralizing nutrients (Hoorman 2011). Protozoa increase organic matter decomposition and turnover rate of essential nutrients and carbon, thereby increasing microbial activity and microbial biomass in soil (Esteban et al. 2006; Geisen et al. 2018). Moreover, protozoa are used as indicators of quality of soil (Madoni 2011).

10.2.3.4 Nematodes

They are the earliest known microfauna and their abundance is next to protozoa. They are called threadworm because of its narrow long body and small enough to fit in most soil pores and soil aggregates. They cause infectious diseases in plants and animals, but because of their significant role in maintaining the soil ecosystem, they are beneficial in soil (Hoorman 2011). They help in maintaining ecosystem and nutrient cycling in soil. They feed on bacteria, fungi, algae, plant, and other nematodes; so depending on the organisms they feed on, nematodes can be classified as bacterivorous, which feed on bacteria; fungivorous, which feed on fungi; algivorous, which feed on algae; omnivorous, which feed on many soil organisms; and predatory nematodes, which feed on nematodes (Tahat et al. 2020). It has been further reviewed that a healthy soil with high organic matter content has a larger and wider range of beneficial nematodes which maintained a good balance of different other microorganisms and bacterial feeders as well as having predatory and omnivorous nematodes. Because of the diverse feeding behaviours and ubiquity of nematodes, they can respond quickly to soil health changes. Stone et al. (2016) also reported the involvement of nematodes in nutrient recycling process and habitat for biodiversity, recommending as a good biological indicator of soil quality assessment. They serve as potential biological pest control agents by consuming disease-causing organisms.

10.3 CORRELATION BETWEEN SOIL PROPERTIES WITH SOIL MICROORGANISMS

Soil is a natural medium compost of organic and inorganic constituents in different stages of organization where different organisms live and die providing organic matter and recycled for plant nutrients. Soil microbial population and its activities are important for deriving a good soil health of its ability to regulate mineralization, organic matter decomposition, nutrients cycling, and soil stability (Rao et al. 2019).

Different nutrients present in organic form have to be mineralized for its availability to other organisms; otherwise, continuation of life could be impossible. The primary particles present in soil should be cemented for good soil structure or aggregate stability. These aggregations of primary particles are done by secretion of polysaccharides and gummy substance by different microorganisms in soil (Bharadwaj 2021). Soil is one of the most important reservoirs of carbon that help to reduce the present global climate change. With the actions of soil microorganisms, unstable organic matters are converted into stable humic substances that stored in the soil and improved the physical properties. Soil health is the capacity of soil to function which depends on the physical, chemical, and biological properties of soil. And for sustaining soil health for better crop production and crop health, the role of microorganisms is well known. In view of its important role in maintaining soil health and sustainability, it is now essential to understand the factors affecting microbial distribution and composition in soil to maintain (Figure 10.2). Many factors regulate organism population in soil ecosystems; however, the most prominent factor that significantly positively correlated with the soil microbial community was found to be the pH of soil, which was reported in many studies. Different microbial communities can survive at some specific ranges of soil pH to avoid competition among them. In general conditions, neutral pH bacterial population is high whereas when the pH of the soil drops, the fungal population increases and when the pH increases above neutral actenomycetes, the population increases. The significant positive correlation between soil pH and microbial diversity exists under neutral soil pH; however, the correlation decreases with either an increase or decrease of soil pH. Different investigations have been reported to find the soil factors influencing the ecological function and distribution of bacteria in soil. Among these, soil pH is one of the

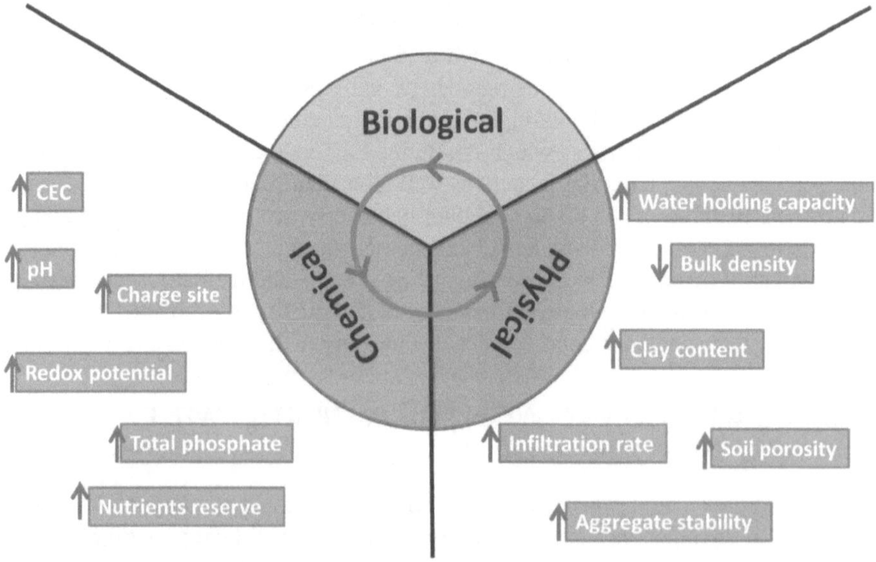

FIGURE 10.2 Important soil indicators (physicochemical) that are responsible for improving biological diversity.

important factors that significantly affect the bacterial population and its diversity when compared with nutrients and organic carbon content in soil (Wang et al. 2019). For proper shape of bacterial communities, its functioning in soil and biogeographic distribution pH plays a major role. The bacterial genera *Actinobacteria, Acidobacteria, Gemmatimonadetes, Chloroflexi, Proteobacteria, Nitrospirae*, and *Gammaproteobacteria* found responded strongly to soil pH followed by climatic variables, particularly mean annual precipitation (Bainard et al. 2016; Wang et al. 2019). In high nutrient content and alkaline soil conditions, there is greater metabolic demand which increases several metabolic and transport pathways (Bahram et al. 2018). The significant influence of pH on soil microorganisms is attributed due to either directly involved in physiological stress tolerance by a different group of microorganisms or indirectly affecting different properties of soil like nutrient cycling, ion exchange reactions, and organic matter decomposition.

Soil microbial activity highly influences nutrients content in soil (Cao et al. 2016). A proper ratio of carbon to different nutrients should maintain for a mineralization reaction to take place. Soil is a medium where different organisms derived their nutrients. Maximum of the nutrients present in soil are primarily derived from the disintegration and decomposition of rocks and minerals (weathering) (Schuessler et al. 2018). Only after consolidated rocks and minerals were weathered to form an unconsolidated soil, different organisms started habitat on it. The same nutrients which are utilized by different organisms have to recycle for the habitat of other organisms. So, organic matter decomposition is an important step for recycling nutrients in soil, and its presence also determined the amount of microbial population. Besides nitrogen and carbon, sulphur is also an important constituent of plant residue which influences different groups of microorganisms (Saha et al. 2018). Microorganisms capable of oxidizing sulphur could be either aerobic or anaerobic. *Thiobacillus*, *Beggiatoa*, *Thothrix*, and *Thioploca* are important S oxidizing bacteria and several fungi and *actenomycetes* such as *Aspergillus*, *Penicillium*, and *Microsporeum* have also been reported to be S oxidizer. Phosphorus in soil occurs in both organic and inorganic form and its availability depends on soil pH. Many bacteria and fungi have the potential to solubilize this P which are bound with cations (Ca, Mg, Fe, and Al) with the change in pH. So, presence of essential nutrients in soil determines the microbial population in soil. Besides the chemical properties, environmental and geographic factors including distance and current environmental factors encompassing climate gradient (rainfall and temperature) and soil type were observed to be highly involved in regulating the microbial activity. The predominant role of pH in the bacterial community is in accordance with the findings of Lan et al. (2017b), where they reported that soil pH and total phosphorous are the primary drivers for the bacterial community. However, the fungal community was reported to be greatly influenced by organic matter and total nitrogen content of the soil.

Effects of abiotic and biotic factors in shaping soil microbial communities (Wang et al. 2020b) reported that abiotic factors such as soil pH, moisture content, soil temperature, nutrients distribution, soil texture, structure, and BD (Figure 10.2) highly influence bacterial communities whereas fungal communities are influenced by biotic factors (Stevens et al. 2020). Soil moisture also appeared to be an important factor managing different soil microbial communities in agroecosystems. Microbial activities are inhibited by low water content in soil due to plasmolysis.

In order to maintain cell integrity, it is important to increase intracellular solutes than extracellular (Manzoni et al. 2012). When soil moistures dry up, substrate supply on soil microbial cells declines and nutrient transportation through mass flow and diffusion also halt (Bremer and Krämer 2019). The soil moisture level at the range of 50–70% of WHC is optimum for maximum activity of aerobic microorganisms (Franzluebbers 1999). In natural soil, the microbial population is also affected by the soil texture. Soil texture is simply the percentage of sand, silt, and clay. With the percentage distributions of these primary particles in soil, the microbial biomass return and turnover varies. The turnover of microbial biomass is more under coarse textured soil than a fine textured but the proportion of carbon and nitrogen in biomass is reversed. Soil texture also plays an important role in the faunal-induced mineralization process. Depending on the texture, soil faunal activities vary, finer the soil there is physical restriction for the fauna to feed upon certain biomass in soil. So, texture with its effect on the turnover and mineralization process in soil plays an important role in maintaining soil habitat and microbial communities. The primary particles present in soil do not remain individual but instead they remain in an organized way which we call soil aggregates. This aggregation of primary particles by physical, chemical, and biological forces to form a stable aggregate is called soil structure (Pepper and Brusseau 2019). A good soil structure possesses enough pores which maintained a proper ratio of solid, liquid, and gas which is essential for microbial population. So, more soil aggregates stability, more diversity of microbial population will find. Another important physical soil property is BD, which signifies the compactness of soil. A higher value of BD means more compaction or lower porosity, which results in low percentage of water and air in soil. So, higher the value, lower will be the microbial population and vice versa.

Organic matter is also a crucial factor in regulating soil microbial abundance and function (Li et al. 2018), where the main parameters are C/N ratio and N content of organic matter. The primary source of organic matter is from plant and during any plant life cycle, they assimilate different essential nutrients which are integral components of its tissue. Different communities of microorganisms residing in soil required essential nutrients for their growth and activities. The carbon and other nutrients required will be assimilated after decomposition of organic matter. So, the microbial population and its activities are influenced with the presence or absence of organic matter. Many researchers have studied different field experiments comparing organic materials (wheat straw, corn straw, wheat root, corn root, pig manure, and cattle manure), in soil and the microbial diversity. Since the microbial abundance in soil depends on the nutrient released from the organic matter to the soil, so the process of mineralization solely depends on the C:N ratio of the organic materials applied. Narrower the C:N ratio, higher will be the mineralization rate and vice versa. The C:N ratio of the organic residue in soil will depict the mineralization rate which indirectly influences the population of the microorganisms. When compared with manure-treated soil and crop residue treatment, there is a significantly higher microbial biomass carbon (MBC) and MBN in the first few months of the manure treatments (Li et al. 2018). But with an increase in time, the crop residue-treated soil shows significantly higher MBC and MBN than the manure-treated soil. These show that organic matter containing wider C:N ratio prolongs mineralization process than

OM with a narrow C:N ratio. The functional diversity of microorganisms in soil with crop residue with wider C:N ratio and low lignin increases, suggesting its long-term effect to increase microbial biomass and can further induce the changes of soil properties to regulate the soil microbial community. Biochar, a carbon-rich material produced during pyrolysis process, is used to enrich soil health. Biochar when applied to soil increased the bacterial population but reduced the fungal community, suggesting the importance of organic carbon for the bacterial community and the carbon-to-nitrogen ratio for fungi (Wang et al. 2020b). It has also been reported that soil pH, soil organic carbon, total nitrogen, and total phosphate were all significantly correlated with bacteria, actinomycetes, fungi, and total microorganisms; however, the impact greatly differs with the difference of the factors and microbial biomass.

There is a positive correlation of organic compounds with the microbial community in soil, where the organic matter favours the growth of microorganisms; therefore, higher organic matter resulted in higher microorganisms (Bhattarai et al. 2015). The organic carbon content in soil is directly proportional to the microbial biomass return. Babur and Dindaroglu (2020) also reviewed the positive effect of soil characteristics including pH, content of moisture and nutrients, temperature, and clay content on soil microbial diversity and activity. Woldemariam and Ghoshal (2020) studied the correlation between physicochemical properties and microbial biomass of soil using different land use systems. Almost 95% of microbial biomass and activities were governed by the physiochemical properties of soil, namely, soil organic carbon, soil total nitrogen, WHC, soil aggregates, porosity, and BD, where the properties were found to be higher in natural forests when compared with agricultural land. Hermans et al. (2017) observed a strong relationship between specific taxa and anthropogenesis-related soil variables including the relative abundances of members of *Pirellulaceae*a and soil pH, members of *Gaiellaceae* and carbon-to-nitrogen ratios, members of *Bradyrhizobium* and the levels of Olsen P (a measure of plant available phosphorus), and members of *Chitinophagaceae* and aluminium concentrations, suggesting their relationships not only show up ecological characteristics of these organisms but also could serve as a biologically relevant indicator of soil conditions. In Table 10.1, different techniques for the study microbial conditions in soil are given, which effectively determine the soil microbial diversity and activity in different soils and the soil quality. All these methods are associated with biological samples but the approaches and techniques differ. Biochemical methods deal with the cellular and subcellular levels which are applied to determine the activities of microorganisms in soil. Most of the biochemical methods deal with the ability of soil microorganisms to decompose different fractions of organic matter with secretion of specific enzymes, whereas molecular methods deal exclusively with nucleic acid and DNA at the molecular level. For determining a true picture of microbial diversity, molecular microbial methods have shown great advantages over the conventional dilution plate and direct microscopic count. In this molecular method, extraction and purification of nucleic acid from soil and DNA hybridization provide a true picture of microbial diversity in a given soil. Many studies have demonstrated the direct relationship of soil microbial health and soil quality; thereby determining the microbial health in soil and the status of soil health could be detected (Bolat et al. 2015; Trivedi et al. 2016; Frąc et al. 2018; Bhowmik et al. 2019).

TABLE 10.1
Different Techniques Used to Determine Microbial Diversity and Function in Soil

Sl.	Analysis	Technique	Parameter	Signification	References
1.	Biochemical and molecular analysis	Enzyme activity and soil DNA estimation using PCR-DGGE	Dehydrogease, fluorescein diacetate (FDA) hydrolysis	Microbial activity and microbial community structure	Gajda et al. (2018)
2.	Molecular analysis	IlluminaMiSeq sequencing analysis	Operational taxonomic units, Chao index, and Shannon index	Microbial composition and diversity	Lan et al. (2017a)
3.	Molecular analysis	IlluminaMiSeq sequencing analysis	Operational taxonomic units	Community structure of soil microorganisms	Stevens et al. (2020)
4.	Biochemical analysis	Counting, respiratory and enzyme activities	CO_2 Production and dehydrogenase activity	Microbial number and activities	Rigobelo and Nahas (2004)
5.	Molecular analysis	Metagenomic analysis	Genome size of the community (community functional gene content)	Microbial community	Johnston et al. (2019)
6.	Biochemical analysis	Chloroform fumigation extraction and basal respiration methods	C and N content and CO_2 production	Microbial biomass and activity	Bolat et al. (2015)
7.	Biochemical analysis	Chloroform-fumigation-extraction and basal respiration methods	Soil microbial biomass carbon and microbial indexes (Cmic/Corg, MR/Cmic) and microbial respiration	Microbial biomass and activity	Babur (2019)
8.	Molecular analysis	Miseq sequencing	Operational taxonomic units and Shannon index	Microbial communities	Wang et al. (2020a)
9.	Molecular analysis	IlluminaHiSeq sequencing	Operational taxonomic units	Bacterial community structure and functions	Wang et al. (2019a)

(Continued)

TABLE 10.1 *(Continued)*
Different Techniques Used to Determine Microbial Diversity and Function in Soil

Sl.	Analysis	Technique	Parameter	Signification	References
10.	Molecular analysis	Pyrosequencing analysis	Chao, Shannon's and Inverse Simpson's indices	diversity and composition of soil bacterial communities	Bainard et al. (2016)
11.	Physical and biochemical analysis	Microcalorimetric measurement Gas chromatography mass spectrometer	Thermal activity and phospholipid fatty acid (PLFA)	Microbial activity and community structure	Cao et al. (2016)
12.	Molecular analysis	Metagenomics and metabarcoding	Operational taxonomic units, latitudinal diversity gradient, antibiotic-resistance genes, phospholipid fatty acids, and carbon-nitrogen content	Microbial diversity and function	Bahram et al. (2018)
13.	Biochemical analysis	Microbial count and enzyme assay	Dehydrogenase, acid and alkaline phosphatase activities	Microbial community	Bhowmik et al. (2019)

10.4 CONCLUSION AND FUTURE PERSPECTIVES

The inevitable processes for a healthy soil including N_2 fixation, phosphate solubilization, and nutrient cycling are performed by soil organisms. They also enhance plant nutrient availability and uptake, plant hormones production, and plant resistance to various environmental stresses, figuring a healthy soil ecosystem. They are the most important components of soil health, which could bring to sustainability of soil and agriculture. The major challenge of agriculture is to achieve sustainability, i.e., providing sufficient food of the growing population without harming the environment could be overcome by giving more emphasis on beneficial soil microorganisms. For example, the use of chemical fertilizers, pesticides, and insecticides for the improvement of soil fertility should be replaced by beneficial microorganisms, which can increase soil fertility without affecting the environment. There are various factors that encourage or inhibit the soil microbial community. Factors like pH, content of organic matter, moisture, temperature, texture, structure, nutrients, and clay encourage the microbial diversity and function; therefore, more research for a better understanding of the factors and mechanisms on soil microbial communities is needed to maintain soil health. Soil microorganisms are the novel indicators of soil health as they respond quickly to any changes in the soil ecosystem than physical or chemical properties. Therefore, by determining the presence or absence of soil microorganisms or alteration of the microbial population, soil health can be evaluated. Many techniques have emerged for the determination of soil microbial diversity and activity including physical, biochemical, and molecular techniques, which provide knowledge on the status of the microbial community.

REFERENCES

Al-Ani, L.K.T., Aguilar-Marcelino, L., Salazar-Vidal, V.E., Becerra, A.G., Raza, W. 2021. Role of useful fungi in agriculture sustainability. In *Recent Trends in Mycological Research* (pp. 1–44). Springer, Cham.

Babur, E. 2019. Effects of parent material on soil microbial biomass carbon and basal respiration within young afforested areas. *Scandinavian Journal of Forest Research* 34(2): 94–101.

Babur, E., Dindaroglu, T. 2020. Seasonal changes of soil organic carbon and microbial biomass carbon in different forest ecosystems. *Environmental Factors Affecting Human Health*, 1:1-21

Bagyaraj, D.J., Ashwin, R. 2017. Can mycorrhizal fungi influence plant diversity and production in an ecosystem. In *Microbes for Restoration of Degraded Ecosystems* (pp. 1–7). NIPA, New Delhi.

Bahram, M., Hildebrand, F., Forslund, S.K. et al. 2018. Structure and function of the global topsoil microbiome. *Nature* 560(7717): 233–237.

Bainard, L.D., Hamel, C., Gan, Y. 2016. Edaphic properties override the influence of crops on the composition of the soil bacterial community in a semiarid agroecosystem. *Applied Soil Ecology* 105: 160–168.

Bharadwaj, A. 2021. Role of microbial extracellular polymeric substances in soil fertility. In *Microbial Polymers* (pp. 341–354). Springer, Singapore.

Bhattarai, A., Bhattarai, B., Pandey, S. 2015. Variation of soil microbial population in different soil horizons. *Journal of Microbiology & Experimentation* 2(2): 2–5.

Bhowmik, A., Kukal, S.S., Saha, D., Sharma, H., Kalia, A., Sharma, S. 2019. Potential indicators of soil health degradation in different land use-based ecosystems in the Shiwaliks of northwestern India. *Sustainability* 11(14): 3908.

Bolat, I., Kara, Ö., Sensoy, H., Yüksel, K. 2015. Influences of Black locust (*Robinia pseudoacacia* L.) afforestation on soil microbial biomass and activity. *iForest-Biogeosciences and Forestry* 9(1): 171.

Bremer, E., Krämer, R. 2019. Responses of microorganisms to osmotic stress. *Annual Review of Microbiology* 73: 313–334.

Bulgariu, L., Bulgariu, D. 2020. Bioremediation of toxic heavy metals using marine algae biomass. In *Green Materials for Wastewater Treatment* (pp. 69–98). Springer, Cham.

Cao, H., Chen, R., Wang, L., Jiang, L., Yang, F., Zheng, S., Wang, G., Lin, X., 2016. Soil pH, total phosphorus, climate and distance are the major factors influencing microbial activity at a regional spatial scale. *Scientific Reports* 6(1): 1–10.

Chinnarajan, R., Raveendran, H.P., Irudayarajan, L., 2021. Ciliated protozoan occurrence and association in the pathogenesis of coral disease. *Microbial Pathogenesis* 162: 105211.

Esteban, G.F., Clarke, K.J., Olmo, J.L., Finlay, B.J. 2006. Soil protozoa – An intensive study of population dynamics and community structure in an upland grassland. *Applied Soil Ecology* 33(2): 137–151.

Fierer, N., Wood, S.A., de Mesquita, C.P.B. 2021. How microbes can, and cannot, be used to assess soil health. *Soil Biology and Biochemistry* 153: 108111.

Finlay, B.J., Black, H.I., Brown, S., Clarke, K.J., Esteban, G.F., Hindle, R.M., Olmo, J.L., Rollett, A., Vickerman, K. 2000. Estimating the growth potential of the soil protozoan community. *Protist* 151(1): 69–80.

Frąc, M., Hannula, S.E., Bełka, M., Jędryczka, M. 2018. Fungal biodiversity and their role in soil health. *Frontiers in Microbiology* 9: 707.

Franzluebbers, A.J. 1999. Microbial activity in response to water-filled pore space of variably eroded southern piedmont soils. *Applied Soil Ecology* 11(1): 91–101.

Gaby, J. C., Buckley, D. H. (2011). A global census of nitrogenase diversity. *Environmental Microbiology* 13(7), 1790–1799.

Gajda, A.M., Czyż, E.A., Dexter, A.R., Furtak, K.M., Grządziel, J., Stanek-Tarkowska, J. 2018. Effects of different soil management practices on soil properties and microbial diversity. *International Agrophysics* 32(1): 81.

Geisen, S., Mitchell, E.A., Adl, S. et al. 2018. Soil protists: A fertile frontier in soil biology research. *FEMS Microbiology Reviews* 42(3): 293–323.

Griffiths, H.M., Ashton, L.A., Parr, C.L., Eggleton, P. 2021. The impact of invertebrate decomposers on plants and soil. *New Phytologist* 231: 2142–2149.

Hayat, R., Ali, S., Amara, U., Khalid, R., Ahmed, I. 2010. Soil beneficial bacteria and their role in plant growth promotion: A review. *Annals of Microbiology* 60(4): 579–598.

Hermans, S.M., Buckley, H.L., Case, B.S., Curran-Cournane, F., Taylor, M., Lear, G. 2017. Bacteria as emerging indicators of soil condition. *Applied and Environmental Microbiology* 83(1): 2826–2816.

Hoorman, J.J. 2011. The role of soil protozoa and nematodes. In: Smith, K.L. (Ed.), *Fact Sheet: Agriculture and Natural Resources*. The Ohio State University Extension, Columbus, Ohio.

Khanipour Roshan, S., Dumack, K., Bonkowski, M., Leinweber, P., Karsten, U., Glaser, K. (2021). Taxonomic and functional diversity of heterotrophic protists (Cercozoa and endomyxa) from biological soil crusts. *Microorganisms* 9(2), 205.

Johnston, E.R., Hatt, J.K., He, Z., Wu, L., Guo, X., Luo, Y., Schuur, E.A., Tiedje, J.M., Zhou, J., Konstantinidis, K.T. 2019. Responses of tundra soil microbial communities to half a decade of experimental warming at two critical depths. *Proceedings of the National Academy of Sciences* 116(30): 15096–15105.

Köhl, J., Kolnaar, R., Ravensberg, W.J. 2019. Mode of action of microbial biological control agents against plant diseases: Relevance beyond efficacy. *Frontiers in Plant Science* 10: 845.

Kulasooriya, S.A., Magana-Arachchi, D.N. 2016. Nitrogen fixing cyanobacteria: Their diversity, ecology and utilisation with special reference to rice cultivation. *Journal of the National Science Foundation of Sri Lanka* 44(2): 111–128.

Kutílek, M., Nielsen, D.R., 2015. Soil is the skin of the planet earth. In *Soil* (pp. 13–19). Springer, Dordrecht.

Kuppardt-Kirmse, A., Chatzinotas, A. (2020). Intraguild predation: predatory networks at the microbial scale. In *The Ecology of Predation at the Microscale* (pp. 65–87). Springer, Cham.

Lan, G., Li, Y., Jatoi, M.T., Tan, Z., Wu, Z., Xie, G. 2017a. Change in soil microbial community compositions and diversity following the conversion of tropical forest to rubber plantations in Xishuangbanan, Southwest China. *Tropical Conservation Science* 10: 1940082917733230.

Lan, G., Li, Y., Wu, Z., Xie, G. 2017b. Soil bacterial diversity impacted by conversion of secondary forest to rubber or eucalyptus plantations: A case study of Hainan Island, South China. *Forest Science* 63(1): 87–93.

Latz, M.A., Jensen, B., Collinge, D.B., Jørgensen, H.J. 2018. Endophytic fungi as biocontrol agents: Elucidating mechanisms in disease suppression. *Plant Ecology & Diversity* 11(5–6): 555–567.

Li, L., Xu, M., Eyakub Ali, M., Zhang, W., Duan, Y., Li, D. 2018. Factors affecting soil microbial biomass and functional diversity with the application of organic amendments in three contrasting cropland soils during a field experiment. *PLoS ONE* 13(9): 0203812.

Madoni, P. 2011. Protozoa in wastewater treatment processes: A minireview. *Italian Journal of Zoology* 78(1): 3–11.

Manzoni, S., Schimel, J.P., Porporato, A. 2012. Responses of soil microbial communities to water stress: Results from a meta-analysis. *Ecology* 93(4): 930–938.

Mishra, V.K., Singh, G., Shukla, R. 2019. Impact of xenobiotics under a changing climate scenario. In *Climate Change and Agricultural Ecosystems* (pp. 133–151). Woodhead Publishing, United Kingdom.

Pankaj, U., Singh, D.N., Singh, G., Verma, R.K. 2019. Microbial inoculants assisted growth of *Chrysopogon zizanioides* promotes phytoremediation of salt affected soil. *Indian Journal of Microbiology* 59(2): 137–146.

Pankaj, U., Singh, D.N., Mishra, P., Gaur, P., Vivekbabu, C.S., Shanker, K., Verma, R.K. 2020. Autochthonous halotolerant plant growth promoting rhizobacteria promote bacoside A yield of *Bacopa monnieri* (L.) Nash and phytoextraction of salt-affected soil. *Pedosphere* 30(5): 671–683.

Pankaj, U., Kurmi, A., Lothe, N.B., Verma, R.K. 2021. Influence of the seedlings emergence and initial growth of *palmarosa* (*Cymbopogon martinii* (Roxb.) Wats. var. Motia Burk) by arbuscular mycorrhizal fungi in soil salinity conditions. *Journal of Applied Research on Medicinal and Aromatic Plants* 24: 100317.

Pepper, I.L., Brusseau, M.L. 2019. Physical-chemical characteristics of soils and the subsurface. In: *Environmental and Pollution Science*, (pp. 9–22). Academic Press.

Rao, D.L.N., Aparna, K., Mohanty, S.R. 2019. Microbiology and biochemistry of soil organic matter, carbon sequestration and soil health. *Indian Journal Fertilizer* 15: 124–138.

Rashid, M.I., Mujawar, L.H., Shahzad, T., Almeelbi, T., Ismail, I.M., Oves, M. 2016. Bacteria and fungi can contribute to nutrients bioavailability and aggregate formation in degraded soils. *Microbiological Research* 183: 26–41.

Rigobelo, E.C., Nahas, E. 2004. Seasonal fluctuations of bacterial population and microbial activity in soils cultivated with *Eucalyptus* and *Pinus*. *Scientia Agricola* 61: 88–93.

Saha, J.K., Selladurai, R., Coumar, M.V., Dotaniya, M.L., Kundu, S., Patra, A.K. 2017. Soil and its role in the ecosystem. In *Soil Pollution – An Emerging Threat to Agriculture* (pp. 11–36). Springer, Singapore.

Saha, B., Saha, S., Roy, P.D., Padhan, D., Pati, S., Hazra, G.C. 2018. Microbial Transformation of Sulphur: An Approach to Combat the Sulphur Deficiencies in Agricultural Soils. In *Role of Rhizospheric Microbes in Soil* (pp. 77–97). Springer, Singapore.

Sahu, N., Vasu, D., Sahu, A., Lal, N., Singh, S.K. 2017. Strength of microbes in nutrient cycling: A key to soil health. In *Agriculturally Important Microbes for Sustainable Agriculture* (pp. 69–86). Springer, Singapore.

Santi, C., Bogusz, D., Franche, C. 2013. Biological nitrogen fixation in non-legume plants. *Annals of Botany* 111(5): 743–767.

Schuessler, J.A., von Blanckenburg, F., Bouchez, J., Uhlig, D., Hewawasam, T. 2018. Nutrient cycling in a tropical montane rainforest under a supply-limited weathering regime traced by elemental mass balances and Mg stable isotopes. *Chemical Geology* 497: 74–87.

Steenhoudt, O., Vanderleyden, J. 2000. *Azospirillum*, a free-living nitrogen-fixing bacterium closely associated with grasses: Genetic, biochemical and ecological aspects. *FEMS Microbiology Reviews* 24(4): 487–506.

Stevens, B.R., Roesch, L., Thiago, P., Russell, J.T., Pepine, C.J., Holbert, R.C., Raizada, M.K., Triplett, E.W. 2020. Depression phenotype identified by using single nucleotide exact amplicon sequence variants of the human gut microbiome. *Molecular Psychiatry* 26(8): 4277–4287.

Stevens, B.M., Sonderegger, D.L., Johnson, N.C. 2020. Biotic and abiotic factors predict the biogeography of soil microbes in the Serengeti. BioRxiv. https://doi.org/10.1101/2020.02.06.936625

Stone, D., Costa, D., Daniell, T.J., Mitchell, S.M., Topp, C.F.E., Griffiths, B.S. 2016. Using nematode communities to test a European scale soil biological monitoring programme for policy development. *Applied Soil Ecology* 97: 78–85.

Suhameena, B., Devi, S.U., Gowri, R.S., Kumar, S.D. 2020. Utilization of *Azospirillum* as a biofertilizer – An overview. *International Journal of Pharmaceutical Sciences Review and Research* 62(2): 141–145.

Tahat, M.M., Alananbeh, K.M., Othman, Y.A., Leskovar, D.I. 2020. Soil health and sustainable agriculture. *Sustainability* 12(12): 4859.

Trivedi, P., Delgado-Baquerizo, M., Anderson, I.C., Singh, B.K. 2016. Response of soil properties and microbial communities to agriculture: Implications for primary productivity and soil health indicators. *Frontiers in Plant Science* 7: 990.

Urra, J., Alkorta, I., Garbisu, C. 2019. Potential benefits and risks for soil health derived from the use of organic amendments in agriculture. *Agronomy* 9(9): 542.

Van Nguyen, T., Pawlowski, K. 2017. Frankia and Actinorhizal Plants: Symbiotic Nitrogen Fixation. In *Rhizotrophs: Plant Growth Promotion to Bioremediation* (pp. 237–261). Springer, Singapore.

Vermeire, M.L., Cornélis, J.T., Van Ranst, E., Bonneville, S., Doetterl, S., Delvaux, B. 2018. Soil microbial populations shift as processes protecting organic matter change during podzolization. *Frontiers in Environmental Science* 6: 70.

Wang, C.Y., Zhou, X., Guo, D., Zhao, J.H., Yan, L., Feng, G.Z., Gao, Q., Yu, H., Zhao, L.P. 2019. Soil pH is the primary factor driving the distribution and function of microorganisms in farmland soils in northeastern China. *Annals of Microbiology* 69(13): 1461–1473.

Wang, J., Lu, J., Zhang, Y., Wu, J. 2020a. Microbial ecology might serve as new indicator for the influence of green tide on the coastal water quality: Assessment the bioturbation of *Ulva prolifera* outbreak on bacterial community in coastal waters. *Ecological Indicators* 113: 106211.

Wang, H., Yan, S., Ren, T., Yuan, Y., Kuang, G., Wang, B., Yun, F., Feng, H., Ji, X., Yuan, X., Liu, G. 2020b. Novel environmental factors affecting microbial responses and physicochemical properties by sequentially applied biochar in black soil. *Environmental Science and Pollution Research* 27(30): 37432–37443.

Woldemariam, W.G., Ghoshal, N. 2020. Microbial community, biomass and physico-chemical properties of soil in dry tropics. https://doi.org/10.21203/rs.3.rs-36821/v1

Xin, X.L., Yang, W.L., Zhu, Q.G., Zhang, X.F., Zhu, A.N., Zhang, J.B. 2018. Abundance and depth stratification of soil arthropods as influenced by tillage regimes in a sandy loam soil. *Soil Use and Management* 34(2): 286–296.

11 Recent Advances in Potentiality of Microorganisms in Promoting Plant Growth and Managing Degraded Land

Yumnam Bijilaxmi Devi
Rani Lakshmi Bai Central Agricultural University, Jhansi, India

Thounaojam Thomas Meetei
Lovely Professional University, Phagwara, India

CONTENTS

DOI: 10.1201/9781003147077-11

11.1 INTRODUCTION

The earth inhabits around 7.41 billion people who occupy 6.38 billion hectares of area, and among them, agriculture is giving livelihood to around 1.3 billion people in all over the world (Gouda et al. 2018). The dependence of people on agriculture also marks the need of improving production and yielding higher incomes. Terrestrial agriculture provides for almost 99.7% of the human population (Alexandratos and Bruinsma 2003). The increase in the need of food has also increased with the increase in human population. In order to achieve these goals, intensive agriculture has played an important role which improvises maximum use of chemicals like fertilizers, insecticides, herbicides, etc., which consequently result in disturbing the environment. In order to sustain the environment for future generations, minimum use of these chemicals should be encouraged and the farming system should follow integrated systems of cultivation. Here, the importance of microorganisms comes into play in promoting plant growth.

Talking about India, out of the total area, 60.6% of the area is used for agricultural purposes (Gouda et al. 2018). Agriculture is the largest source of livelihood in India. Around 70% of the population of India rely on agriculture, and in rural India, almost 80% of the population is agriculture dependent with most of the farmers being small and marginal. So, improving the production has been an ever-ending target. India has also adopted intensive agriculture directives to meet the demands of the increasing population and its dependence. Climatic conditions also play a vital role in agricultural productivity and in order to make up food demand, the resources are being over exploited, which consequently affected the sustainability of environment, atmosphere and the ecosystem (Lehmann and Kleber 2015). Substantial pollution has made changes in exchangeable bases and pH, thus reducing the productivity of crops by making nutrients unavailable (Gupta et al. 2015). Increasing the land used for agriculture is not feasible as land is a restricted source and land has been degraded by 20–25% worldwide, and each year, 5–10 Mha land may be degraded (Abhilash et al. 2016). Under these circumstances, proper usage of microorganisms as plant growth promoters, biofertilzers and biopesticides will help in sustaining environment without compromising crop's productivity (Kamkar 2016; Pérez et al. 2016; Suhag 2016). Leguminous plants and microbial symbiotic relationship help in atmospheric nitrogen fixation and bio-mineralization, thus having great potential to improve soil fertility and quality (Herrera Paredes and Lebeis 2016; Rosenberg and Zilber-Rosenberg 2016; Agler et al. 2016). Plant growth-promoting rhizobacteria (PGPR) can best explain the association of microorganisms with plants and microorganisms' role in promoting plant growth either in terms of synergistic or antagonistic interaction (Bhardwaj et al. 2014) and these rhizobacteria greatly influence soil properties which will consequently help in improving plant growth promotion. Therefore, exploration of PGPR has already been started in various parts of the world, thus revitalizing the quality of soil (Fasciglione et al. 2015).

Marginal land may be referred agricultural areas which are unsuitable for crop productivity due to low fertility caused by various soil properties and adverse climate (Fernández et al. 2020). Marginal land can be divided into abandoned farmland, degraded land, wasteland and idle land depending on soil nutritional, chemical and physical properties (Usmani 2020). Human activities form degraded lands due to their irrational behavior to natural resources. Most of the degraded types of land has been triggered and intensified by human activities like soil erosion, soil contamination, soil acidification, sheet erosion, silting, salinization, alkalization, urbanization, etc. Land degradation has been the consequences of increasing human population on earth. In order to meet the human needs, natural resources have been exploited to such an extent that land degradation has fasten up its pace and caused decline in plant growth and development. Soil is an important natural resource which is vital and non-replenishable asset for food and nutritional security (Rojas et al. 2016). Land degradation can simply be defined as a long-term destruction of the ecosystem functions due to exploitation of natural resources, which cannot be retrieved to its original ability to support life unaided (Van Vuuren et al. 2012). Land degradation has caused a drastic reduction in biological and economic ability to produce from a land resource. Degraded lands are exaggerated by human activities followed by natural processes and consequently interwoven with climate change and consequently loss of biodiversity (Kust et al. 2017). Intensive cropping, erosion, acidity of soil, water logging, salinity/alkalinity, avalanche and anthropogenic activities like mining industrial effluents, etc., are some of the major aspects which are triggering land degradation, and mismanagement of these factors is aggravating the consequences faced by plants, soil and environment as a whole (Rojas et al. 2016). Urbanization has also been one of the major causes in degradation of land by reducing the area for agricultural purposes and making it unfit for production (Ceccarelli et al. 2013). Soil fertility also favors good soil health that will help in retrieving the degradation, which we can achieve by preventing the indiscriminate use of chemicals and fertilizers (Chasek et al. 2019).

Use of microorganisms in amending these shortcomings from degraded lands can be an important aspect as the native microorganisms has already been adjusted and growing against the harsh environment due to degradation. They have acquainted themselves by producing growth-regulating hormones, enzymes and exudates which will help the plants grow in these areas. If these beneficial properties of microbes can be explored from certain degraded lands, characterize them and reapplied to the same soil after mass multiplication of the screened beneficial microorganisms. The approach of managing degraded lands can be achieved with the use of a direct or indirect approach (Goswami et al. 2016). A direct approach may comprise direct application of compounds helpful in promoting plant growth like biofertilizers. An indirect approach may include strengthening the rhizospheric microorganisms as they possess plant growth-promoting traits; this can be done through amendment processes like the addition of organic matter, liming and drainage to provide optimum conditions for growth of these rhizospheric microorganisms (Tripathi et al. 2012).

11.2 PLANT GROWTH-PROMOTING TRAITS PROVIDED BY MICROORGANISMS

Microorganisms have always been in a symbiotic relationship with plants and directly/indirectly impact on development and growth; specifically, the microorganisms residing in the rhizosphere of plant roots have various useful effects on the growth of host plants by providing various services (Figure 11.1) like nitrogen fixation, phosphate solubilization and mobilization, potash solubilization, etc. (Raza et al. 2016). The bacteria residing in the rhizosphere and promoting growth of plants are generally referred to as PGPR (Gouda et al. 2018). The practice of using these microorganisms in defending the plants from diseases and adverse climatic conditions is eco-friendly and sustainable (Akhtar et al. 2012). The crops that have been deliberated for the relationship with microorganisms include maize, tomato and barley wheat (Gray and Smith 2005). And this relationship has been exploited commercially and scientifically for sustainable agriculture in the new era of agriculture (Gonzalez et al. 2015).

PGPR have been active in providing various mechanisms like nitrogen fixation, phosphate solubilization (Ahemad and Khan 2012), production of siderophores (Sayyed et al. 2019), indole 3 acetic acid (Yousef 2018), 1-amino-cyclopropane-1-carboxylate (ACC) deaminase (Zafar-ul-Hye et al. 2019) and hydrogen cyanide production (Guo et al. 2020). These PGPR have also been reported to act as degrader of pollutants, thus reducing the effect of pollution in soil and the overall environment (Xie et al. 2016).

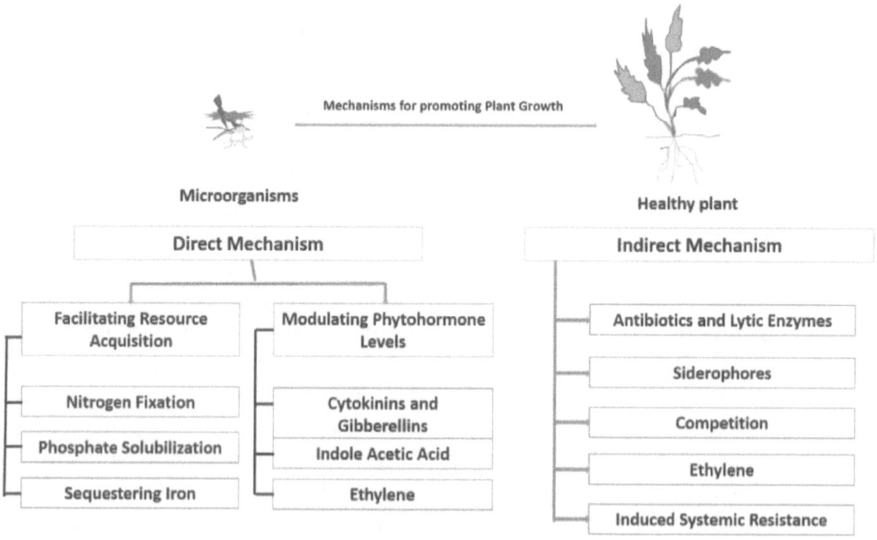

FIGURE 11.1 Mechanisms of promoting plant growth by microbial application in degraded lands (Glick 2012).

11.2.1 Mechanisms of Plant Growth Promotion by Microorganisms

There are some direct and indirect mechanisms followed by microorganisms to promote plant growth. Direct mechanisms may include atmospheric nitrogen fixation, solubilizing insoluble phosphorus, sequestering iron under direct nutrient acquisition and modulating phytohormones (Bhattacharyya and Jha 2012). Indirect mechanisms include benefits to plants by inducing systemic resistance, producing antibiotics against pathogens, etc. (Goswami et al. 2016).

11.2.1.1 Biological Nitrogen Fixation

Nitrogen is a macronutrient which is required by plants in large quantity, but in nature, most of the nitrogen is present in elemental form of N_2. Biological nitrogen fixation is the main process by which atmospheric N_2 is converted to ammonia which can further be converted to nitrate; both these forms of nitrogen can be used by plants as a nitrogen source (Huebert et al. 1999).

11.2.1.2 Phosphate Solubilization

Phosphorus is one of the macro-nutrients which is also an important nutrient in plant growth and development. Large amounts of phosphorus are present in soil but in an insoluble form; therefore, plants are not able to take up these phosphates as a nutrient source (Rehman et al. 2020). Microorganisms have the ability to solubilize this insoluble phosphorus into a soluble form of phosphorus which can be easily taken up by the plants for their growth. The mechanism underneath may be due to acidification, chelation and enzymatic action (Rodríguez and Fraga 1999).

11.2.1.3 Production of Phytohormones

Phytohormones are the hormones produced by plants for their various functions in responding against the adverse climate. Phytohormones include auxins, gibberellins, ethylene, cytokinin and abscissic acid (ABA). Some of the microorganisms have the ability to produce some of these hormones, which will have great influence on the growth of plants (Wong et al. 2015).

11.2.1.4 Antibiotic Production

Antibiotics are compounds produced for competing the harmful microorganisms in the environment. Microorganisms produce these compounds which will help the plants to compete with pathogens in the soil (Tariq et al. 2017).

11.2.1.5 Enzyme Production

PGPR also secrete lytic enzymes which are responsible for eliminating pathogens like *Fusarium oxyporum*, *Sclerotium rolfsii*, *Rhizoctonia solani*, etc. Cellusases, chitinases, proteases and lipases are some of the enzymes produced by the microorganisms. These help in strengthening plants in harsh climate (Kenneth et al. 2019).

11.2.1.6 Siderophore Production

Iron is one of the immobile nutrients in soil. Availability of iron is low in soil but with the help of iron chelators, iron is brought near the plant roots for its uptake.

PGPR have the ability to produce low-mass iron chelators. These are known as siderophores which have high affinity toward iron (Qi and Zhao 2013).

11.3 BENEFICIAL PLANT-MICROORGANISM INTERACTIONS IN THE RHIZOSPHERE REGION

The ecosystem service provided by rhizosphere is of supreme importance as they directly influence the evolution and growth of the vegetation by carrying out carbon and water cycling, trapping of nutrients, uptake and storage of carbon, etc. (Adl 2016). Land degradation along with global climate change has brought stress on soil environment; for surviving these stress conditions, plants need some physiological adaptation mechanisms which are provided by rhizospheric microorganisms in a direct or indirect way (Ahkami et al. 2017). Some of the mechanisms provided by microorganisms include producing plant hormones which improve root formation and growth; thus, the roots will be able to withstand the stress. The auxin which is a phytohormone is the main controller of growth in every aspect of plants; the growth in root hair length, increase in lateral root primordial and its response to gravity have been found in many studies (Ishida et al. 2008; Péret et al. 2009; Overvoorde et al. 2010). Other phytohormones like gibberellic acids, ABA, cytokinins and ethylene all play an important role in fighting abiotic stress in the soil (Werner et al. 2003; Negi et al. 2008; Xiong et al. 2006).

Phytohormones are produced by plants in response to stress; rhizosphere microorganisms are also capable of improving root biomass, increasing productivity and carbon storage under abiotic stress conditions (Coleman-Derr and Tringe 2014). Plant growth-promoting microorganisms remain in close relation with rhizosphere (Figure 11.2). Some of examples of microorganisms, which interact with plants,

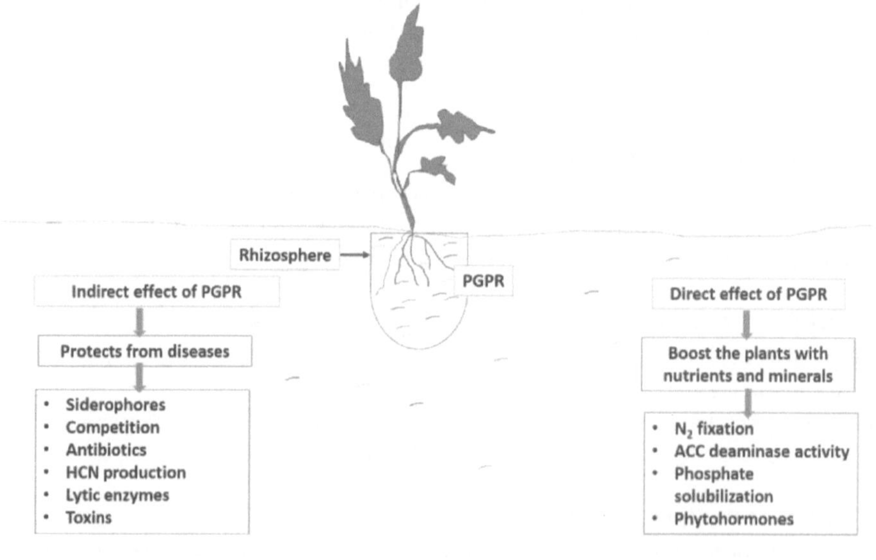

FIGURE 11.2 Direct and indirect growth-promoting microbial activities in the plant rhizospheric zone.

improve the N_2 fixation, phosphate solubilization, etc. These interactions benefit crop productivity and ability to tolerate stress (Marulanda et al. 2009; Tank and Saraf 2010; Marasco et al. 2013). There has been a record of colonizing and promoting growth of switch grass under diverse conditions (Kim et al. 2012). These species of plants are found to be improving bioenergy crop production and supporting sustainable feedstock production system.

A PGPR strain of *Pseudomonas fluorescens*, i.e. FPT9601-T5, has been found to be under functional categories such as basic metabolism, stress response, signal transduction, etc., and is associated with auxin- and ethylene-responsive genes (Wang et al. 2005). Another strain PAL5 of *Gluconacetobacter diazotrophicus* has been found to activate ABA depending on signaling genes which can promote drought resistance in sugar cane (Vargas et al. 2014). *Bacillus subtilis* strain JS improved photosynthesis pathway and consequently enhanced the growth of plants (Kim et al. 2015). *Bacillus amyloliquefaciens* strain KPS46 has the ability to stimulate certain extracellular protein elicitors for boosting plant growth and promoting induced systemic resistance in soybean plants (Buensanteai et al. 2008).

11.4 PLANT GROWTH-PROMOTING MICROORGANISMS IN DEGRADED LANDS

Human activities form degraded lands due to their irrational behavior to natural resources. Most of the degraded type of land has been triggered and intensified by human activities like soil erosion, soil contamination, soil acidification, sheet erosion, silting, salinization, alkalization, urbanization, etc. Soil is an important natural resource that serves as a habitat for microorganisms and plants which is non-replenishable; so its use has not to be jeopardized. Managing degraded land is one of the important factors to sustain the soil resources. Alternatively, some of PGPR have the ability to promote plant growth in these degraded lands like biological nitrogen fixation, phosphate solubilization, phytohormones production, etc. (Table 11.1). The use of these beneficial microorganisms should be brought under focus as they are non-hazardous to environment along with providing plant growth-promoting traits.

11.5 RECENT ADVANCES IN MICROBIAL PLANT GROWTH PROMOTION

Removal/Detoxification of organic and inorganic pollutants from soil has also been termed as a function of microorganisms, which will reduce the stress in the environment near the plant. Bacteria use various intra- and extra-mechanisms to detoxify the adverse effects of heavy metals in their tissues. The mechanisms used by PGPR for tolerating and detoxifying the pollutants like heavy metals may also vary from bacterial species to species and also for different metals. Some microorganisms can detoxify Zn by forming complexation of organic acids (Choudhury and Srivastava 2001). Phyto-stimulation is also one of the important direct mechanisms that provides beneficial effects to plants (Khalid et al. 2006). These phytohormones improve plant growth by increasing cell division, root elongation and development, stress

TABLE 11.1
Various Plant Growth-Promoting Traits Produced by Different Plant Growth-Promoting Rhizobacteria (PGPR)

PGPR Traits	Useful in Types of Degraded Land	Microorganisms Responsible	References
Biological nitrogen fixation	Nutrient-depleted soil Eroded soil	*Rhizobium*, *Azospirillum*, *Azotobacter*, *Azoarcus*, *Bacillus polymyxa*, *Gluconoacetobacter*, *Burkholderia* and *Herbaspirillum*	Chen and Chen (2004)
Indole acetic production (IAA)	Stress environment like low moisture content	*Bacillus* sp. *Serratia* sp.	Lwin et al. (2012)
Antibiotic production like 2,4-diacetyl phloroglucinol, phenazine-1-carboxyclic acid, phenazine-1-carboxamide, pyoluteorin, pyrrolnitrin, oomycin A	Soil with pathogens causing diseases	*Fluorescent pseudomonads* *Pseudomonas fluorescens* *Bacillus* sp.	Fernando et al. (2005)
Mineral phosphate solubilization	Nutrient-depleted soil, eroded soil	*Bacillus* sp. *Pseudomonas striata* *B. polymyxa*	Gupta and Gopal (2008)
Sidephore production	Salt-affected soil	*Brevibacillus brevis* *Enterobacter* sp. *Pseudomonas* sp. *Azospirillum brasilense*	Gupta and Gopal (2008)
Gibberellic acid production	Stress environment	*Bacillus megaterium* *P. fluorescens* *Azotobacter chroococcum*	Lenin and Jayanthi (2012)
Inducing systemic resistance against *Fusarium oxysporum*	Soil with pathogens causing diseases	*P. fluorescens* *Serratia mercescens*	Raupach et al. (1996)

tolerance, etc.; the physiological process of the plants is highly facilitated by these hormones (Nadeem et al. 2013). Some of the PGPR have also shown effectiveness against stress environment due to salinity and drought. It may due to the mechanism followed by the bacteria to produce exopolysaccharides which protects them against the unfavorable conditions (Sandhya et al. 2009).

11.6 FACTORS AFFECTING MICROORGANISM GROWTH IN DEGRADED SOIL AND THEIR MANAGEMENT THROUGH PLANT-MICROORGANISM INTERACTIONS

Microbial growth is affected by numerous kinds of degraded lands but certain plant-microorganism interactions may help in managing these degraded soils in optimizing the productivity and conserving the soil from loss due to erosion, acidity, salinity, etc.

11.6.1 Temperature

Temperature plays a critical part in plants as well as microbial growth. Screening microorganisms with capacity to withstand extreme temperatures will help in managing the degraded soil with the use of these microorganisms. Some findings have confirmed the ability of rhizospheric microorganisms in tolerating temperature up to 60°C which is an important finding in benefit to microorganisms (Rodriguez et al. 2008) and some microorganisms improve the side effects caused by low-temperature stress (Barka et al. 2006; Dimkpa et al. 2009). *Bacillus phytofirmans* strain PsJN inoculated in grapevine was found to be accumulating higher levels of carbohydrates, proline and phenols and also showed tolerance up to 4°C temperature (Barka et al. 2006).

11.6.2 Moisture

Abiotic strains are the chief limiting features in agricultural productivity. Moisture is one of the important abiotic components required for plant growth. Excess or lack of moisture both hampers the plants, microorganisms and plant-microorganism interaction. Findings prove that bacterial communities modulate the moisture availability, altering soil chemistry and plant phenotypes, thus helping in fighting the drought condition (Naylor and Coleman-Derr 2018). It has been found that low growth rates are associated with air-dried soil, but with increasing moisture content, growth rates increase gradually and remain stable in moist soils. Respiration and growth of bacteria are correlated with different soil moisture and rewetting the soil resulted in better plant-microorganism interaction (Iovieno and Bååth 2008).

11.6.3 Plant Types

Plant community dynamics affect microbial growth in soil. Long-rooted plants improve soil structure of the soil, thus improving microbial growth. Short-rooted plants like grasses protect the soil from direct heating of the soil by rainfall, thus reducing the degradation process. Pine-type trees release more acidity in soil and broad leaf trees conserve bases near their root rhizosphere, therefore affecting microbial growth in the soil. This terrestrial plant community affects soil microbial ecology which will consequently result in alterations in ecosystem function (Ho et al. 2017). Microorganism mediated positive and negative feedback on the entire plant ecosystem and helped in improving plant-plant and plant-microorganism interactions (Bever et al. 2010). Symbiotic association of microorganism and plants is affected by plant phenotype and ecology.

11.6.4 Nutrients in Soil

There are 17 essential macro and micronutrients for the growth of plants and optimum yield production (Miransari 2013). These elements are also required by microorganisms for their development and functioning in the soil. The deficiency or toxicity of these elements will hamper the plant-microorganism interaction and their functionality in yield production and managing land degradation. Nutrient availability is the main criteria for plants to grow; microorganisms play a significant part in increasing fertility in soil. Many of the mechanisms followed by microorganisms like production of ACC deaminase, N_2 fixation, phosphate solubilization, siderophore production, etc. The optimum level of nutrients in the soil will control plant and microbial growth; however, microorganisms have better tolerance ability toward nutrient stress as they undergo many of the processes which helps in improving nutrient availability (Richardson et al. 2009; Hirel et al. 2011; Marschner 2012).

11.6.5 Soil Types

Soil types like clayey, sandy, alluvial, etc., determines the soil health as it is the property that will improve soil physical structure. Soil structure will consequently control all other important physical and chemical properties of soil which will affect both plant as well as microorganism growth. Addition of organic matter will surely help improve the soil type of a particular land. Soil with high cation exchange capacity, water-holding capacity and low bulk density will be suitable for proper microbial habitats and this type of soil will be found where more organic matter is added and decomposed to humus. Rest, all types of soil will have some adaptations to the prior habitat; which is why, selecting site-specific microbial inoculants is the need of the hour. Bacteria like *Bacillus* and *Pseudomonas* prevail in alkaline soil whereas fungi like *Aspergillus* prefer acidic conditions. Much variation with soil type cannot be done with even microorganisms; however, screening of microorganisms adapted to particular soil type and can be used to promote growth of the plant in degraded land due to adverse soil type (Girvan et al. 2003).

11.6.6 Heavy Metals

Heavy metals like zinc, arsenic, cadmium, etc., challenge the plants and microbial growth by hampering their normal growth mechanisms. The increased concentration of these metals causes toxicity to plants and microorganisms; however, certain microorganisms have the capacity to tolerate the outcome of these heavy metals. The use of microorganisms improves the phytoremediation, most specifically phytoremediation assisted by microorganism (Glick 2003; Jamil et al. 2014). PGPR like *Rhizobium*, *Mesorhizobium*, *Azotobacter*, *Pseudomonas*, *Bacillus* spp., etc., have shown the ability to protect its host plant from any detrimental effect of heavy metal toxicity (Shinwari et al. 2015). The symbiotic bacteria rhizobium has been found to be sensitive to heavy metals and has the ability to adapt in elevated metal concentrations (Giller et al. 2009). Now, the genetically engineered microorganisms have the capacity to degrade pollutants and bioremediation may be explored for managing these degraded soils (Boricha and Fulekar 2009).

11.7 CONCLUSION AND FUTURE PROSPECTS

Excess exploitation of natural resources has no doubt created a major impact on our land and caused degradation of the same. But the increasing human population and need to sustain the environment and management of degraded lands are of utmost importance. Exploring microbial communities especially plant growth-promoting microorganisms from rhizospheric soils have been found to be meeting the exquisite need of the hour by providing nutrients to plants and maintaining the environment and protecting it from further degradation. PGPR enhances plant growth and remediates and also manages the degraded lands and water sources reduce loss of nutrients from the soil. PGPR have been found to be improving crop tolerance toward harsh conditions and maintaining a balance nutrient cycling thus improving soil fertility. Using microbial communities will help in exploring multidisciplinary research which will include biotechnology, nanotechnology, agro-biotechnology, etc. The agriculture as well as wastelands requires a boost to manage themselves in the recent decades and this can be achieved using microorganisms. The use of live components in managing is good in most cases but the challenge they face in the field are the climatic conditions and competition offered by other microorganisms. These microbial communities that have been screened and selected for the purpose of managing degraded lands should also be able to adjust in harsh climatic conditions and possess some antibiotic properties to fight against pathogens. A fundamental knowledge of physiological traits of rhizospheric microorganisms is a must in order to select a suitable species of microorganism. The development of microbial inoculum for application on degraded lands should focus on the microorganisms of such climates as these will possess the ability to grow in these conditions more efficiently than introducing any new microbial inoculum to a new place. Efforts at the molecular level should also be done to provide enough genome database information regarding their proteomic characterization. At present, selection of site-specific inoculum is the need of the hour depending on the suitability of microbial inoculants in diverse conditions.

REFERENCES

Abhilash, P.C., Tripathi, V., Edrisi, S.A., Dubey, R.K., Bakshi, M., Dubey, P.K., Singh, H.B., Ebbs, S.D. 2016. Sustainability of crop production from polluted lands. *Energy, Ecology and Environment* 1(1): 54–65.

Adl, S. 2016. Rhizosphere, food security, and climate change: a critical role for plant-soil research. *Rhizosphere* 1(1): 1–3.

Agler, M.T., Ruhe, J., Kroll, S., Morhenn, C., Kim, S.T., Weigel, D., Kemen, E.M. 2016. Microbial hub taxa link host and abiotic factors to plant microbiome variation. *PLoS Biology* 14(1): e1002352.

Ahemad, M., Khan, M.S. 2012. Evaluation of plant growth-promoting activities of rhizobacterium *Pseudomonas putida* under herbicide stress. *Annals of Microbiology* 62(4): 1531–1540.

Ahkami, A.H., White III, R.A., Handakumbura, P.P., Jansson, C. 2017. Rhizosphere engineering: enhancing sustainable plant ecosystem productivity. *Rhizosphere* 3: 233–243.

Akhtar, N., Qureshi, M.A., Iqbal, A., Ahmad, M.J., Khan, K.H. 2012. Influence of *Azotobacter* and IAA on symbiotic performance of *Rhizobium* and yield parameters of lentil. *Journal of Agricultural Research* 50: 361–372.

Alexandratos, N., Bruinsma, J. 2003. In: Bruinsma, J. (Ed.), World Agriculture: Towards 2015/2030 an FAO Perspective. Earthscan Publications, London, pp. 1–28.

Barka, E.A., Nowak, J., Clément, C. 2006. Enhancement of chilling resistance of inoculated grapevine plantlets with a plant growth-promoting rhizobacterium, *Burkholderia phytofirmans* strain PsJN. *Applied and Environmental Microbiology* 72(11): 7246–7252.

Bever, J.D., Dickie, I.A., Facelli, E., Facelli, J.M., Klironomos, J., Moora, M., Rillig, M.C., Stock, W.D., Tibbett, M., Zobel, M. 2010. Rooting theories of plant community ecology in microbial interactions. *Trends in Ecology & Evolution* 25(8): 468–478.

Bhardwaj, D., Ansari, M.W., Sahoo, R.K., Tuteja, N. 2014. Biofertilizers function as key player in sustainable agriculture by improving soil fertility, plant tolerance and crop productivity. *Microbial Cell Factories* 13(1): 1–10.

Bhattacharyya, P.N., Jha, D.K. 2012. Plant growth-promoting rhizobacteria (PGPR): emergence in agriculture. *World Journal of Microbiology and Biotechnology* 28(4): 1327–1350.

Boricha, H., Fulekar, M.H. 2009. *Pseudomonas plecoglossicida* as a novel organism for the bioremediation of cypermethrin. *Biology and Medicine* 1(4): 1–10.

Buensanteai, N., Athinuwat, D., Chatnaparat, T., Yuen, G.Y., Prathuangwong, S. 2008. Extracellular proteome of *Bacillus amyloliquefaciens* KPS46 and its effect on enhanced growth promotion and induced resistance against bacterial pustule on soybean plant. *Agriculture and Natural Resources* 42(5): 13–26.

Ceccarelli, T., Bajocco, S., Luigi, P.L., Luca, S.L. 2013. Urbanisation and land take of high-quality agricultural soils-exploring long-term land use changes and land capability in Northern Italy. *International Journal of Environmental Research* 8(1): 181–192.

Chasek, P., Akhtar-Schuster, M., Orr, B.J., Luise, A., Ratsimba, H.R., Safriel, U. 2019. Land degradation neutrality: the science-policy interface from the UNCCD to national implementation. *Environmental Science & Policy* 92: 182–190.

Chen, W.X., Chen, W.F. 2004. Exertion of biological nitrogen fixation in order to reducing the consumption of chemical nitrogenous fertilizer. *Review of China Agricultural Science and Technology* 6(6): 3–6.

Choudhury, R., Srivastava, S. 2001. Zinc resistance mechanisms in bacteria. *Current Science* 10: 768–775.

Coleman-Derr, D., Tringe, S.G. 2014. Building the crops of tomorrow: advantages of symbiont-based approaches to improving abiotic stress tolerance. *Frontiers in Microbiology* 5: 283.

Dimkpa, C., Weinand, T., Asch, F. 2009. Plant–rhizobacteria interactions alleviate abiotic stress conditions. *Plant, Cell & Environment* 32(12): 1682–1694.

Fasciglione, G., Casanovas, E.M., Quillehauquy, V., Yommi, A.K., Goni, M.G., Roura, S.I., Barassi, C.A. 2015. *Azospirillum* inoculation effects on growth, product quality and storage life of lettuce plants grown under salt stress. *Scientia Horticulturae* 195: 154–162.

Fernández, M.J., Barro, R., Pérez, J., Ciria, P. 2020. Production and composition of biomass from short rotation coppice in marginal land: a 9-year study. *Biomass and Bioenergy* 134: 105478.

Fernando, W.D., Nakkeeran, S., Zhang, Y. 2005. Biosynthesis of antibiotics by PGPR and its relation in biocontrol of plant diseases. PGPR: Biocontrol and Biofertilization. Springer, Dordrecht, pp. 67–109.

Giller, K.E., Witter, E., McGrath, S.P. 2009. Heavy metals and soil microbes. *Soil Biology and Biochemistry* 41(10): 2031–2037.

Girvan, M.S., Bullimore, J., Pretty, J.N., Osborn, A.M., Ball, A.S. 2003. Soil type is the primary determinant of the composition of the total and active bacterial communities in arable soils. *Applied and Environmental Microbiology* 69(3): 1800–1809.

Glick, B.R. 2003. Phytoremediation: synergistic use of plants and bacteria to clean up the environment. *Biotechnology Advances* 21(5): 383–393.

Glick, B.R. 2012. Plant growth-promoting bacteria: mechanisms and applications. *Scientifica*. https://doi.org/10.6064/2012/963401.

Gonzalez, A.J., Larraburu, E.E., Llorente, B.E. 2015. *Azospirillum brasilense* increased salt tolerance of jojoba during in vitro rooting. *Industrial Crops and Products* 76: 41–48.

Goswami, D., Thakker, J.N., Dhandhukia, P.C. 2016. Portraying mechanics of plant growth promoting rhizobacteria (PGPR): a review. *Cogent Food & Agriculture* 2(1): 1127500.

Gouda, S., Kerry, R.G., Das, G., Paramithiotis, S., Shin, H.S., Patra, J.K. 2018. Revitalization of plant growth promoting rhizobacteria for sustainable development in agriculture. *Microbiological Research* 206: 131–140.

Gray, E.J., Smith, D.L. 2005. Intracellular and extracellular PGPR: commonalities and distinctions in the plant–bacterium signaling processes. *Soil Biology and Biochemistry* 37(3): 395–412.

Guo, D.J., Singh, R.K., Singh, P., Li, D.P. et al. 2020. Complete genome sequence of *Enterobacter roggenkampii* ED5, a nitrogen fixing plant growth promoting endophytic bacterium with biocontrol and stress tolerance properties, isolated from sugarcane root. *Frontiers in Microbiology* 11: 2270.

Gupta, A., Gopal, M. 2008. Siderophore production by plant growth promoting rhizobacteria. *Indian Journal of Agricultural Research* 42(2): 153–156.

Gupta, G., Parihar, S.S., Ahirwar, N.K., Snehi, S.K., Singh, V. 2015. Plant growth promoting rhizobacteria (PGPR): current and future prospects for development of sustainable agriculture. *Journal of Microbial and Biochemical Technology* 7(2), 96–102.

Herrera Paredes, S., Lebeis, S.L. 2016. Giving back to the community: microbial mechanisms of plant–soil interactions. *Functional Ecology 30*(7), 1043–1052.

Hirel, B., Tétu, T., Lea, P.J., Dubois, F. 2011. Improving nitrogen use efficiency in crops for sustainable agriculture. *Sustainability* 3(9): 1452–1485.

Ho, Y.N., Mathew, D.C., Huang, C.C. 2017. Plant-microbe ecology: interactions of plants and symbiotic microbial communities. In: Yousaf, Z. (Ed.), Plant Ecology – Traditional Approaches to Recent Trends, IntechOpen, pp. 93–119. doi: 10.5772/intechopen.69088.

Huebert, B., Vitousek, P., Sutton, J., Elias, T., Heath, J., Coeppicus, S., Howell, S., Blomquist, B. 1999. Volcano fixes nitrogen into plant-available forms. *Biogeochemistry* 47(1): 111–118.

Iovieno, P., Bååth, E. 2008. Effect of drying and rewetting on bacterial growth rates in soil. *FEMS Microbiology Ecology* 65(3): 400–407.

Ishida, T., Kurata, T., Okada, K., Wada, T. 2008. A genetic regulatory network in the development of trichomes and root hairs. *Annual Review of Plant Biology* 59: 365–386.

Jamil, M., Zeb, S., Anees, M., Roohi, A., Ahmed, I., ur Rehman, S., Rha, E.S. 2014. Role of *Bacillus licheniformis* in phytoremediation of nickel contaminated soil cultivated with rice. *International Journal of Phytoremediation* 16(6): 554–571.

Kamkar, B. 2016. Sustainable development principles for agricultural activities. *Advances in Plants & Agriculture Research* 3(5): 1–2.

Kenneth, O.C., Nwadibe, E.C., Kalu, A.U., Unah, U.V. 2019. Plant growth promoting rhizobacteria (PGPR): a novel agent for sustainable food production. *American Journal of Agricultural and Biological Sciences* 14: 35–54.

Khalid, A., Arshad, M., Zahir, Z.A. 2006. Phytohormones: microbial production an application. In: Biological Approaches to Sustainable Soil System, CRC Press, Boca Raton, FL, pp. 207–220.

Kim, S., Lowman, S., Hou, G., Nowak, J., Flinn, B., Mei, C. 2012. Growth promotion and colonization of switchgrass (*Panicum virgatum*) cv. Alamo by bacterial endophyte *Burkholderia phytofirmans* strain PsJN. *Biotechnology for Biofuels* 5(1): 1–10.

Kim, J.S., Lee, J., Seo, S.G., Lee, C., Woo, S.Y., Kim, S.H. 2015. Gene expression profile affected by volatiles of new plant growth promoting rhizobacteria, *Bacillus subtilis* strain JS, in tobacco. *Genes & Genomics* 37(4): 387–397.

Kust, G., Andreeva, O., Cowie, A. 2017. Land degradation neutrality: concept development, practical applications and assessment. *Journal of Environmental Management* 195: 16–24.

Lehmann, J., Kleber, M. 2015. The contentious nature of soil organic matter. *Nature* 528(7580): 60–68.

Lenin, G., Jayanthi, M. 2012. Indole acetic acid, gibberellic acid and siderophore production by PGPR isolates from rhizospheric soils of *Catharanthus roseus*. *International Journal of Pharmaceutical and Biological Science Archive* 3: 933–938.

Lwin, K.M., Myint, M.M., Tar, T., Aung, W.Z.M. 2012. Isolation of plant hormone indole-3-acetic acid (IAA) producing rhizobacteria and study on their effects on maize seedling. *Engineering Journal* 16(5): 137–144.

Marasco, R., Rolli, E., Vigani, G., Borin, S., Sorlini, C., Ouzari, H., Zocchi, G., Daffonchio, D. 2013. Are drought-resistance promoting bacteria cross-compatible with different plant models?. *Plant Signaling & Behavior* 8(10): e26741.

Marschner, H. 2012. Marschner's Mineral Nutrition of Higher Plants. Academic Press, Cambridge, MA.

Marulanda, A., Barea, J.M., Azcón, R. 2009. Stimulation of plant growth and drought tolerance by native microorganisms (AM fungi and bacteria) from dry environments: mechanisms related to bacterial effectiveness. *Journal of Plant Growth Regulation* 28(2): 115–124.

Miransari, M. 2013. Soil microbes and the availability of soil nutrients. *Acta Physiologiae Plantarum* 35(11): 3075–3084.

Nadeem, S.M., Naveed, M., Zahir, Z.A., Asghar, H.N. 2013. Plant–microbe interactions for sustainable agriculture: fundamentals and recent advances. Plant Microbe Symbiosis: Fundamentals and Advances, Springer, New Delhi, pp. 51–103.

Naylor, D., Coleman-Derr, D. 2018. Drought stress and root-associated bacterial communities. *Frontiers in Plant Science* 8: 2223.

Negi, S., Ivanchenko, M.G., Muday, G.K. 2008. Ethylene regulates lateral root formation and auxin transport in *Arabidopsis thaliana. Plant Journal* 55(2): 175–187.

Overvoorde, P., Fukaki, H., Beeckman, T. 2010. Auxin control of root development. *Cold Spring Harbor Perspectives in Biology* 2(6): a001537.

Péret, B., De Rybel, B., Casimiro, I., Benková, E., Swarup, R., Laplaze, L., Beeckman, T., Bennett, M.J. 2009. Arabidopsis lateral root development: an emerging story. *Trends in Plant Science* 14(7): 399–408.

Pérez, Y.M., Charest, C., Dalpé, Y., Séguin, S., Wang, X., Khanizadeh, S. 2016. Effect of inoculation with arbuscular mycorrhizal fungi on selected spring wheat lines. *Sustainable Agriculture Research* 5 526: 2017–2645.

Qi, W., Zhao, L. 2013. Study of the siderophore-producing *Trichoderma asperellum* Q1 on cucumber growth promotion under salt stress. *Journal of Basic Microbiology* 53(4): 355–364.

Raupach, G.S., Liu, L., Murphy, J.F., Tuzun, S., Kloepper, J.W. 1996. Induced systemic resistance in cucumber and tomato against cucumber mosaic cucumovirus using plant growth-promoting rhizobacteria (PGPR). *Plant Disease* 80(8): 891–894.

Raza, W., Yousaf, S., Rajer, F.U. 2016. Plant growth promoting activity of volatile organic compounds produced by biocontrol strains. *Science Letters* 4(1): 40–43.

Rehman, F.U., Kalsoom, M., Adnan, M., Toor, M., Zulfiqar, A. 2020. Plant growth promoting rhizobacteria and their mechanisms involved in agricultural crop production: a review. *SunText Review of Biotechnology* 1(2): 1–6.

Richardson, A.E., Barea, J.M., McNeill, A.M., Prigent-Combaret, C. 2009. Acquisition of phosphorus and nitrogen in the rhizosphere and plant growth promotion by microorganisms. *Plant and Soil* 321(1): 305–339.

Rodríguez, H., Fraga, R. 1999. Phosphate solubilizing bacteria and their role in plant growth promotion. *Biotechnology Advances* 17(4–5): 319–339.

Rodriguez, R.J., Henson, J., Van Volkenburgh, E., Hoy, M., Wright, L., Beckwith, F., Kim, Y.O., Redman, R.S. 2008. Stress tolerance in plants via habitat-adapted symbiosis. *The ISME Journal* 2(4): 404–416.

Rojas, R.V., Achouri, M., Maroulis, J., Caon, L. 2016. Healthy soils: a prerequisite for sustainable food security. *Environmental Earth Sciences* 75(3): 180.

Rosenberg, E., Zilber-Rosenberg, I. 2016. Microbes drive evolution of animals and plants: the hologenome concept. *MBio* 7(2): e01395-15.

Sandhya, V., Ali, SK.Z., Grover, M., Reddy, G., Venkateswarlu, B. 2009. Alleviation of drought stress effects in sunflower seedlings by the exopolysaccharides producing *Pseudomonas putida* strain GAP-P45. *Biology and Fertility of Soils* 46(1): 17–26.

Sayyed, R.Z., Seifi, S., Patel, P.R., Shaikh, S.S., Jadhav, H.P., El Enshasy, H. 2019. Siderophore production in groundnut rhizosphere isolate, *Achromobacter* sp. RZS2 influenced by physicochemical factors and metal ions. *Environmental Sustainability* 2(2): 117–124.

Shinwari, K.I., Shah, A.U., Afridi, M.I., Zeeshan, M., Hussain, H., Hussain, J., Ahmad, O., Jamil, M. 2015. Application of plant growth promoting rhizobacteria in bioremediation of heavy metal polluted soil. *Asian Journal of Multidisciplinary Studies* 3(4): 179–185.

Suhag, M. 2016. Potential of biofertilizers to replace chemical fertilizers. *International Journal of Advanced Research in Science, Engineering and Technology* 3(5): 163–167.

Tank, N., Saraf, M. 2010. Salinity-resistant plant growth promoting rhizobacteria ameliorates sodium chloride stress on tomato plants. *Journal of Plant Interactions* 5(1): 51–58.

Tariq, M., Noman, M., Ahmed, T., Hameed, A., Manzoor, N., Zafar, M. 2017. Antagonistic features displayed by plant growth promoting rhizobacteria (PGPR): A review. *Journal of Plant Science and Phytopathology* 1(1): 38–43.

Tripathi, D.K., Singh, V.P., Kumar, D., Chauhan, D.K. 2012. Impact of exogenous silicon addition on chromium uptake, growth, mineral elements, oxidative stress, antioxidant capacity, and leaf and root structures in rice seedlings exposed to hexavalent chromium. *Acta Physiologiae Plantarum* 34(1): 279–289.

Usmani, R.A. 2020. Potential for energy and biofuel from biomass in India. *Renewable Energy* 155: 921–930.

Van Vuuren, D.P., Kok, M.T., Girod, B., Lucas, P.L., de Vries, B. 2012. Scenarios in global environmental assessments: key characteristics and lessons for future use. *Global Environmental Change* 22(4), 884–895.

Vargas, L., Santa Brigida, A.B., Mota Filho, J.P. et al. 2014. Drought tolerance conferred to sugarcane by association with *Gluconacetobacter diazotrophicus*: a transcriptomic view of hormone pathways. *PLoS ONE* 9(12): e114744.

Wang, Y., Ohara, Y., Nakayashiki, H., Tosa, Y., Mayama, S. 2005. Microarray analysis of the gene expression profile induced by the endophytic plant growth-promoting rhizobacteria, *Pseudomonas fluorescens* FPT9601-T5 in Arabidopsis. *Molecular Plant-Microbe Interactions* 18(5): 385–396.

Werner, T., Motyka, V., Laucou, V., Smets, R., Van Onckelen, H., Schmülling, T. 2003. Cytokinin-deficient transgenic arabidopsis plants show multiple developmental alterations indicating opposite functions of cytokinins in the regulation of shoot and root meristem activity. *Plant Cell* 15(11): 2532–2550.

Wong, W.S., Tan, S.N., Ge, L., Chen, X., Yong, J.W.H. 2015. The importance of phytohormones and microbes in biofertilizers. In: Bacterial Metabolites in Sustainable Agroecosystem, Springer, Cham, pp. 105–158.

Xie, J., Shi, H., Du, Z., Wang, T., Liu, X., Chen, S. 2016. Comparative genomic and functional analysis reveal conservation of plant growth promoting traits in *Paenibacillus polymyx*a and its closely related species. *Scientific Reports* 6(1): 1–12.

Xiong, L., Wang, R.G., Mao, G., Koczan, J.M. 2006. Identification of drought tolerance determinants by genetic analysis of root response to drought stress and abscisic acid. *Plant Physiology* 142(3): 1065–1074.

Yousef, N.M. 2018. Capability of plant growth-promoting rhizobacteria (PGPR) for producing indole acetic acid (IAA) under extreme conditions. *European Journal of Biological Research* 8(4): 174–182.

Zafar-ul-Hye, M., Danish, S., Abbas, M., Ahmad, M., Munir, T.M. 2019. ACC deaminase producing PGPR *Bacillus amyloliquefaciens* and *Agrobacterium fabrum* along with biochar improve wheat productivity under drought stress. *Agronomy* 9(7): 343.

12 Productivity Losses through Bio-Deterioration in Water-Deficit Harvested Sugarcane

Varucha Misra, A.K. Mall, and Santeshwari
ICAR–Indian Institute of Sugarcane Research
Lucknow, India

CONTENTS

12.1 INTRODUCTION

Sugarcane (*Saccharum* sp.) is a plant mainly used for sugar extraction and ethanol, besides producer of other products like paper and fodder for dairy animals, because of its high forage production, good nutritional value of protein and mineral contents as well as low production cost of dry matter. Sugar prices are achieving high peaks over the last 30 years, due to increased pressure on water, energy, and many other factors. In the field of agricultural abiotic stresses, water deficit plays a vital role in cane productivity. Water scarcity and deficiency are becoming crucial issues for production, and high yield for any crop. This is further accentuating with

DOI: 10.1201/9781003147077-12

the changes in climatic conditions. Shrivastava et al. (2016) stated that half of the world faced the problem of water deficit in crop production as water tables are falling down throughout the continents of the world. It has been anticipated that by the year 2025 more than half the world's population will be living in regions suffering from a shortage of water (Rockstrom 1999), and crops growing in these areas will also be facing severe losses in crop yield and production. Sugarcane covers an area of 41.7 lakh ha in India as it is a cash crop of the country which makes the country stand second in largest production in the world. Kumar et al. (2019) reported that its long growth cycle makes it more prone to major changes in the environmental condition and constrains its growth and productivity under such conditions. Environmental constraints give rise to many abiotic stresses or a combination of multiple abiotic stresses, which causes a strong and negative impact on the sugarcane growth cycle. Direct or indirect relationship arose up by abiotic stresses with the crop. Among several abiotic stresses, water deficit plays a major role in sugarcane perspective as it is known as a water-loving crop (Misra et al. 2020a). Even though sugarcane is an amazing crop and has a tendency to tolerate water-deficit conditions through its compensatory ability, still its growth and development can be affected negatively by this situation (Shrivastava et al. 2017) as a response towards physiological and biochemical processes (Garcia et al. 2020). Being a semi-perennial crop, sugarcane faces seasonal water-deficit conditions wherein photosynthesis, carbohydrate accumulation, change in antioxidant metabolism, and impairment of plant growth and sucrose yield are known to be highly affected (Azcon et al. 1991; Anjos et al. 2010). Related to these conditions, post-harvest sucrose loss, i.e., sucrose deterioration, is a vital problem after harvesting of canes. Several other factors like time lag, milling, varietal changes, etc., are also responsible for these losses to occur at a more pronounced rate (Misra et al. 2016b; Misra et al., 2022c). During the process of handling after harvesting of canes, 20–30% of sucrose deterioration is often seen. Canes exposed to water-deficit conditions are known to have low sucrose content which deteriorates with increasing time after harvest (Misra et al. 2016b). The loss in sucrose content in harvested canes under water-deficit conditions causes a huge decrease in the recovery of sugar (Misra et al. 2020a; Misra et al. 2020b).

Reduction in post-harvest deterioration of canes is an important aspect which can be attained by curtailing the cut to crush period, particularly in the case of canes in water-deficit conditions. The time interval between crushing and clarification is approximately 15–20 min, but the level of microbial contamination of the juice is usually extremely high, typical viable counts being 10^8–10^9 cells/mL juice. During the process of deterioration, metabolic conversion of stored sucrose takes place, resulting in less economic product through the involvement of enzymes and microorganisms. Sucrose deterioration occurs not only in the extracted juice but even in cane stalks that are harvested and left open in the fields. Primary and secondary losses are the two determinants of post-harvest quality losses. Direct association of microorganism is known to be seen in secondary losses of harvested sugarcane that involves dextran, alcohol and acid formation. No use of control measures and no biocide application cause the population of microorganisms to increase rapidly. During the process of liming and sulphitation, almost all microorganisms are removed but the disturbance in temperature causes certain bacteria like thermophilic ones to proliferate. The sucrose deterioration is majorly due to the inversion process followed by deterioration by bacterial activities, enzymatic action and other biological processes. The total loss due to all these mentioned reasons may begin from 0.5% to 4–5% of the total sugars reaching the mills. Of the total sucrose loss, the highest loss contributed by 62% by microbial activity, 25% by free enzymatic activity and 13% by chemical inversion.

This chapter highlights the microorganisms and their interaction with cane staling, factors responsible for microbial activity in harvested canes and the role of microorganisms during sugar processing of water-deficit canes.

12.2 MICROORGANISMS AND THEIR INTERACTION WITH SUGARCANE STALING

Microorganisms play a crucial role in sugarcane staling, leading to low sugar recovery (Figure 12.1). In burnt canes, 15 microorganisms have been reported whereas in green canes about 50 microorganisms have been known to cause a reduction in sugar recovery (Solomon et al. 2006). Changes in enzymatic activity, chemical and respiratory losses are seen at higher rates along with growth and proliferation of microorganisms with increasing staling conditions. In harvested canes, a number of microorganisms is known to invade the inside of the stalk through the cut/damaged parts of sugarcane (Figure 12.1). These are *Leuconostoc*, yeast, *Aerobacter*, *Streptomyces*, *Xanthomonas*; *Saccharomyces, Pichia* and *Torula* belong to the yeast category that was prominently seen in stale canes. Kulkarni and Kulkarni (1987) found that *Actinomyces*, *Penicillium* and *Streptomyces* are certain microorganisms which are found on harvested canes and their population increases with increasing time after harvest. This adds up to sucrose losses in harvested sugarcanes. El-Tabey Shehata (1960) and Tilbury (1970) illustrated that per gram of sugarcane, mould and yeast loads were 10^4 and 10^3, respectively. Stale sugarcanes have the highest number of *Leuconostoc* bacteria as these bacteria are mainly responsible for occurrence of bio-deterioration in harvested and stale canes. There are two major species of this

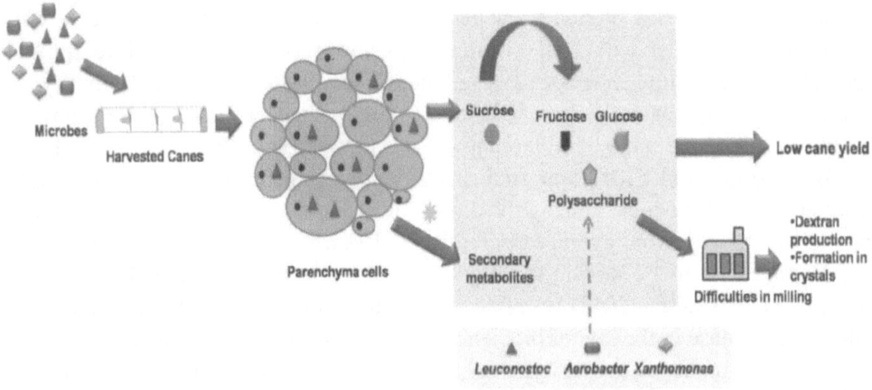

FIGURE 12.1 Microbial action on stale sugarcanes under water-deficit conditions.

bacterium that harm the sucrose content in sugarcanes. These are *Leuconostoc mesenteroides* and *Leuconostoc dextranicum*. Endophytic microbial flora like *Vibrio*, *Bacillus*, etc., has also been seen on sugarcane stalks and these were enhanced by several folds due to cane staling, which is accountable for the poor quality of cane juice.

The cut/damaged sugarcanes from where the microorganisms invade in sugarcane stalk reach to parenchymatous tissues for proliferation and deteriorate sucrose content either to reducing sugars (fructose or glucose) or by inverting sugars or polysaccharides. Secondary metabolites are also produced by microbial invasion in harvested sugarcanes and their production increases with increase in staling time of canes. The polysaccharide production in harvested canes is mainly due to *L. dextranicum* in which the enzyme dextrasucrase plays a crucial role, though *Xanthomonas* and *Aerobacter* are other microorganisms that produce dextran in harvested canes. Rise in dextran content causes difficulties in sugar processing like shape of crystal alteration, etc.

12.3 FACTORS RESPONSIBLE FOR MICROBIAL ACTIVITY IN HARVESTED SUGARCANE

The invasion and proliferation of microorganisms in the harvested sugarcanes are governed by several factors. Depending on the presence or absence of these factors, the rate of augmentation of sucrose losses in harvested sugarcane is also dependent (Duncan and Colmer 1964). Higher the invasion of microorganisms inside the harvested sugarcanes, more is the sucrose losses. Following are the important factors that influence the microbial activity in harvested sugarcane.

12.3.1 Physical Damage to Sugarcane and Its Duration

The physical damage mostly occurs at the time of harvesting. Mechanical harvesting though had many benefits but often use of mechanical cutting system influences many factors that are not in support of the crop. One such is the interaction of the machine and plant during the process of harvesting. Mello and Harris (2003) showed that mechanical cutting of sugarcane causes damage and shock to root-stocks. This is due to the contact with base-cutting discs and noticeable losses like splinter application onto culms. The speed of mechanical harvesters causes different types of damage in sugarcane. Fragmented, breaks and peripheral damage are the three types of physical damage of sugarcane (Kroes and Harris 1994; Mello and Harris 2003). Noronha et al. (2011) reported that there is a significant variation in peripheral damage on rootstocks during the day (42.6%) and evening (35.8%) by mechanical harvesting when the average speed was 6.2 km/h. Fragmentation damage has been reported at 36.4% during the day while 31.2% in the evening (Martins et al. 2019). Manhães et al. (2013) revealed that damage due to mechanical harvesting causes an increase in the rate of incidence of diseases and fungi in sugarcanes as the injuries aid in pathogen invasion. Another reason for physical damage on sugarcane is freeze cracks on the stalks. Freeze cracks are seen when temperatures are low. The severity of physical damage on sugarcane stalks depends on temperature and duration of freeze (Eggleston et al. 2004). Moderate freezing conditions damage the lateral buds and internodes at the upper end of the sugarcane stalk, which becomes the entry site for microorganisms particularly *Leuconostoc*, while severe freezing causes death of all the aboveground tissues resulting in cracked openings in the stalks. This serves as the open access entrance gate for the microorganism invasion (Irvine 1980). Microbial attack is more favourable when freeze is accompanied with warm and wet weather. The invasion of microorganisms is easier in dead and moribund tissues of sugarcane stalk that are exposed to freezing temperatures. The entry of microorganisms is favoured at different temperatures depending on the stalk damage. Dead lateral buds facilitate microbial invasion at –4.4°C and freeze cracks at –5.6°C (Eggleston et al. 2004). Furthermore, it is known that as the period after damage increases prior to crushing of harvested canes, more will be the microbial invasion and proliferation. This leads to heavier deterioration in sucrose content resulting in reduced sugar recovery (Misra et al. 2017a).

12.3.2 Time Delay After Harvest and Before Milling

This is the most important and crucial factor on which microbial activity in harvested sugarcane is dependent. In the current system of cane transportation, a delay of 3–5 days is a common or normal process and at times this may increase up to 7 days (Solomon 2009). Cane transportation from fields to mills is also known to be responsible for the delay in time after harvest. The transportation of harvested canes to mills varies from bullock and camel carts to highly mechanized machines on roads and rail (Solomon 2000). Deterioration in sucrose content is directly proportional to time delay after harvest up to milling of canes (Misra et al. 2016a). Freshly harvested canes transported to mills are at relatively lesser risk of deterioration than ones which are delivered late to the mills. Storage conditions, damage percentage and even shape of containers are other factors involved in deteriorating the sucrose content and in microbial activity after harvest in canes (Solomon 2009).

This delay in time after harvest and before milling gives sufficient time to the microorganisms for entering through cut ends and proliferating in the sucrose-rich parenchymatous tissues (Solomon 2000). After harvest, the juice that oozes out from the cut ends gives a luxurious feast environment for growth and proliferation of microorganisms (Bevan and Bond 1971). Chains, grab loaders, push pile rakes, etc., which hold the harvested sugarcane in mills can even cause injury to harvested canes and these injuries also serve as entry site for microorganisms which further aggravate with mud and high temperatures, particularly for *Leuconostoc* populations.

12.3.3 Storage Condition of Harvested Canes

Storage conditions of harvested canes are another factor on which microbial growth and population are dependent (Misra et al. 2017b, 2018, 2019). It is not only considered how canes are piled and stored in farmers' fields but also in mill yards (Muller and Von 1900; Lee 1923; Misra et al. 2017a). It has been reported that in Louisiana, when harvested canes were cut late and left overnight in transportation vehicles a favourable environment was created for higher proliferation of *Leuconostoc* and also caused more production of dextran content (Hommes 1926; Young 1955; Misra et al. 2017a). The moisture content of sugarcane stalks and their storage method (such as open storage or in proper aerated piles) have also been revealed as a factor for cane deterioration (Wold 1946; Misra et al. 2017a). Clean, dry and healthy canes have relatively lesser deterioration than canes with injury, insect pest infestation, dirty, mud clinging and wet (Solomon 2009).

12.3.4 Atmospheric Conditions

Microbial activity plays a crucial role with the changing atmospheric conditions. Microbial growth and proliferation have been reported to be relatively higher in stale canes. Heavy rainfall and wet and warm weather are favourable conditions for microbial growth which causes losses in harvested sugarcanes. The muddy environment created by heavy rainfall causes *Leuconostoc* and other microorganisms to grow quickly in harvested sugarcanes. Muddy canes block the aeration in harvested sugarcanes which is the reason behind favourable conditions for these bacterial growths. These bacteria are soilborne and so are closer to the sugarcane tissues. Mechanical harvesting of wet canes also causes higher sucrose deterioration (Wold 1946). Juice obtained from these harvested canes had relatively higher bacteria, yeasts and moulds. In a gram of sugarcane, the bacterial load was found to be 10^9, while the yeast load was 10^6 and the mould load was 10^6 (Mayeux 1960; Misra et al. 2017a). The microorganisms may be soilborne or originate through decaying plant material. In bacterial growth, the most important bacteria are *Leuconostoc* which proliferates at a faster rate in heavy rainfall and humid climate. Canes possessing these bacteria after harvest had been reported to increase dextran production. Alvarez and Cardenley (1988) revealed the importance of temperature on cane deterioration. During late crushing periods of sugarcane which begin from April to June, a rise in inversion rate in harvested canes has been reported (Solomon et al. 1997). Association of internodal invertases with microbial activity had also been illustrated and showed that both aspects increased rapidly with the rise in reducing sugars. This in turn causes low sugar recovery. High temperatures during the night do not generate the dextran content in stored canes (Solomon 2000).

12.4 EFFECT OF WATER DEFICITS ON SUGARCANE GROWTH AND DEVELOPMENT

Several morphological and physiological alterations are known in sugarcane grown under water-deficit conditions (Mall et al. 2019; Mall and Misra 2017; Misra et al. 2020a, 2020b). The first and foremost impact of water-deficit conditions can be seen in sugarcane leaves where water level hampers the progression of this stress. Mild stress conditions have relatively higher water status than severe water-deficit stress conditions. Chen et al. (1995) revealed that only relative water content and free-water levels were reduced whereas the bound water content was enhanced in water-deficit conditions. This was clearly seen in Co 6304 and Yuetang 57-423 varieties (water-deficit sensitive cultivars), where water content was worse after

30 (June–August) and 50 days (October–November) exposing to both magnitudes of water-deficit stress, viz., moderate and severe stress, but in water-deficit tolerant varieties (Guitang 11 and NCo 310), the impact of water-deficit stress on water content was not affected as much. Uprety et al. (1999) further reported that canes exposed to water-deficit stress have decreased water potential and relative water content. The thickness of the sugarcane stalk is dependent on the magnitude of the water-deficit condition. In canes with thinner stalks under water-deficit conditions, height of the mother shoot, leaf coloration, tiller number and rate of leaf emergence were altered and affected relatively higher than the canes with thicker stalks and taller mother shoots (Misra et al. 2022b). Water-deficit conditions are also known to affect tillering rate and ratio of tiller/root weight. It is known that thicker stalk canes are more prone to strong impact of water-deficit conditions than thinner stalks (Lal et al. 1968). Negi et al. (1971) illustrated that stomatal conductance, photosynthesis, photochemical electron transport, rate of transpiration and photo-assimilate partitioning are some of the physiological traits that were influenced under water-deficit conditions. Sairam (1994) showed decreased nitrate reductase activities in canes exposed to such stress. Nutrient uptake was also affected and reduced their content under water-deficit conditions. Growth hormones are also known to be reduced under this stress condition and has influenced numerous metabolic activities. During water-deficit conditions, solute accumulation such as malondialdehyde was enhanced and permeability of plasma membrane was elevated (Gao et al. 2006). Reduction in moisture content in soil causes a strong impact on the yield of sugarcane. The yield losses vary in the duration of the water-deficit stress conditions. After 75–90 days of this stress, cane yield loss was reported to be 0.39 mt/day, while after 90–105 days, this loss was of 0.256 mt/day which is relatively less in comparison to 75–90 days of stress (Anonymous 1959).

12.5 OCCURRENCE AND ROLE OF MICROORGANISMS DURING SUGAR PROCESSING OF WATER-DEFICIT CANES

There are a number of microorganisms that affect sugar processing by hindering sugar production and their rate increases when canes are exposed to water-deficit conditions. Moulds are mostly seen at the centrifugation stage in sugar mills; however, this microorganism is also seen during the crystallization and drying stages (Kumar and Singh 2012). The mean amount of moulds and yeasts in cane juice were 6.26 and 5.20 log cfu/mL, respectively (Silva et al. 2016). The occurrence of moulds was also seen on raw sugars and a number of species have been identified by Kopoloffs (1920). The major moulds causing sucrose losses are *Aspergillus niger* and *Aspergillus sydowii. Aspergillus glacus* is another mould that possesses the capability for cane deterioration (Sherwood and Hines 1950). *Penicillium* spp. may also damage the sugar product (Kumar and Singh 2012). Yeasts are the other microorganisms which have been revealed to enter cane stalks through longitudinals split in the sheathed leaf base. James (1993) illustrated that osmophilic yeasts are responsible for sucrose deterioration due to cane staling. *S. rouxii* and *S. mallis* are the two species of yeasts that were found in raw sugars though yeast invasion in sugarcanes is known to be the least responsible for post-harvest sucrose losses (Scarr and Rose 1966).

Other species of *Zygosaccharomyces* and species of *Pichia*, *Candida*, *Dekkeromyces* and *Endomycopsis* have been also been reported to be causing difficulties in sugar processing (Tilbury 1968). An increase in yeast population (about 10^6 yeasts per g of cane) has been revealed when hygiene levels in sugar mills were not practised properly (Tilbury 1980). The presence of yeast in sugarcane juice at mills results in several problems like souring of canes. Xerophilic yeasts are more prominently found in storage tanks of mills and even on wetty sugar residues (Tilbury 1980). Yeasts cause ethanol production from sucrose content by fermentation method (Reis et al. 2004). *Leuconostoc* bacteria (*L. mesenteroides*, *L. dextranicum*, *L. lactis*) are the main microorganism entering the water-deficit harvested sugarcanes. Besides harvested cut ends, corners, gutters, pipelines and the mixed juice tank are other sites of infection of this microorganism in sugar mills. These bacteria are solely responsible for higher production of dextran in water-deficit conditions. Dextran production in stale canes under water-deficit conditions increases the problem of crystallization, filtration and clarification. Alteration in crystals of sugar is the other major problem in sugar mills which influences the quality of the final product (Solomon 2009; Misra et al. 2020; 2020c).

12.6 MANAGEMENT OF MICROORGANISMS IN WATER-DEFICIT HARVESTED CANES

Management of microorganisms in water-deficit harvested canes is an important measure (Misra and Ansari 2022; Misra et al. 2022a). Neat and clean harvested canes and taking care of hygiene are the first management step for preventing the bio-deterioration in harvested canes. The harvested canes should be devoid of mud which may cling to the stalks after harvest and trash should be removed. The loading of stalks in truck should also be taken care of as fracture/damage makes more entrance sites for microorganisms to invade inside which further augment the losses occurring due to it. At sugar mills, clarification stage in sugar processing should be followed by extraction stage immediately as it involves the application of lime for maintenance of juice pH to 8.0 and high temperature of 80–100°C. The combination of lime and high temperature causes microbial degradation and is known for almost complete reduction in microbial proliferation. Contrastingly, Chen and Chou (1993) illustrated that though microbial proliferation is almost completely reduced, dextran production which has been already produced has not had much of an effect on this combination. Moreover, mesophilic or thermophilic spores are not affected by the environmental condition of this stage. Another stage during sugar processing is the diffuser stage where high temperatures are maintained, which help in managing the microbial growth as well as its proliferation. Klaushofer et al. (1998) showed that equipment cleaning and hygiene maintenance is another approach by which the accumulation of slime-producing microorganisms could be controlled. Chemical usage of the harvested sugarcanes grown under water-deficit conditions is helpful in managing the growth of microorganisms as well as their invasion. A formulation of benzalkonium chloride and sodium metasiliscate is one such example (Misra et al. 2020a). Polmax ESR, dithiocarbamate-based biocides and Polmax Supreme are some other examples (Solomon 2009).

12.7 CONCLUSION AND FUTURE PROSPECTS

Water-deficit conditions are a common and frequently occurring problems due to climate change. Sugarcane crops are severely affected by this abiotic stress and its severity is dependent on the magnitude of the stress. Sucrose content in sugarcane stalks is the best feast for microorganisms. Microorganisms invade sugarcane stalks through cut-harvested ends to increase their population and growth so that they can consume sucrose content and deteriorate it to reduce sugars and other secondary metabolites. Mucoid-producing microorganisms have a higher probability to be seen in harvested water-deficit conditions like *Leuconostoc*, moulds, etc., though these are also present in normal harvested canes. These microorganisms have the capability of producing dextran and acids. Microbial incidence further creates difficulties in sugar processing. Physical damage in canes, time delay after harvest and prior to milling, atmospheric and storage conditions are some of the factors which determine the microbial invasion in sugarcane harvested stalks, in general basis. Complete control of microorganisms for entering into harvested canes has been known to be not possible in many studies of ancient times, but with advancement in research, microbial invasion in harvested stalks could be minimized to a certain extent when precautions are adopted during cane harvesting and processing. Cleaning harvested canes and loading in a proper manner with aeration, help in reducing microbial growth. Water-deficit canes should be prioritized to be milled earlier so that sucrose deterioration losses can be seen. Chemical application on harvested canes has also been shown to have positive results in controlling the microbial invasion in harvested water-deficit sugarcanes. Future approaches to manage microorganisms' growth and proliferation are needed for water-deficit stress so that bio-deterioration losses can also be reduced to some extent.

REFERENCES

Alvarez, J.F., Cardenley, H. 1988. Practical aspects of the control of dextran at Atlantic Sugar Association. *International Sugar Journal* 90: 182–184.

Anjos, E.C.T., Cavalcante, Goncalves, D.M.C., Pedrosa, E.M.R., Santos, V.F., Malia, L.C. 2010. Interaction between mycorrhizal fungus (*Scutellospora heterogama*) and the root knot nematode (*Meloidogyne incognita*) on sweet passion fruit (*Passiflora alata*). *Brazilian Archives of Biology and Technology* 53: 801–809.

Anonymous. 1959. Survey of sugarcane research and development in India: Andhra Pradesh. New Delhi, India: Indian Central Sugarcane Committee, p. 126.

Azcon, R. Rubio, R, Barea, J.M. 1991. Selective interactions between different species of Mycorrhizal fungi and *Rhizobium meliloti* strains, and their effects on growth, N_2 fixation (N^{15}) and nutrition of *Medicago sativa* L. *New Phytology* 117: 399–404.

Bevan, D., Bond, J. 1971. Micro-organisms in field and mill – A preliminary study. In: *Proceeding Qd. Society of Sugar Cane Technology Thirty-eighth Conference*, pp. 137–143.

Chen, R.-K., Zhang, M.-Q., Lu, Y.-B. 1995. Studies on the effect of water deficits on physiology in sugarcane (*Saccharum officinarum* L.). *Sugarcane*. http://en.cnki.com.cn/Article_en/CJFDTOTAL-GZZZ501.000.htm (accessed 09.07.2016).

Chen, J.C.P. and Chou, C.C. 1993. Microbiological control in sugar manufacturing and refining. In: Chen, J.C.P. and Chou, C.C. (eds), Sugarcane handbook. New York: John Wiley and Sons Inc.

Czernicki, M., Von, O.F. 1900. Arcch. Java-Suikerind. *Jaarg* 8: 597–657.
Duncan, C.L., Colmer, C.L. 1964. Coliforms associated with sugarcane plants and juices. *Applied Microbiology* 12(2): 173–177.
Eggleston, G., Legendre, B., Tew, T. 2004. Indicators of freeze damaged sugarcane varieties which can predict processing problem. *Food Chemistry* 87(1): 119–133.
El-Tabey Shehata, A.M. 1960. Yeasts isolated form sugar cane and its juice during the production of aguardente de cana. *Applied Microbiology* 8(2): 73–76.
Gao, F., Masek, J., Schwaller, M., Hall, F. 2006. On the blending of the Landsat and MODIS surface reflectance: Predicting daily Landsat surface reflectance. *IEEE Transactions on Geoscience and Remote Sensing* 44(8): 2207–2218.
Garcia, F.H.S. Hemdonca, A.M.D.C. Rodrigues, M. et al. 2020. Water deficit tolerance in sugarcane is dependent on the accumulation of sugar in the leaf. *Annual Applied Biology* 176: 65–74.
Hommes, F. 1926. Archteuitgang van reit op het emplacement. *Archiv Suikerindus Nederland, Indie* 34: 545–551.
Irvine, J.E. 1980. Field origins of dextran and other substances affecting sucrose crystallization. In: *Proceedings of the 1980 technical session on cane sugar refining research*, October 19–21, 1980, New Orleans, LA, pp. 116–120.
James, G. 1993. Sugarcane, 2nd ed. Blackwell Science Ltd., UK, p. 224.
Klaushofer, H., Clarke, M.A., Rein, P.W., Mauch, W. 1998. Microbiology. In: Van der Poel P.W., Schiweck, H., Schwartz T. (eds), Sugar technology – Beet and cane sugar manufacture. Berlin: Verlag, pp. 993–1007.
Kopoloffs. 1920. c.f. Panda, H. 2011. The complete book on sugarcane processing and by-products of molasses (with analysis of sugar, syrup and molasses), p. 503.
Kroes, S., Harris, H.D. 1994. Effects of cane harvester basecutter parameters on the quality of cut. *Proceedings of Australian Society of Sugar Cane Technologists* 16: 169–177.
Kulkarni, V.M., Kulkarni, S.S. 1987. Microbiology of sugar manufacturing process. *Proceedings of the Deccan Sugar Technologists Association* 37: M37–M43.
Kumar, A., Singh, R. 2012. Microorganism in sugarcane and sugar processing: Impact and remedial actions to eliminate its effect on recoverable sugar. In: *Proceedings of the 71st Annual Convention of Sugar Technologists Association of India*, pp. 50–61.
Kumar, D., Malik, N. Sengar, R.S. 2019. Physio-biochemical insights into sugarcane genotypes under water stress. *Journal Biological Rhythm Research* 52(1): 1–4.
Lal, K.N., Mehrotra, D.N., Tandon, J.N. 1968. Growth behavior, root extension and juice characters of sugarcane in relation to nutrient deficiency and drought resistance. *Indian Journal of Agriculture Science* 38(5): 790–804.
Lee, H.A. 1923. The deterioration of cut cane in Pampanga. *Sugar Cane Plant News* 4: 7–15.
Mall, A.K., Misra, V. 2017. Biotechnological approaches: Sustaining sugarcane productivity and yield. In: Bhore, S., Marimutchu, K., Ravichandran, M. (eds), Biotechnology for sustainability Achievements, challenges and perspectives. Malaysia: AIMST University, pp. 387–398 (ISBN: 978-967-14475-3-6; eISBN: 978-967-14475-2-9).
Mall, A.K., Misra, V., Kumar, A., Pathak, A.D. 2019. Drought: Status, losses and management in sugarcane crop. In: Bharti, P.K., Ray, J. (eds), Farm and development: Technological perspectives. New Delhi: Discovery Publishing House Pvt. Ltd., pp. 165–187.
Manhães, C.M.C., Garcia, R.F., Júnior, D.R., Francelino, F.M.A., Júnior, J.F.S.A., Francelino, H.O. 2013. Perdasquantitativas e danosàssoqueirasnacolheita de cana-de-açúcar no Norte Fluminense. *Revista Vértices* 15: 3.
Martins, M.B., Testa, J.V.P., Drudi, F.S., Sandi, J., Ramos, C.R.G., Lancas, K.P. 2019. Interference of speed cutting height and damage to rootstock in mechanical harvesting of sugarcane. *Australian Journal of Crop Science* 13(8): 1305–1308.
Mayeux, P.A. 1960. *Some studies on the microbial flora of sugar cane*. M.Sc. Thesis, Louisiana State University, Baton Rouge.

Mello, R.C., Harris, H. 2003. Desempenho de cortadores de base paracolhedoras de cana-de-açúcar com lâminasserrilhadas e inclinadas. *Agriambi* 7: 2.

Misra, V., Solomon, S., Singh, P., Prajapati, C.P., Ansari, M.I. 2016a. Effect of water logging on post-harvest sugarcane deterioration. *Agrica* 5(2): 119–132.

Misra, V., Solomon, S. Ansari, M.I. 2016b. Impact of drought on post-harvest quality of sugarcane crop. *Advances in Life Sciences* 20(5): 9496–9505.

Misra, V., Mall, A.K., Pathak, A.D., Solomon, S., Kishor, R. 2017a. Micro-organisms affecting post-harvest sucrose losses in sugarcane. *International Journal of Current Microbiology and Applied Sciences* 6(7): 2554–2566.

Misra, V., Mall, A.K., Shrivastava, A.K., Solomon, S., Pathak, A.D., Hasan, S., Ansari, M.I. 2017b. Efficacy of antibacterial activity of plants against *Leuconostoc* spp. – a potent cause for post-harvest sucrose loss in sugarcane. *Journal of Food, Agriculture and Environment* 15(3–4): 73–77.

Misra, V., Mall, A.K., Tripathi, S., Shrivastava, A.K., Pathak, A.D. 2018. Assessment of comparative characterization of *Leuconostoc* spp. in sugarcane grown soil and its crop. *International Journal of Agriculture and Statistical Sciences* 14(2): 569–573.

Misra, V., Mall, A.K., Shrivastava, A.K., Solomon, S., Shukla, S.P., Ansari, M.I. 2019. Assessment of *Leuconostoc* spp. invasion in standing sugarcane with cracks internode. *Journal of Environmental Biology* 40(3): 316–321.

Misra, V., Solomon, S., Hashem, A., Abd_Allah, E.F., Al-Arjani, A.F., Mall, A.K., Prajapati, C.P., Ansari, M.I. 2020a. Minimization of post-harvest sucrose losses in drought affected sugarcane using chemical formulation. *Saudi Journal of Biological Sciences* 27(1): 309–317.

Misra, V., Solomon, S., Hassan, A., Allah, E.F.E., Al-Arjani, A.F., Mall, A.K., Prajapati, C.P., Ansari, M.I. 2020b. Morphological assessment of water stressed sugarcane: A comparison of drought and water logging canes. *Saudi Journal of Biological Sciences* 27(5): 1228–1236.

Misra, V., Solomon, S., Mall, A.K., Prajapati, C.P., Ansari, M.I. 2020c. Impact of chemical treatments on *Leuconostoc* bacteria from harvested stored/stale cane. *Biotechnology Reports* 27: e00501.

Misra, V., Ansari, M.I. 2022. Water logging tolerance and crop productivity. In: Ansari, S.K., Ansari, M.I., Husen, A. (eds), Augmenting crop productivity in stress environment. Springer, Singapore, pp. 161–175.

Misra, V., Mall, A.K., Ansari, M.I. 2022a. Role of effective management of harvested crop to increase productivity under stress environment. In: Ansari, S.K., Ansari, M.I., Husen, A. (eds), Augmenting crop productivity in stress environment. Springer, Singapore, pp. 223–238.

Misra, V., Solomon, S., Mall, A.K., Abid, M., Khan, M.M.A.A, Ansari, M.I. 2022b. Drought stress and sustainable sugarcane production. In: Arora, N.K., Bouizgarne B. (eds), Microbial biotechnology for sustainable agriculture, Vol. 1, pp. 353–368.

Negi, O.P., Naithani, S.P., Poddar, S. 1971. Root studies of outstanding sugarcane varieties of Bihar, India. *Proceeding of 14th Congress International Society of Sugarcane Technology* 1: 733–738.

Noronha, R.H.F., Silva, R.P., Chioderoli, C.A., Santos, E.P., Cassia, M.T. 2011. Controleestatísticoaplicadoaoprocesso de colheitamecanizadadiurna e noturna de cana-de-açúcar. *Bragantia* 70: 4.

Reis, R.A., Siqueira, G.R., Bernardes, T.F. 2004. Experiência da UNESP Jaboticabalnaensilagem da cana-de-açúcar. I ReuniãoTécnicasobreSilagem.com Cana-de-açúcar. EMBRAPA Pecuária Sudeste, p. 13.

Rockstrom, J. 1999. On farm green water estimates as a tool for increased food production in water scarce regions. *Physics and Chemistry of the Earth (B)* 24(4): 375–383.

Sairam, R.K. 1994. Effect of moisture stress on physiological activities of two contrasting wheat genotypes. *Indian Journal Experimental Biology* 32: 594–597.

Scarr, M.P., Rose, D. 1966. Study of osmophilic producing invertase. *Journal of General Microbiology* 45: 9–16.

Sherwood, I.R., Hines, W.J. 1950. In: *Proceeding of International Society of Sugar Cane Technologists, 7th Congress*, I, pp. 591–607.

Shrivastava, A.K., Srivastava, T.K., Srivastava, A.K., Misra, V., Shrivastava, S., Singh, V.K., Shukla, S.P. 2016. Climate change induced abiotic stresses affecting sugarcane and their mitigation. Lucknow: IISR, p. 108.

Shrivastava, A.K., Misra, V., Srivastava, S., Shukla, S.P., Pathak, A.D. 2017. Contributions of sugarcane researches in development of biology and agriculture: A historical perspective. *Indian Journal of Sugarcane Technology* 32(1): 41–44.

Silva, M.D.O., Freire, F.J., Kuklinsky-Sobral, J., Oliveira, C.A.D., Freire, M.B.G.D.S., Apolinario, V.X.D.O. 2016. Bacteria associated with sugarcane in Northeastern Brazil. *African Journal of Microbiology Research* 10(37): 1586–1594.

Solomon, S. 2000. Post-harvest cane deterioration and its milling sequences. *Sugar Tech* 2(1–2): 1–18.

Solomon, S. 2009. Post-harvest deterioration of sugarcane. *Sugar Tech* 11(2): 109–123.

Solomon, S., Shrivastava, A.K., Srivastava, B.L., Madan, V.K. 1997. Pre-milling sugar losses and their management in sugarcane. Technical Bulletin No. 37. Lucknow: Indian Institute of Sugarcane Research, pp. 1–217.

Solomon, S., Banerji, R., Shrivastava, A.K., Singh, P., Singh, I., Verma, M., Prajapati, C. P., Sawnani, A. 2006. Post-harvest deterioration of sugarcane and chemical methods to minimize sucrose losses. *Sugar Tech* 8(1): 74–78.

Tilbury, R.H. 1968. Bio-deterioration of harvested sugarcane. In: Walters, A.H., Elphick, J.J. (eds), Biodeterioration of materials: Microbiological and allied aspects. Amsterdam: Elsevier, pp. 717–730.

Tilbury, R.H. 1970. Biodeterioration of harvested sugarcane in Jamaica. Ph.D. Thesis, University of Aston, Birmingham, England.

Tilbury, R.H. 1980. Xerotolerant (osmophilic) yeasts. In: Skinner, F.A., Passmore, S.M., Davenport, R.R. (eds), Biology and activities of yeasts. London: Academic Press, pp. 153–179.

Uprety, K.K., Murti, G.S.R., Bhatt, R.M. 1999. Response of French bean cultivars to water deficits: Changes in endogenous hormones, praline and chlorophyll. *Biology Plant* 40: 381–388.

Wold, R.L. 1946. Cane deterioration in a storage pile. *Hawaiian Plant Records* 50(1): 5–10.

Young, D.P. 1955. Review of foliage desiccant experiments for sugarcane in Hawaii. *Reports of Hawaiian Sugar Technology* 14: 96.

13 Bio-Inoculants
The Potential Microorganisms for Restoration of Degraded Soil

Shubha Trivedi and Mukesh Srivastava
Rani Lakshmi Bai Central Agricultural University, Jhansi, India
Sonika Pandey
ICAR–Indian Institute of Pulses Research, Kanpur, India
Sanat Kumar Dwibedi
Odisha University of Agriculture and Technology
Bhubaneswar, India

CONTENTS

DOI: 10.1201/9781003147077-13

13.1 INTRODUCTION

The global population at present is estimated to be around 7 billion (Godfray et al. 2010; Glick 2015), and if it rises at the same pace, then it will reach its carrying capacity of around 9–10 billion by 2050. Hence, the greatest challenge before us in the future is to find a way to feed the ever-growing human population (Glick 2015; Dwibedi et al. 2017). As the demand for food will then increase by 70%, more land will need to be brought under cultivation. Technologies of intensive multi-cropping protected high-density farming and integrated farming systems have been yielding significantly but still further horizontal expansion to satiate the future needs can never be overruled fully. Intensive mono-cropping with a few selected extractive crops demands excessive inputs and depletes soil nutrient reserve readily that ultimately disturb the harmonious balance of nutrients causing soil deterioration and bio-physicochemical transformations.

The ecosystem of the earth has been continuously disturbed due to natural and/or anthropogenic reasons. Such degraded, damaged or destroyed ecosystem can be restored through various remedial measures; among them, the role of microorganisms is very important due to their unique characteristics. They are ubiquitous and freely available in nature in abundant quantities and in all types of ecosystems such as in hot springs, deep oceans, earth's crust, etc. Microorganisms play a great role in nutrient mobilization, transformation, conservation and their availability across the biosphere.

In the present scenario, around 40% of the terrestrial vegetation has been disturbed through many anthropogenic activities such as overexploitation, overgrazing, deforestation, and industrial, agricultural and urban use. Due to the burgeoning pressure of the human and livestock population on land for mitigating their demands for food, fuel, feed and fodder, more and more areas are brought under intensive farming. Injudicious application of synthetic pesticides, fertilizers, hormones, probiotics and monoculture has deteriorated the soil quality. Soil health is manifested by the balanced presence of four major components, viz., minerals, air, water and organic matter. Any significant variation in their concentration leads to serious degradation of soil and loss of soil quality. Soil degradation is a major threat to sustainable crop production and global food security. There is a dire need to develop environmentally friendly technologies for the restoration of degraded lands to bridge the gap between the present and future needs.

The role of microorganisms in remediating the exogenous toxic heavy metals, polycyclic aromatic hydrocarbons, allelopathic and many such xenobiotic and hazardous chemicals is now universally acclaimed. They disintegrate harmful chemicals and toxic substances into relatively less harmful resources. Each and every food web in the biosphere is directly or indirectly influenced by microorganisms. Soil degradation has tremendous effects on microbial abundance and distribution. Microbial diversity is affected by many anthropogenic activities contributing to the environmental pollution and habitat destruction. Degraded soils mostly have depleted organic matter, minerals and nutrients (Srivastava et al. 1989). The basic step in restoration process is cycling of nutrients without which the nutrients will be lost and plants will not be able to complete their life cycle successfully. During the initial stages of soil restoration, substrate supply and microbial population size have major decisive role. Species composition and richness can be a powerful tool for the success of any restoration process. In degraded soils, it is mandatory to maintain a green vegetation cover (Jha and Singh 1992).

Oldeman (1990) divided degraded land into three categories, viz., lightly degraded (9%), moderately degraded (10%) and strongly degraded (4%). Light and moderately degraded soils are suitable for agriculture farming and their restoration is possible through change in farm management practices. In such soils, organic matter and nutrient layers are depleted, which lead to the loss of soil fertility, structure and reduce the water-holding capacity (Montgomery 2007). Excessive use of synthetic pesticides results in soil degradation. Consistent efforts are thus necessary to find alternative, environmentally friendly and innovative options to reduce the use of chemical fertilizers and pesticides. Microorganisms such as bacteria and fungi provide essential primary nutrients (N, P and K) and micronutrients (Mg and Fe) for plant growth. Nitrogen-fixing bacteria and mycorrhizal fungi can add 5–20% of the total N required by a natural grassland annually (Rashid et al. 2016). Free-living and symbiotic bacteria also enhance nitrogen fixation and produce auxins, cytokines and gibberellins.

Bacteria and fungi are well known for their role in improving the soil structure by making it porous. Fungal cells release mucilage that helps in making the soil porous while bacteria release exopolysaccharides that form organo-mineral complexes helping in soil particle aggregation. Soil structure is not only influenced by the mineral constituents of the soil but also by its porosity that render it fertile. Around 40% of the total agriculture land in the world is already degraded and another 24% is under the threat of degradation. Thus, it is high time for finding solutions to such stupendous ecological aberrations and taking stringent ameliorative measures for sustainable management that can maintain soil fertility and increase crop production (Bai et al. 2015). In this chapter, attempts have been made to elucidate the important microbial taxa that play a crucial role in land restoration.

13.2 TYPES OF BIO-INOCULANTS INVOLVED IN THE RESTORATION OF DEGRADED LAND

Different types of microorganisms play important roles in restoration of degraded lands and their spatiotemporal variations occur depending on type and community.

13.2.1 Soil Microorganisms

The soil microbial population helps in the stabilization and structural composition of soil. Soils with active microbial communities have stable aggregation, while at lower activity, the soils have poor aggregation and are compact. Changing microbial communities by inoculation can help in the successful restoration of degraded lands (Wubs et al. 2016). Soil microbial activity declines with the disturbance of the soil layer. Microorganisms produce antibiotics and hormones which help in plant root development. Vesicular-arbuscular mycorrhiza (VAM) supplies phosphorus to leguminous crops. Nematodes, mites and other micro-fauna in soil recycle organic matter that supports in sustenance of the bacteria, fungi and protozoa, and through mineralization, the nutrients are made available to the plants. It has been observed that when degraded lands are introduced with bacterial and fungal feeding nematodes, they are restored in lesser time as compared to the lands without nematodes.

13.2.1.1 Bacteria

Bacteria play an important role in soil organic matter decomposition. Rhizobia bacteria take nitrogen from air and convert it into ammonia. Free living and symbiotic bacteria enhance plant growth by providing nitrogen and phosphorus to plants. Different species of *Rhizobia*, members of the family Rhizobiaceae, are found in root nodules of legume crops in symbiotic association. These bacteria take up nitrogen from air and convert it into ammonia, thereby enhancing plant growth directly by fixing nitrogen for plant use and increasing soil fertility and stability (Singh et al. 2016).

13.2.1.2 Actinomycetes

Actinomycetes with characteristics of both bacteria and fungi give a unique odor to the soil. They are often believed to be the missing evolutionary link between the latter two but the commonalities are more with bacteria than fungi. Apart from their significance in many therapeutic medicines, they take part in organic matter and complex polymer decomposition, nutrient recycling, growth inhibition of several plant pathogens in the rhizosphere, and production of many extracellular enzymes that are beneficial to crop plants.

13.2.1.3 Algae

Algae grow in the top soils with proper sunlight and moisture. Their number in soil ranges from 100 to 10,000 g^{-1} of soil (García-Garibay et al. 2014). Being autotrophic in nature, algae prepare their food by photosynthesis; thus, they release huge amount of oxygen into the soil facilitating submerged aeration. They play a major role in maintaining soil fertility, specifically in the tropical soils. Upon death, they add organic matter which in turn increases organic carbon in the soil. They bind the soil particles and prevent soil erosion that helps in increasing water retention capacity of the soil for longer duration. Cyanobacteria, often called blue-green algae, are presumed to be the first colonizers of terrestrial ecosystems. They are the ubiquitous components of biocrust communities that take part in fixing C and N, and synthesizing exopolysaccharides, increasing soil fertility and water retention through improvement in soil structure and stability (Chamizo et al. 2018).

13.2.1.4 Protozoa

Protozoa are colourless, single-celled animal-like organisms abundantly found in the surface soil at a population range of 10,000–100,000 g^{-1} (Cutler et al. 1923). Most of them derive their nutrition by ingesting soil bacteria, thereby maintaining microbial equilibrium in the soil. Due to an incipient dormancy phase in their life cycle, they can withstand adverse soil conditions.

13.2.1.5 Nematodes

Nematodes are the miniature look-alike of worms which play a major role at several levels of the food web in soil. They usually feed on plant parts and algae at the first level, on bacteria and fungi at the second level and on other nematodes at the higher levels. Their tremendous contribution towards soil fertility due to their significant participation in many functions at different levels of the food web in soil renders them to suitably treat and act as indicators of soil quality (Christopher 2017). Like protozoa, nematodes also play an important role in mineralization and release of plant nutrients in the soil in easily available forms. For example, ammonium is released into the soil with their feeding on bacteria or fungi.

13.2.1.6 Arbuscular Mycorrhizal Fungi

Arbuscular mycorrhizal fungi (AMF) have a close symbiotic association between plant root and soil fungi that facilitate the uptake of phosphorous. They stabilize the soil and enhance plant growth and are important in agriculture. For the restoration of any degraded land, the first step is to determine the status of mycorrhiza in the degraded land. The soil structure is strengthened by mycorrhizal fungi both physically and chemically (Naheeda et al. 2019). They produce humic compounds and organic glue called glomalin that bind the soil particles into water-stable aggregates (Sutton and Sheppard 1976). Better soil aggregates allow water, nutrients, roots and soil microorganisms to move freely within the soil. They help in mitigating plant stresses and are also easily found in harsh environments.

13.2.1.7 Plant Growth-Promoting Rhizobacteria (PGPR)

In land restoration, plant growth-promoting rhizobacteria (PGPR) play an important role in plant and soil interactions. They remain in the rhizosphere and actively participate in plant growth-promoting activities (Kloepper et al. 1980). Various rhizobacteria such as *Achromobacter*, *Arthrobacter*, *Azotobacter*, *Azospirillum*, *Bacillus*, *Enterobacter*, *Pseudomonas* and *Serratia* (Gray and Smith 2005) as well as *Streptomyces* sp. have been found to be beneficial to plants as well as soil (Dimkpa et al. 2009). PGPR secrete several enzymes, siderophores, nitric oxide, organic acids, antibiotics and metabolites which are responsible for increasing mineral uptake, nitrogen fixation and phyto-hormone production, destruction of harmful microorganism, and tolerance against abiotic stresses. Extensive research is required for exploring the remediation potential of these microorganisms in restoration of degraded land.

13.2.1.8 Biofertilizers

Biofertilizers are the most precious gift by nature, and for nature, they are the main component of nutrient management, productivity and sustainability. These biofertilizers (algae, fungi and PGPRs) are the renewable and eco-friendly sources of plant nutrients. Inorganic fertilizers directly supply nutrients to the soil and plants while biofertilzers are living organisms which help to increase the process of mineral uptake by the plants (Jiang et al. 2017; Luginbuehl et al. 2017). Biofertilizers help in reducing environmental pollution on the one hand, while increasing the nutrient availability and improving the bio-physicochemical properties of the soil. They increase root proliferation and solubilize complex nutrients of soil. They are cost-effective and better substitutes for inorganic fertilizers.

13.2.1.9 Viruses

Soil viruses play an important role in influencing the soil ecology as they have the unique ability to transfer genes from different hosts. They act as a potential cause of microbial mortality that ultimately influence the soil nutrient and gas concentration regulating the microbial population (Azam et al. 1983). Soil harbours many novel viruses that represent a large reservoir of genetic diversity (Christopher 2017) and are abundantly present in areas with sparse bacterial populations.

13.2.2 Earthworms

Earthworms are the intestines of the earth as they maintain soil fertility. They constitute a significant portion of the macro-fauna biomass, mostly in the humid tropics. They modify the soil structure by their biogenic properties like casts, pellets, galleries, etc. Anecic species of earthworms are known as ecological engineers as they build permanent burrows into the deeper soil surface and bring organic matter from soil into their burrows for food. Epigeic species, on the other hand, build non-permanent soil structures and mainly feed on litter and humus. In nature, for restoration, they work together or individually depending on the weather conditions. They affect the nutrient cycling of soil through their burrowing activities. The soil becomes porous and thus increases nutrient cycling. Their casts are rich in C and N, thus increasing the ratio of C and N by 1.5 and 1.3 factors (Bhadauria and Saxena 2010).

13.3 ROLE OF SOIL MICROORGANISMS IN DEGRADED SOILS

Soil microorganisms have the ability to decompose organic matter and enhance soil fertility. Soil microbial population contains bacteria, fungi and micro-fauna (Dwivedi and Soni 2011). Soil biota contains a large number of microorganisms that naturally live in the soil and perform a large number of activities which make the soil healthy (Figure 13.1). Microorganisms also play a major role in the destruction of hazardous chemicals like nitrous oxide, methane and other toxic compounds (Doran and Linn 1993). Microorganisms help in the solubilization of minerals like phosphate by phosphate-solubilizing microorganism and affect the weathering process of soil, thus builds soil structure and aggregation. They produce gummy polysaccharides and mucilage that help to cement soil aggregates (Bhadauria and Saxena 2010) and perform biochemical reactions for cycling of elements and development of healthy soil structure. Among the microorganisms, fungi help to stabilize soil structure because their hyphal branches aggregate like a hairnet and act as a thread of soil fabric. A microorganism improves the soil organic carbon, which enhances the soil fertility and water-retaining capacity (Arias et al. 2005). This helps to reduce greenhouse gas-induced climate change.

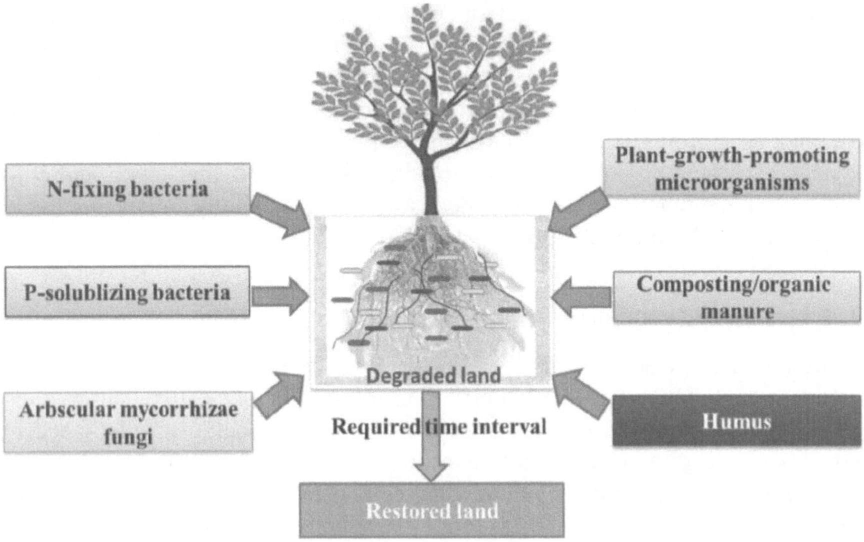

FIGURE 13.1 Role of various microorganisms in the process of degraded land restoration.

13.3.1 Soil Microbial Biomass

The soil microbial biomass (SMB) consists of the total mass of bacteria, fungi and micro-fauna in the soil. All natural organic materials must enter the soil through the above organisms at least once as they get degraded from which they come (Patra 1994). It acts as both source and sink for plant nutrients in available form and also transforming into other forms in soil. This pool is a direct measure of C and N in soil and is closely associated with the soil organic matter (Anderson and Domsch 1989; Smith and Paul 1990). The soil microorganisms dissociate complex organic substrates for deriving energy and biomass carbon leaving behind the excess inorganic nutrients in the soil. Looking at the huge nutrient pool in soil organic matter, the SMB represents a relatively small-standing stock of nutrients. Thus, SMB is the living component of soil organic matter excluding macro-fauna and plant roots.

13.3.2 Conditions for Microbial Activity under Stress

Healthy soils inhabit an enormous microbial population and substantial microbial biomass. Favourable conditions with better available carbon sources, moisture, aeration, temperature, acidity/alkalinity and available inorganic nutrients such as nitrogen and phosphorous facilitate microbial activity. Under the limited availability of such substrates, the microbial population are stressed and they do not receive adequate supply of readily available substrates for higher rates of growth. All soil organisms except some bacteria need food, water and oxygen for their living. They eat a carbon-based food for getting nutrients (Figure 13.1) and require a moist habitat and oxygen in soil. Such typical needs of the soil organisms render their availability mostly restricted to the top few centimetres of the soil. Nearly about 75% of soil organisms are concentrated within the top five centimetres of soil, whereas agricultural activities such as ploughing, hoeing and earthing limit their growth due to frequent disturbance in their habitat. Unfortunately, many traditional and conventional agricultural practices such as burning of stubble, total killer application for complete clearance of vegetation, injudicious fertilizer application and conventional tillage practices have degraded soils through salinity, acidification and desertification (Ozturk et al. 2004).

13.3.3 Land Remediation

During the last few decades, most of our agricultural lands are under threat due to modern exploitative farming but changing farming practices like conservation agriculture and precision farming are helping to create healthier soils in recent years. By creating the right environment for the microorganisms, the soil fertility gets restored automatically and that can be boosted up with the inoculation of beneficial bacterial and mycorrhizal fungi to promote healthier soils and increase the crop yield (Figure 13.1). Research on such microbial remediation is still in their infant stage that needs further refinement for getting a wider range of benefits based upon improved soil biological fertility.

13.4 ROLE OF SOIL MICROBIAL BIOMASS AND ORGANIC MATTER UNDER DEGRADED LAND

Soil is a natural medium in which the microorganisms live, multiply and finally die. The fundamental functionality of the soil ecosystem could be restored through an effective strategy to catalyse the natural processes and cycles. Its four basic components, viz., minerals, air, water and organic matter when removed as in open cast mining, affect the natural life cycle of the soil biota. Organic matter and mineral nutrients decrease drastically due to surface mining (Srivastava and Singh 1991; Soni and Rawat 2005). Mine spoils are not suitable for both plants and microorganisms due to unfavourable pH, low organic matter content and drought-like conditions arising out of coarse texture and soil compaction (Agarwal et al. 1993). Soil salinity, acidity, poor water-holding capacity, improper nutrient supply and accelerated soil erosion are the few other limiting factors for re-vegetation of the mine spoils (Jha and Singh 1991).

Mine spoils alter the bio-physicochemical properties of the soil that affect plant as well as microbial growth. Restoration of mine spoils primarily requires to establish and maintain a vegetation cover (Rimmer 1982) while its sustainability depends on appropriate nutrient recycling. Plant roots release significant quantities of photo-assimilated carbon into the rhizosphere zone (Norton et al. 1989). Dutta and Agarwal (2000) reported significant positive effects of *Grevellia pteridifolia* plantation over *Cassia siamea* under coal mine spoil with higher microbial biomass, C, N and P content. A gradual increase of organic carbon and microbial biomass has been reported in reclaimed mine spoils (Grahm and Haynes 2004; Sourkova et al. 2005). The role of SMB has been well studied in the agricultural systems and even to some extent in forest ecosystem. The SMB and their activities are dependent on the quality, quantity and turnover of detrital organic matter in the forest floor (Barbhuiya et al. 2008). If the nutrients are lost or immobilized by microorganisms, then the plants suffer from nutrient deficiency. Increase in soil organic carbon and SMB may increase the functionality and stability of soil ecosystems (Degens et al. 2001; Lynch et al. 2004). Microbial structural and functional diversity spontaneously help restore the mine spoils (Chodak et al. 2009). The rate of mineralization of soil organic matter depends on substrate supply and microbial population size (Ross et al. 1990). The ratio of microbial biomass carbon and total carbon is proposed to be the best measure of the success of any reclamation efforts (Insam and Domsch 1988; Insam and Haselwandter 1989). The microbial species composition and biomass estimation could be a powerful tool for prediction of smooth reclamation process (Smith and Paul 1990). Poor vegetation and land degradation are the prime determinants of the marginal SMB (Moussa et al. 2007). Higher SMB is an indicator of better soil fertility as microbial biomass is a potential source of plant nutrients. The progress of the restoration process in degraded areas can be ascertained through microbial dynamics. SMB, their respiration rate and metabolic quotients, which have been used to assess the restoration process of degraded lands (Grahm and Haynes 2004; Frouz and Novakova 2005; Sourkova et al. 2005). Although the role of soil microorganisms in biogeochemical cycles, energy transfer and formation of soil is well understood, information on optimum levels of SMB for the development of degraded land is lacking.

The conversion of forests into other land uses results in remarkable degradation in soil nutrients and SMB. Microbial biomass in the dry tropical environment is a very sensitive indicator of land use change (Srivastava and Singh 1991). Singh et al. (1989) considered SMB as both source and sink in dry tropical environment and microbial immobilization may be the main source for plant nutrients that may lead to nutrient conservation. The function of SMB is to conserve nutrients in biologically active and available forms during dry period. Microbial fraction are related to the biochemical parameters such as enzyme activities and labile organic matter that could be a sensitive indicator and early predictor of the (Jenkinson and Ladd 1981; Stout et al. 1981; O'Brien 1984; Jenkinson 1990) ecological stress and its recovery (Dick 1997) in the soil mostly due to increase in the carbon and energy sources from plant root exudates and plant biomass remaining after their death (Rutigliano et al. 2004; Singh et al. 2004). SMB is dependent on the type of plant species as both quality and quantity of soil organic matter are important for microbial activity (Jia et al. 2005). Evidence of the elevated microbial biomass C and N with the plantation of *Caragana micropylla* Lam. in sandy soil have been reported by Cao et al. (2008). An effective rehabilitation plantation strategy can restore degraded land satisfactorily in natural forest areas undergoing massive degradation due to deforestation (Chodak et al. 2009) as enhanced nutrient status of degraded mine spoils has been reported through appropriate plantation programmes (Dutta and Agarwal 2002). Some studies on mine land reclamation, estimate a period of 20 years or longer to recover through microbial inoculation and afforestation/reforestation (Insam and Domsch 1988; Sawada 1996; Anderson et al. 2002; Mummey et al. 2002a, b).

Microbial biomass is directly linked with the degree of immobilization or mineralization of C and N in the soil. Higher mineralization of nutrients reflects reduction in SMB, while microbial biomass increase may indicate immobilization of nutrients (Gill et al. 1986). Large pool of organic matter may produce a significant amount of NH_4^+ that favours plant growth (Finzi et al. 1998). The successional dynamics of microbial biomass C and N are linked to the soil organic C and N contents in the soil during secondary succession of the forest (Arunachalam and Pandey 2003; Jia et al. 2005). However, in semi-arid mine lands, the microbial biomass takes much longer to recover without a significant noticeable difference between the reclaimed land and undisturbed sites (Anderson et al. 2008). An impact study of the inorganic fertilizers on SMB in paddy fields indicated a larger correlation with soil C than P and N in the soil (Zhong and Cai 2007), which clearly suggests taking measures for upgradation of the soil organic carbon with the help of biofertilizer application in degraded lands. Impact of seasonal changes and soil moisture status on SMBs indicated higher SMBs during the summer and lower due to grazing of microorganisms by the soil fauna during rainy season at higher moisture levels (Srivastava 1992). The population of microorganism-feeding fauna such as amoebae, flagellates, nematodes and microarthropods increase (Dash and Guru 1980) that accelerate the faunal predation.

13.5 MEASUREMENT OF RESTORATION SUCCESS

Restoration success in a degraded ecosystem can be measured with two approaches; first, to return to the condition target in an ecosystem, and second, to increase the efficiency of the ecosystem with respect to its function. Another approach to detect the absence or presence of ecto and endo mycorrhizae is the best option that affects plant growth and soil succession after distortion. The basic objective of restoration is to develop a self-sustaining ecosystem (Society for Ecological Restoration International Science and Policy Working Group 2004). By taking measure of the soil microbial communities, the restoration process can be monitored. Microbial community composition and their functionality also differ with the geographical locations. For restoration of an ecosystem, the following attributes should be taken into account:

a. Microbial diversity and community structure should be similar to the indigenous species and microbial inoculation should preferably be from the local strains.
b. Elimination of the potential threats such as predators.
c. Ensuring the presence of sufficient carbon and organic source for microbial multiplication.
d. Microbial population should be self-sustainable.

13.6 CONCLUSION AND FUTURE PROSPECTS

For enhancement of agriculture production to meet the future food demands, horizontal expansion of crop areas through reclamation of degraded lands is the call of the day. More and more land should be brought under cultivation through various processes of land restoration. Among the various options, soil microorganisms play an important role for ecological restoration but unfortunately not much attention has yet been paid to explore such processes deeper. Use of algae, fungi, bacteria, mycorrhiza, protozoa, nematodes, PGPR, biofertilizers, etc., has a tremendous contribution in the transformation of degraded land structures. The best way of reclamation is to select a combination of such site-specific bio-ameliorants in view of the geographical location, season, vegetation type and abundance, characteristics of the land degrading elements, and moisture levels. Preference should be given for the use of stress-tolerant microorganisms with suitable plant hosts and organic compost. Research evidence supports the significance of management practices in enhancing the soil biological activity, soil health and of course increasing crop health under degraded land. The future task for researchers should be to explore microbial strains with high nutrient fixing and drought tolerance ability.

REFERENCES

Agarwal, M., Singh, J., Jha, A.K., Singh, J.S. 1993. Coal-based environmental problems in a low rainfall tropical region. In: R.F. Keefer & K.S. Sajwan (eds.), Trace Elements in Coal Combustion Residues. Lewis Publishers, Boca Raton, FL, 27–57.

Anderson, L.T.H., Domsch, K.H. 1989. Ratios of microbial biomass carbon to total organic-C in arable soils. *Soil Biology & Biochemistry* 21: 471–479. https://doi.org/10.1016/0038-0717(89)90117-X.

Anderson, J.D., Stahl, P.D., Mummey, D.L. 2002. Indicators of soil recovery. In: Proceedings of the western soil science 7th annual meeting, Colorado State University, Fort Collins.

Anderson, J.D., Ingram, L.J., Stahl, P.D. 2008. Influence of reclamation management practices on microbial biomass carbon accumulation in semiarid mined lands of Wyoming. *Applied Soil Ecology* 40: 387–397. https://doi.org/10.1016/j.apsoil.2008.06.008

Arias, M.E., González-Pérez, J.A., González-Vila, F.J., Ball, A.S. 2005. Soil health – A new challenge for microbiologists and chemists. *International Microbiology* 8: 13–21.

Arunachalam, A., Pandey, H.N. 2003. Microbial C, N and P along a weeding regime in a valley cultivation system of northeast India. *Tropical Ecology* 44: 147–154.

Azam, F., Fenchel, T., Field, J.G., Gray, J.S., Meyer-Reil, L.A., Thingstad, F. 1983. The ecological role of water-column microbes in the sea. *Marine Ecology Progress Series* 10: 257–263.

Bai, Z.G., Dent, D.L., Olsson, L., Tengberg, A. Tucker, C., Yengoh, G. 2015. A longer, closer, look at land degradation. *Agriculture for Development* 24: 3–9. ISSN 1759-0604.

Barbhuiya, A.R., Arunachalamm, A., Pandey, H.N., Khan, M.L., Arunachalam, K. 2008. Effects of disturbance on fine roots and soil microbial biomass C, N and P in a tropical rainforest ecosystem of Northeast India. *Current Science* 94: 5–10.

Bhadauria, T., Saxena, K.G. 2010. Role of earthworms in soil fertility maintenance through the production of biogenic structures: Status, trends, and advances in earthworm research and vermitechnology. *Applied and Environmental Soil Science*. https://doi.org/10.1155/2010/816073.

Cao, C., Jiang, D., Teng, X., Jiang, Y., Liang, W., Cui, Z. 2008. Soil chemical and microbiological properties along a chronosequence of *Caragana microphlla* Lam. plantations in the Horqin Sandy Land of Northeast China. *Applied Soil Ecology* 40: 78–85. https://doi.org/10.1016/j.apsoil.2008.03.008

Chamizo, S., Mugnai, G., Rossi, F., Certini, G., De Philippis, R. 2018. Cyanobacteria inoculation improves soil stability and fertility on different textured soils: Gaining insights for applicability in soil restoration. *Frontiers of Environmental Science* 6: 49. https://doi.org/10.3389/fenvs.2018.00049

Chodak, M., Pietrzykowski, M., Niklinska, M. 2009. Development of microbial properties in a chronosequence of sandy mine soils. *Applied Soil Ecology* 41: 259–268. https://doi.org/10.1016/j.apsoil.2008.11.009

Christopher, J. 2017. Living Soils: The Role of Microorganisms in Soil Health. Northern Australia and Land Care Research Programme Published by Future Directions International Pvt. Ltd., Birdwood Parade, Dalkeith, WA.

Cutler, D.W., Lettice, M.C., Sandon, H. 1923. A quantitative investigation of the bacterial and protozoan populations of the soil, with an account of the protozoan fauna. *Philosophical Transactions of the Royal Society London B* 211: 382–390. https://doi.org/10.1098/rstb.1923.0007

Dash, M.C., Guru, B.C. 1980. Distribution and seasonal variation in number of *Testacea* (protozoa) in some Indian soils. *Pedobiologia* 20: 325–342. https://doi.org/10.1007/s00300-005-0006-4

Degens, F.B.P., Schipper, L.A., Sparling, G.P., Duncan, L.C. 2001. Is the microbial community in soil with reduced catabolic diversity less resistant to stress or disturbance? *Soil Biology and Biochemistry* 33: 1143–1153.

Dick, R.P. 1997. Soil enzyme activities as integrative indicators of soil health. In: C.E. Pankhurst, B.M. Dube & V.V.S.R. Gupta (eds.), Biological Indicators of Soil Health. CAB International, Wallingford, 121–156.

Dimkpa, C.O., Merten, D., Svatoš, A., Büchel, G., Kothe, E. 2009. Metal-induced oxidative stress impacting plant growth in contaminated soil is alleviated by microbial siderophores. *Soil Biology and Biochemistry* 41: 154–162. https://doi.org/10.1016/j.soilbio.2008.10.010

Doran, J.W., Linn, D.M. 1993. Microbial ecology of conservation and management. In: L. Hatfield & B.A. Stewart (eds.), Advances in Soil Science, J. Lewis Publishers, Boca Raton, FL.

Dutta, R.K., Agarwal, M. 2000. Reclamation of mine spoils: A need for coal industry. In: A. Kumar & P.K. Goel (eds.), Industry, Environment and Pollution. Technoscience Publications, Jaipur, India, pp. 239–250.

Dutta, R.K., Agarwal, M. 2002. Effect of tree plantation on the soil characteristics and microbial activity of coal spoil land. *Tropical Ecology* 43: 315–324.

Dwivedi, V., Soni, P. 2011. A review on the role of soil microbial biomass in eco-restoration of degraded ecosystem with special reference to mining areas. *Journal of Applied and Natural Science* 3: 151–158. https://doi.org/10.31018/jans.v3i1.173

Dwibedi, S.K., Sahu, S.K., Patnaik, R.K., Tarai, R.K., Dash, A. 2017. Effects of varying levels of fly ash and vermicompost amendment on floristic composition of weeds in rice nursery. *International Journal of Current Microbiology and Applied Sciences* 6(12): 3565–3579. https://doi.org/10.20546/ijcmas.2017.612.414

Finzi, A.C., Breeman, N.V., Canham, C.D. 1998. Canopy tree-soil interactions within temperate forests: Species effects on soil carbon and nitrogen. *Ecological Applications* 8: 440–446. https://doi.org/10.2307/2641083

Frouz, J., Novakova, A. 2005. Development of soil microbial properties in topsoil layer during spontaneous succession in heaps after brown coal mining in relation to humus microstructure development. *Geoderma* 129: 54–64. https://doi.org/10.1016/j.geoderma.2004.12.033

García-Garibay, M., Gómez-Ruiz, L., Cruz, A., Bárzana, E. 2014. Single cell protein: The algae. In: C.A. Batt & M.L. Tortorello (eds.), Encyclopedia of Food Microbiology, 2nd ed. Academic Press, Oxford, UK, 425–430.

Gill, W.B., Cannon, K.R., Robertson, J.A., Cook, F.D. 1986. Dynamics of soil microbial biomass and water soluble organic C in Breton L after 50 years of cropping of two rotations. *Canadian Journal of Soil Science* 66: 1–19. https://doi.org/10.4141/cjss86-001

Glick, B.R. 2015. Introduction to plant growth-promoting bacteria. Beneficial Plant-Bacterial Interactions. Springer, Cham, Switzerland, 1–28.

Godfray, H.C.J., Beddington, J.R., Crute, I.R. et al. 2010. Food security: The challenge of feeding 9 billion people. *Science* 327: 812–818. https://doi.org/10.1126/science.1185383

Grahm, M.H., Haynes, R.J. 2004. Organic matter status and the size, activity and metabolic diversity of soil microflora as indicators of the success of rehabilitation of sand dunes. *Biology and Fertility of Soils* 39: 429–437.

Gray, E.J., Smith, D.L. 2005. Intracellular and extracellular PGPR: Commonalities and distinctions in the plant–bacterium signaling processes. *Soil Biology and Biochemistry* 37: 395–412. https://doi.org/10.1016/j.soilbio.2004.08.030

Insam, H., Domsch, K.H. 1988. Relationship between soil organic carbon and microbial biomass on chronosequences of reclamation sites. *Microbial Ecology* 15: 177–188.

Insam, H., Haselwandter, I. 1989. Metabolic quotient of the soil microflora in relation to plant succession. *Oecologia* 79: 171–178. https://doi.org/10.1007/BF00388474

Jenkinson, D.S. 1990. The turnover of organic carbon and nitrogen in soil. *Philosophical Transactions of Royal Society (London) Series B* 329: 361–368.

Jenkinson, D.S., Ladd, J.N. 1981. Microbial biomass in soil: Measurement and turnover. In: E.A. Paul & J.N. Ladd (eds.), Soil Biochemistry, Vol. 5. Marcel Dekker, Inc., New York and Basel, 415–471.

Jha, A.K., Singh, J.S. 1991. Spoil characteristics and vegetation development of an age series of mine spoils in a dry tropical environment. *Vegetatio* 97: 63–76. https://doi.org/10.1007/BF00033902

Jha, A.K., Singh, J.S. 1992. Rehabilitation of mine spoils. In: J.S. Singh (ed.), Restoration of Degraded Land: Concept and Strategies. Rastogi Publications, Meerut, India, 211–254.

Jia, G.M., Cao, J., Wang, C., Wang, G. 2005. Microbial biomass and nutrients in soil at the different stages of secondary forest succession in Ziwulin, northwest China. *Forest Ecology and Management* 217: 117–125. https://doi.org/10.1016/j.foreco.2005.05.055

Jiang, Y.N., Wang, W.X., Xie, Q.J., Liu, N., Liu, L.X., Wang, D.P. 2017. Plants transfer lipids to sustain colonization by mutualistic mycorrhizal and parasitic fungi. *Science* 356: 1172–1175. https://doi.org/10.1126/science.aam9970

Kloepper, J., Leong, J., Teintze, M. et al. 1980. Enhanced plant growth by siderophores produced by plant growth-promoting rhizobacteria. *Nature* 286: 885–886. https://doi.org/10.1038/286885a0

Luginbuehl, L.H., Menard, G.N., Kurup, S., Van, H., Radhakrishnan, G.V., Breakspear, A. 2017. Fatty acids in arbuscular mycorrhizal fungi are synthesized by the host plant. *Science* 356: 1175–1178. https://doi.org/10.1126/science.aan0081

Lynch, J.M., Benedetti, A., Insam, H., Nuti, M.P., Smalla, K., Torsvik, V., Nannipieri, P. 2004. Microbial diversity in soil: Ecological theories, the contribution of molecular techniques and the impact of transgenic plants and transgenic microorganisms. *Biology and Fertility of Soils* 40: 363–385. https://doi.org/10.1007/s00374-004-0784-9

Montgomery, D.R. 2007. Soil erosion and agricultural sustainability. *Proceedings of National Academy of Science* 104(33): 13268–13272.

Moussa, A.S., Rensburg, L., Kellner, K., Bationo, A. 2007. Soil microbial biomass in semi-arid communal sandy rangelands in the Western Bophirima district, South Africa. *Applied Ecology and Environmental Research* 5(1): 43–56.

Mummey, D.L., Stahl, P.D., Buyer, J.S. 2002a. Microbial biomarkers as an indicator of ecosystem recovery following surface mine reclamation. *Applied Soil Ecology* 21: 251–259. https://doi.org/10.1016/S0929-1393(02)00090-2

Mummey, D.L., Stahl, P.D., Buyer, J.S. 2002b. Soil microbiological properties 20 year after surface mine reclamation: Spatial analysis of reclaimed and undisturbed sites. *Soil Biology & Biochemistry* 34: 1717–1725. doi: 10.1016/S0038-0717(02)00158-X

Naheeda, B., Cheng, Q., Ahanger, A., Raza, S., Khan, I. Ashraf, M., Ahmed, N., Zhang, L. 2019. Role of arbuscular mycorrhizal fungi in plant growth regulation: Implications in abiotic stress tolerance. *Frontiers of Plant Science* 10: 1068. https://doi.org/10.3389/fpls.2019.01068

Norton, J.M., Smith, J.L., Firestone, M.K. 1989. Carbon flow in the rhizosphere of *ponderosa* pine seedlings. *Soil Biology & Biochemistry* 22: 449–455.

O'Brien, B.J. 1984. Soil organic carbon fluxes and turnover rates estimated from radiocarbon enrichments. *Soil Biology & Biochemistry* 16: 115–120. https://doi.org/10.1016/0038-0717(84)90100-7

Oldeman, R.A.A. 1990. Forests: Elements of Silvology. Springer Verlag, Berlin, Heidelberg, New York, London, Paris, Tokyo, Hong Kong, Barcelona, 624.

Ozturk, M., Ozcelik, H., Sakcali, M.S., Guvensen, A. 2004. Land degradation problems in the Euphrates basin. *Turkey Environmental News* 10: 7–9.

Patra, D.D. 1994. Soil microbial biomass. In: S.C. Bhandari & L.L. Somani (eds.), Ecology and Biology of Soil Organisms. Agrotech Publishing Academy, Udaipur, 82.

Rashid, M.I., Mujawar, L.H., Shahzad, T., Almeelbi, T., Ismail, I.M.I., Oves, M. 2016. Bacteria and fungi can contribute to nutrients bioavailability and aggregate formation in degraded soils. *Microbiological Research* 183: 26–41. https://doi.org/10.1111/j.1365-2389.1982.tb01790.x

Rimmer, D.L. 1982. Soil physical conditions on reclaimed spoil heaps. *Journal of Soil Science* 33: 567–579. https://doi.org/10.1111/j.1365-2389.1982.tb01790.x

Ross, D.J., Hart, P.B.S., Sparling, G.P., August, A. 1990. Soil restoration under pasture after top soil removal: Some factors influencing C and N mineralization and measurements of microbial biomass. *Plant and Soil* 127: 49–59.

Rutigliano, F.A., Ascoli, R.D., Santo, A.V. 2004. Soil microbial metabolism and nutrient status in the Mediterranean area as affected by plant cover. *Soil Biology and Biochemistry* 36: 1719–1729. https://doi.org/10.1016/j.soilbio.2004.04.029

Sawada, Y. 1996. Indices of microbial biomass and activity to assess mine site rehabilitation. In: Proceedings of the combined meetings of 3rd international and 21st annual Minerals Council of Australia.

Singh, J.S., Raghubanshi, A.S., Singh, R.S., Srivastava, S.C. 1989. Microbial biomass act as a source of plant nutrients in dry tropical forest and Savanna. *Nature* 388: 499–500. https://doi.org/10.1038/338499a0

Singh, A.N., Raghubanshi, A.S., Singh, J.S. 2004. Comparative performance and restoration potential of two *Albizia* species planted on mine spoil in a dry tropical region, India. *Ecological Engineering* 22: 123–140. https://doi.org/10.1016/j.ecoleng.2004.04.001

Singh, A., Vaish, B., Singh, R.P. 2016. Eco-restoration of degraded lands through microbial biomass: An ecological engineer. *Acta Biomedica Scientia* 3(1): 133–135.

Smith, J.L., Paul, E.A. 1990. The significance of soil microbial biomass estimation. In: J.M. Bollag & G. Stotzky (eds.), Soil Biochemistry, Vol. 6. Routledge, Taylor and Francis, New York,357–396.

Society for Ecological Restoration International Science (SER) and Policy Working Group. 2004. The SER international primer on ecological restoration. Society for Ecological Restoration International, Tucson, Arizona. Available from http://www.ser.org.

Soni, P., Rawat, L. 2005. Eco-restoration concept – Its application for restoration of biodiversity in mined lands. In: P. Soni, V. Chandra & S.D. Sharma (eds.), Mining Scenario and Eco-restoration Strategies. Jyoti Publications and Distribution, Mohit Nagar, Dehradun, 67–84.

Sourkova, M., Frouz, J., Fettweis, U., Bens, O., Huttl, R.F., Santrukova, H. 2005. Soil development and properties of microbial biomass succession in reclaimed post mining sites near Sokolov (Czech Republic) and near Cottbus (Germany). *Geoderma* 73–80. https://doi.org/10.1016/j.geoderma.2004.12.032

Srivastava, S.C. 1992. Microbial C, N and P in dry tropical soils: Seasonal changes and influence of soil moisture. *Soil Biology & Biochemistry* 24: 711–714.

Srivastava, S.C., Singh, J.S. 1991. Microbial C, N and P in dry tropical forest soils: Effect of alternate land uses and nutrient flux. *Soil Biology & Biochemistry* 23: 117–124. https://doi.org/10.1016/0038-0717(91)90122-z

Srivastava, S.C., Jha, A.K., Singh, J.S. 1989. Changes with time in soil biomass C, N and P of mine spoils in a dry tropical environment. *Canadian Journal of Soil Science* 69: 849–855. https://doi.org/10.4141/cjss89-085

Stout, J.D., Goh, K.M., Rafter, T.M. 1981. Chemistry and turnover of naturally occurring resistant organic compounds in soil. In: E.A. Paul & J.N. Ladd (eds.), Soil Biochemistry, Vol. 5. Marcel Dekker, New York, 1–73.

Sutton, J.C., Sheppard, B.R. 1976. Aggregation of sanddune soil by endomycorrhizal fungi. *Canadian Journal of Botany* 54: 326–333. https://doi.org/10.1139/b76-030

Wubs, E.R., Putten, W.H., Bosch, M., Bezemer, T.M. 2016. Soil inoculation steers restoration of terrestrial ecosystems. *Nature Plants* 2: 16107. https://doi.org/10.1038/nplants.2016.107.

Zhong, W.H., Cai, Z.C. 2007. Long-term effects of inorganic fertilizers on microbial biomass and community functional diversity in a paddy soil derived from quaternary red clay. *Applied Soil Ecology* 36: 84–91. https://doi.org/10.1016/j.apsoil.2006.12.001

Index

Note: Page numbers followed by *f* indicate figures and *t* indicate tables.

T

V

W

Y